Yannis Hadzigeorgiou

Imaginative Science Education

The Central Role of Imagination in Science Education

 Springer

Yannis Hadzigeorgiou
University of the Aegean
Rhodes, Greece

ISBN 978-3-319-80590-0   ISBN 978-3-319-29526-8   (eBook)
DOI 10.1007/978-3-319-29526-8

© Springer International Publishing Switzerland 2016
Softcover reprint of the hardcover 1st edition 2016
This work is subject to copyright. All rights are reserved by the Publisher, whether the whole or part of the material is concerned, specifically the rights of translation, reprinting, reuse of illustrations, recitation, broadcasting, reproduction on microfilms or in any other physical way, and transmission or information storage and retrieval, electronic adaptation, computer software, or by similar or dissimilar methodology now known or hereafter developed.
The use of general descriptive names, registered names, trademarks, service marks, etc. in this publication does not imply, even in the absence of a specific statement, that such names are exempt from the relevant protective laws and regulations and therefore free for general use.
The publisher, the authors and the editors are safe to assume that the advice and information in this book are believed to be true and accurate at the date of publication. Neither the publisher nor the authors or the editors give a warranty, express or implied, with respect to the material contained herein or for any errors or omissions that may have been made.

Printed on acid-free paper

This Springer imprint is published by Springer Nature
The registered company is Springer International Publishing AG Switzerland

To the memory of my father whose creative imagination was always an inspiration to me

Foreword

This book *Imaginative Science Education* (ISE) represents an effort that provides new and exciting ways to view school science! It illustrates how students can use their imaginations in ways not found in typical classrooms. As such, the book emphasizes many actions required of students, and in this sense, it is one significant effort which uses students directly for promoting changes in science education. And we hope these changes will also result in major changes in education.

Other science educators have also visualized changes in teaching that also include the way students learn. This visualization includes six science 'domains' that describe science and how science should become central in K-16 education. Two such examples are the identification and use of 'science process skills' and 'science concepts'. Concepts continue to focus on the information often included in textbooks. But on the other hand, processes are not likely to be found in them. The 'science process skills' are observing, classifying, measuring, communicating, inferring, and predicting. These are considered crucial for 'doing' science. However, 'science concepts' are central to science instruction in typical classroom teaching, as they include information about the natural world as well as ideas about explanations accepted as 'known'. Two other domains are 'creativity' and 'attitude' which are known as the 'enabling' domains. They call for experiences that promote visualization, open-ended questioning, problem solving, and positive attitudes concerning science study.

The largest domain is 'applications', where most people live and work. This domain is where students learn to transfer and effectively use what they have learned in science for use in their own lives outside school classrooms. It is important because it also involves students personally using both concepts and process skills. The sixth domain, 'worldview', focuses on philosophy, history, and sociology of science and defines it as a specialized human activity.

All six domains demonstrate major efforts used to assess real student learning of science. ISE, while it addresses all six domains, emphasizes the domains of 'creativity' and 'attitude' by giving primacy to the role of creative imagination in the learning process. For it becomes abundantly clear in the first two chapters that imagination, emotional engagement, and learning are interrelated. However, all the imaginative approaches described in the respective chapters of the book provide opportunities for emotional engagement and real learning. The empirical evidence cited in each chapter, along with the examples illustrating instructional practices, help make the relationship between emotions, imagination, and learning quite apparent.

The National Science Teachers Association (NSTA) has defined science as including three actions. It starts with persons exploring things found in the natural world. It then moves to their seeking explanations of the objects and events encountered. Lastly, it includes seeking evidence to support the explanations. These explorations of the natural world exemplify the real 'doing' of science. These three actions/features engage students in using their imaginations to support their ideas as actions instead of 'learning' things to recite and merely repeating things they are told to do.

It is quite unfortunate that for many the real 'doing' of science is ignored! Students are typically asked to remember what they read as evidence for learning. And teachers often fail to use students' imaginations and creativity in their science classrooms, despite the fact that the 'doing' of science requires much imagination and creativity. This is why ISE deserves attention by the science education community. Even though the examples used to illustrate the ISE approaches are drawn from physical science, all science areas can profit from the implementation of ISE.

The Next Generation Science Standards (NGSS) identify practices, crosscutting concepts, and core ideas as ways to promote major changes in science teaching. But where do we find imagination and personal creativity by students? Typical teaching continues to promote the reading of textbooks, teachers giving lectures, and recipes for use in laboratories. But we do not want students merely reading textbooks and remembering what is needed for testing! Too often, the "testing" has nothing to do with real student learning and the actual 'doing' of science. ISE, by its very nature, is not something that can be taught as 'topics' to remember for testing!

With regard to personal involvement and students' attitudes towards science, the research shows that the longer students study science in typical K-12 classrooms, the worse their 'attitudes' become regarding science! The same is true about student creativity and imaginative skills, as well as their questioning skills. They all diminish with typical classroom teaching. ISE can help with this problem, as all imaginative approaches discussed in the book help engage students in actually 'doing' science.

As we move forward with all four areas of STEM (science, technology, engineering, and mathematics), more basic changes in science teaching are needed. Inquiry and problem-solving deserve particular attention. The most important, however, is to attract students to pursue careers in STEM, and here the potential of ISE needs to be acknowledged. Both the theory and the research studies cited in each chapter make it evident that ISE approaches facilitate student engagement with ideas and also promote scientific inquiry and problem solving.

It must be emphasized that ISE encourages experiences of real science both inside and outside the classroom. Science starts with children even before they start to attend school, and this learning needs to continue beyond elementary, middle, high school, and college. ISE demonstrates new ideas that work in everyday science teaching around the world. It encourages the engaging of the imaginations of students and makes learning more meaningful! ISE also sparks the imagination and creativity in teachers who want to teach differently and break away from the traditional classroom teaching. In short, ISE has the potential to change teaching and also lead to real student learning!

University of Iowa Robert E. Yager
Iowa City, IA, USA

Acknowledgement

I am indebted and wish to express my sincerest thanks to the following colleagues who stimulated my thinking about the topics covered in this book: Greg Stefanich (University of Northern Iowa, USA), Stephen Klassen and Catherine Froese-Klassen (University of Winnipeg, Canada), and Roland Schulz (Simon Fraser University, Canada).

My special thanks are reserved for Kieran Egan (Simon Fraser University), whose work on imagination in education has been invaluable and quite inspirational, and Robert Yager (University of Iowa), who honoured me by writing a foreword for the book.

I also wish to thank the Lawrence Hall of Science, University of California, for kindly granting me permission to include in the book a poem written by Lincoln Bergan (from *GEMS Teachers' Guide*).

Contents

1 **Imaginative Thinking in Science and Science Education** 1
   1.1 Imagination and Its Role in Thinking 3
   1.2 Imagination and the Nature of Science 8
   1.3 Imagination and Science Learning 15
      1.3.1 The Role of the Emotional Imagination 16
      1.3.2 The Role of the Unfamiliar, the Strange,
            the Paradoxical, and the Mysterious 17
      1.3.3 The Role of Thought Experiments 23
      1.3.4 Teaching and Learning Possibilities 24
   1.4 Concluding Comments: The Need for Engaging
       the Imagination in Science Education 31

2 **Engagement and Aesthetic Experience in Science Education** 33
   2.1 The Problem of Engaging Students in Science 35
   2.2 Engagement Through Cognitive Disequilibrium 39
   2.3 The Notion of Aesthetic Experience 43
   2.4 What the Research Shows 49
   2.5 Implications for Science Education 53
   2.6 Concluding Comments: The Need for Aesthetic Approaches
       to Teaching/Learning Science 57

3 **Teaching for Romantic Understanding** 59
   3.1 The Notion of Romantic Science 61
   3.2 The Notion of Romantic Understanding 67
   3.3 Romantic Understanding in Science Education 71
   3.4 Fostering Romantic Understanding 72
      3.4.1 Humanization of Meaning 73
      3.4.2 Associating with Heroic Elements and Qualities 73
      3.4.3 The Extremes of Reality and Human Experience 74
      3.4.4 A Sense of Wonder 75
      3.4.5 Contesting of Conventional Ideas and All Kinds
            of Conventions 76

	3.5 Evaluating Romantic Understanding	77
	3.6 What the Research Shows	79
	3.7 Concluding Comments: The Need for a Romantic Understanding of School Science	80
4	**Narrative Thinking and Storytelling in Science Education**	**83**
	4.1 Narrative Thinking	84
	4.2 Storytelling: A Tool for Understanding the World	89
	4.3 Storytelling in Science Education	93
	4.4 The Characteristics of a Science Story	96
	4.5 The Functions/Purposes of a Science Story	100
	4.5.1 Storytelling as a Means to Humanize Scientific Knowledge	102
	4.5.2 Storytelling for Introducing Ideas from the Nature of Science	104
	4.5.3 Storytelling as a Means to Introduce Scientific Inquiry	107
	4.5.4 Storytelling as a Means for the Development of Romantic Understanding	111
	4.5.5 Storytelling as a Means to Introduce Thought Experiments	112
	4.5.6 Storytelling as a Means for Raising Environmental Awareness	114
	4.6 Storytelling in Science Teacher Education	115
	4.7 What the Research Shows	116
	4.8 Concluding Comments: The Need for Storytelling in Science Education	118
5	**Creative Science Education**	**121**
	5.1 The Notion of Creativity	122
	5.2 Creativity in Science	124
	5.3 Creativity in Science Education	128
	5.4 What the Research Shows	132
	5.5 Fostering Creativity in the Science Classroom	135
	5.6 Concluding Comments: The Need for Creative Thinking in Science Education	139
6	**'Wonder-Full' Science Education**	**143**
	6.1 The Nature of Wonder	144
	6.2 Wonder in Science	151
	6.3 Wonder in Science Education: What the Research Shows	155
	6.3.1 Wonder as a Prerequisite for Engaging Students in School Science	157
	6.3.2 Wonder as a Source of Students' Questions	158
	6.3.3 Wonder as a Prerequisite for Significant Learning	160
	6.4 Implications for Science Education	162
	6.4.1 Wonder Is Evoked	163

	6.4.2 Wonder vs Fun	165
	6.4.3 The Centrality of Questions	166
	6.4.4 The Role of Language	169
	6.4.5 The Role of Content Knowledge	171
	6.4.6 Emphasis on Phenomenological Approaches	171
6.5	Sources of Wonder in School Science: Possibilities and Opportunities	174
	6.5.1 The Ideas of Science	174
	6.5.2 Everyday Ordinary, Familiar Objects and Phenomena	176
	6.5.3 Mysterious Situations and Phenomena	177
	6.5.4 Spectacular/Terrible Phenomena	178
	6.5.5 Unexpected Interconnections (of Phenomena, Entities, Ideas, and Human Life)	179
	6.5.6 The Immensity/Vastness and Smallness of Physical Reality	179
	6.5.7 Amazing, Surprising, Incredible Facts About Natural Entities and Phenomena	181
6.6	Concluding Comments: The Need for Wonder in Science Education	182

7 'Artistic' Science Education ... 185
- 7.1 The Shift from a Positivist/Inductivist Conception of Science ... 186
- 7.2 The Relationship of Art and Science ... 188
- 7.3 Aesthetics and Beauty in Science ... 191
- 7.4 Aesthetics and Beauty in Science Education ... 193
- 7.5 The Pedagogical Importance of Art ... 194
- 7.6 Teaching/Learning Possibilities ... 196
  - 7.6.1 Visual Arts ... 197
  - 7.6.2 Drama/Role Play ... 199
  - 7.6.3 Music ... 203
  - 7.6.4 Poetry ... 204
  - 7.6.5 Integrating the Arts ... 206
- 7.7 What the Research Shows ... 207
- 7.8 Concluding Comments: The Need for Integrating Art and Science ... 214

Erratum ... E1

Conclusion: Towards an 'Imaginative Future' of Science Education ... 217

Appendices ... 223
- Appendix A ... 223
- Appendix B ... 238
- Appendix C ... 242
- Appendix D ... 245

Appendix E 250
Appendix F 251
Appendix G 252
Appendix H 256
Appendix I 258

References 259

Introduction

What is imaginative science education? As the words suggest, it is an approach to science education that gives a more prominent and, I will argue, appropriate role to the imagination in the context of teaching and learning school science. Although imaginative science education can be an innovative approach to teaching and learning school science, in the sense that it goes beyond the traditional/standard, and often times mainstream, teaching/learning models, it is important to stress that what make imaginative science education truly imaginative are the opportunities students are offered for imaginative/creative thinking.

But why do we need an imaginative science education (ISE)? Although for some scientists (e.g., Einstein, Feynman, Holton, Sagan) it sounds a truism to claim that imagination is central to science, in science education, with some dominant paradigms driving science education pedagogy (e.g., conceptual change, inquiry, sociocultural perspective), imagination, at best, has not received enough attention or, at worst, has been totally ignored by science educators (see Hadzigeorgiou, 2005a; Hadzigeorgiou & Fotinos, 2007; Kind & Kind, 2007; Schulz, 2009). Even inquiry science, which requires imaginative/creative thinking, has been implemented in a step-by-step fashion, and more often than not, it was a teacher-controlled process (Hadzigeorgiou, Kabouropoulou, & Fokialis, 2012).

Of course, important changes in the area of the philosophy of science over the last three decades have necessitated changes in science education. The most important changes in the epistemology of science concern (a) the shift from a positivist and empirical view of science to one in which the personal values and beliefs of the scientists influence the standards, which assess the adequacy of the proposed theories (Duschl, 1994; see also Duschl Schweingruber, & Shouse, 2007), and (b) the reconsideration of what can be called the 'scientific method' and the view of science as a rational activity (Chalmers, 1990, 1999; Duschl, 1994; Jenkins, 1996; Trefil, 2003). And the most important change in science education was the shift from the academic, dogmatic, and authoritarian approach to teaching science. However, despite such shift, the role of imagination in the teaching and learning of science has not received the attention it deserves (Hadzigeorgiou, 2005a; Kind & Kind, 2007).

Although the academic tradition of science education has been criticized for a number of reasons (e.g., there has been an emphasis on academic content, students have perceived school science as such content, students have seen science as being irrelevant to their life) (Aikenhead, 2003; Hadzigeorgiou, 2005b; Hadzigeorgiou & Konsolas, 2001), the fact that it (academic tradition) has contributed to making school science unpopular among students when it comes to thinking about careers in science and science related areas (Ceci et al., 2014) is perhaps something that needs particular attention by the science education community.

Attracting students, both males and females, even all those who are considered 'outsiders' in the world of science, is quite crucial if we really wish to address a perennial educational problem, namely, how to engage *all* students in science and certainly how to attract the next generation of scientists and engineers. Yet, what has to be recognized is that making school science attractive is not an easy task. And this not only because student engagement in science, in general, is quite problematic (see Chap. 2), but also because students do not think that science is important to them personally, despite the fact that they view science as an important subject (Jenkins & Nelson, 2005; see also Schmidt, 2011). Considering also the fact that among those who do study science physics is not a popular subject, as students prefer to study biology and environmental issues (Jenkins, 2007; Williams et al., 2003), the goal of making science attractive to students presents a greater challenge for both science educators and science teachers than seems often to be recognized.

It has been pointed out that a great challenge for science education is not how to sugar-coat content knowledge, in order to make it attractive to the students, but mainly how to make it meaningful to them and thus encourage involvement with such content knowledge, both inside and outside the school classroom (Hadzigeorgiou, 1997, 2005a, 2005b; Pugh & Girod, 2007). And science education has to respond to this great challenge; otherwise, 'Science for All' will remain just a slogan. Imaginative engagement with science, through ideas, experiments, situations, and activities that capture the students' imagination, appears to be a promising strategy, as regards student engagement with science. After all, the actual content of science is full of wonder, potentially, but science education seems not to routinely enable students to experience this wonder in everyday classrooms.

Imaginative engagement, as Dewey (1998) had argued, is central to human thought (see Chap. 1), and more recently, Egan's (1997, 1999, 2005) work stressed the role of the imagination as a prerequisite for adequate thinking and learning. As he pointed out, 'Engaging the imagination is not a sugar-coated adjunct to learning; it is the very heart of learning' (Egan, 2005, p. 36). Egan's work, in fact, has clearly pointed to the reconceptualization of education as a process whereby students recapitulate the various kinds of understanding, as these have appeared in our cultural history. Inherent in these forms of understanding, and especially for the education of students of the age range 7–15 (see Chap. 3), is the central role of the imagination. It has been an 'interesting omission' that imagination had not been listed as a characteristic of good teaching by Porter and Brophy (1988).

However, what should be also noted is that research based on the life stories of scientists provides evidence that imaginative skills are not necessarily developed by formal schooling. Shepard (1988) has pointed out that, 'their development occurs before, outside of or perhaps in spite of such schooling – apparently through active but largely solitary interaction with physical objects of one's world' (p. 181). So it appears that the development of imagination in the context of formal education is a real challenge.

Fortunately science is a subject that can provide students with ample opportunities for imaginative engagement. More specifically, science can offer opportunities for a creative approach to inquiry, for exciting science stories, for dramatization, and even opportunities for integrating art and science. And, more fortunately, science is that subject that can inspire students by making them feel the mystery and wonder inherent in natural phenomena, even in its very ideas (Hadzigeorgiou, 1999, 2005a; 2012; Hadzigeorgiou & Garganourakis, 2010). The fact that science has the potential to foster a sense of wonder, and thus inspire students, boys and girls alike, to consider a career in science and/or engineering (Hadzigeorgiou & Garganourakis, 2010), is something to consider in regard to the potential of ISE.

The empirical evidence for the role of the imagination in the context of learning through storytelling – as has been reported by Bruner (1986, 1990) and more recently by some science educators (e.g., Hadzigeorgiou et al., 2010; Hadzigeorgiou, Klassen, & Froese-Klassen, 2012; Klassen, 2009; Stinner, 1995) – as well as through dramatization (e.g., Darlington, 2010; McGregor, 2012; Stinner & Teichmann, 2003) and through a sense of wonder (e.g., Hadzigeorgiou, 2012) is more than encouraging, although, as has already been emphasized, capturing the students' imagination is a real challenge. What should be noted though here is that Egan's (1986, 1997) arguments for the importance of imagination and storytelling in education (see Egan's *The Educated Mind* for a more complete account of the ideas that form the basis of his educational theory) have been fully supported by the empirical evidence of some studies, despite some arguments that can be raised in regard to the design of these studies (see Chap. 4).

Regardless of these arguments, however, ISE is in line with the nature of science, the teaching of which has been considered central to science education (Duschl & Grandy, 2013; Lederman, 2004; Lederman & Abd-El-Khalick, 1998; McComas, 1996, 2002). For it goes without saying that any approach to science education should reflect, in some way, the nature of science – something Dewey showed was both sensible and necessary for adequate learning. Thus, imagination is needed not only because scientists deal, by and large, with unobservable entities, but also because they design experiments, they interpret observational and experimental data, and they propose theories that explain phenomena. However, what should be also admitted is that science is not just logic and mathematics, not just experiments and interpretation of results, but also beauty and, as Richard Dawkins (1998) suggests, 'poetry'. How scientists go from a list of observational data to knowledge and understanding is both remarkable and astonishing. This process is so complex, since not only reason but also intuition, imagination, aesthetics, even serendipity play an important role. In particular, the role of the imagination in science is extremely crucial as this is reflected in the notion of scientific creativity (Hadzigeorgiou et al., 2012; Kind & Kind, 2007).

The fact that curriculum documents worldwide make explicit reference to creative thinking, as a worthwhile aim of education, reflects the great importance we attach to creativity. The *Creative Little Scientists* project (2011–2015), which aimed to foster creativity in early childhood in several European countries, is evidence of the priority given to creative imagination, especially in early years (Cremin et al., 2015). Yet, at the same time, many studies have reported that students do not appreciate the imaginative/creative thinking required in doing science, and they do not view science in general as an imaginative/creative endeavour (see Schmidt, 2011).

Those who agree with Jean Piaget's belief that 'The principal goal of education is to create people who are capable of doing things, not simply of repeating what other generations have done – people who are creative, inventive, and discoverers' (cited in Ginsberg and Opper, 1969, p. 5) can more readily consider the importance of the imagination in education. More recently, however, this importance emerges from the writings of various experts on creativity.

If, as we enter a new era, creativity is becoming increasingly important (McLean, 2009; Pink, 2005; see also Murphy et al., 2010), so much so that 'our future is now closely tied to human creativity' (Csikszentmihalyi, 1996, p. 6), and if creativity should be one of the five cognitive abilities that leaders of the future should seek to cultivate (Gardner, 2010), the role of imagination needs to be seriously considered in the context of education, especially in our attempt to reconceive/re-imagine education and schooling (Egan, 2010).

If the world, as we know it today, is the result or the product of the creative thinking of a few individuals, and if progress in any human endeavour and field of study is due exclusively to the development of new ideas and new ways of seeing reality, then it makes sense to make creative thinking a curricular goal. Science is one of the disciplines that can make a contribution to the achievement of this goal. And it is for this reason that imaginative thinking, as a fundamental ingredient of creative thinking (Gaut, 2003; Sternberg, 2006), should be considered central to science education.

Despite the shift from the academic, authoritarian tradition of science education, which (shift) has necessitated reforms aimed at making science education more student-centred and more socially oriented (see Aikenhead, 2003; Hadzigeorgiou, 2005b; Hadzigeorgiou & Konsolas, 2001; Hodson, 1993, 2003, 2004; Tytler, 2008), one question remains pressing and legitimate: Have these approaches considered the imaginative element inherent in the nature of science? And speaking of the nature of science, what Pugh and Girod (2007) point out and the question they ask are worth considering:

> *Science education is still largely focused around standards, conceptual change, and inquiry. These are some of the dominant paradigms that shape current science education pedagogy. What if art and aesthetics were to join the ranks of these dominant paradigms?* (Pugh & Girod, 2007, p. 10)

Certainly, the aesthetic element of science, which reflects the very nature of science, as a form of inquiry, can be considered by science educators in designing curriculum and instruction. In fact, art and aesthetics appear to offer an alternative or even a complementary way to science teaching and learning, in line with recent changes in science and science education. However, in line with Pugh and Girod's (2007) thinking, a more general question can and should be asked: What if imagination was given its proper place in the science curriculum and students were provided with opportunities for imaginative/creative thinking? Egan's (1999) view – that too much emphasis on logico-mathematical thinking and the neglect of the imagination in the early stages of education may very well be one of the reasons of students' subsequent failure in science and mathematics – should be given more serious thought by all those involved in the field of science education. In fact, science educators and science teachers must consider the role of the imagination in learning, which is supported by advances in science and technology as well as by the arts (Esbin, 2007/2008) and also by brain-based research (Caine & Caine, 2001; Caine et al., 2005).

It is the purpose of this short book to provide an overview of the role of imagination in science education and also discuss some imaginative approaches to science teaching and learning. The first two chapters (i.e., Chap. 1 'The Role of Imagination in Science and Science Education' and Chap. 2 'Engagement and Aesthetic Experiences in Science Education') propose a justification of the imagination in the context of science and science education, while the chapters that follow – 'Teaching for Romantic Understanding' (Chap. 3), 'Narrative Thinking and Storytelling in Science Education' (Chap. 4), 'Creative Science Education' (Chap. 5), 'Wonder-full' Science Education (Chap. 6), and 'Artistic' Science (Chap. 7) – provide an in-depth excursion into the ideas forming the theoretical framework behind imaginative approaches and also offer some practical ideas for the classroom. What needs to be emphasized is that such imaginative approaches to teaching science are not simply creative/innovative approaches that have the potential to motivate student learning; they are approaches based on recent developments in the epistemology of science, which stress the personal element of science, which can be best described as imaginative and aesthetic (see Chaps. 1, 4, and 7) and which, according to the evidence cited in the respective chapters, can provide teachers with first-hand experience that science can really be an exciting subject and not 'a world' that is beyond most students' grasp.

Due to the nature of the ideas discussed in these chapters, and the fact that they are all related to imagination, some repetition of ideas, and perhaps some overlapping between the chapters, inevitably occurs. Indeed, narrative thinking, wonder, creativity, romantic understanding, and aesthetic experience are all related to one another and to imagination. For example, the idea of wonder is directly linked to romantic understanding, but it can be infused into storytelling, which, in turn, can help foster romantic understanding. By the same token, a creative activity, and more specifically an activity involving some art form, can be emotionally engaging and also help evoke a sense of wonder. These experiences (i.e., from wonder, from romantic understanding, from storytelling, from a creative activity) can help students develop an anticipation towards learning, become emotionally involved with their object of study, and thus have more fulfilling experiences, which can be of the sort that Dewey (1934) called 'aesthetic'.

However, I do not think that some repetition of the aforementioned ideas is a problem, as each chapter provides a specific focus for them. Moreover, in some chapters, these ideas are simply mentioned, while their in-depth discussion takes place in those chapters that deal specifically with them. Thus, I chose to discuss narrative thinking and storytelling (Chap. 4) and 'romantic understanding' (Chap. 3) in separate chapters in order to highlight their difference and to provide more space for discussing their nature as well as their specific implications. By the same token, I chose to discuss in separate chapters creativity in science education (Chap. 5) and art and science connections (Chap. 7), as I deemed appropriate that an understanding of the relationship between art and science, in the context of science education, presupposes a good grasp of the notion of creativity in general and that of creativity in science in particular.

Even though the topic of imagination in the context of education has not received the attention that creativity has – most books on imagination fall, more or less, within the area of philosophy – the work of educational theorist Kieran Egan (Egan, 1988, 1997, 2005, 2010) and that of some scholars who wrote about the role of imagination in education (see Egan & Nadaner, 1988; Egan, Stout, & Takaya, 2007; Egan, Cant, & Judson, 2014) is quite illuminating and 'practically' helpful. This book on ISE complements such work and, I believe, can help student teachers, and also science teachers and science educators, to approach the idea of imagination in the context of school science. I have extensively quoted famous scholars and scientists, who have talked about the centrality of the imagination in their work and in the field of science in general, even though an analysis of these quotations is beyond the scope of this book. However, I believe that significant ideas need to be brought forth and thus provide food for thought to those interested in the topic of ISE.

In his *A Vision for What Science Education Should Be Like for the First 25 Years of a New Millennium*, Robert Yager (2000) outlined a broader perspective for school science education, which goes well beyond conceptual understanding and science process skills. His vision is captured in six domains, which refer to 'science content knowledge', 'science process skills', 'applications', 'world-view', 'imagery and creativity', and 'feelings and values'. I believe that an ISE can help science educators and science teachers address the last two neglected domains and thus provide students with more opportunities for aesthetic/transformative experiences.

Happy and constructive reading.

Vancouver, BC, Canada　　　　　　　　　　　　　　　　　　　Yannis Hadzigeorgiou
Rafina, Greece
July 2015

Chapter 1
Imaginative Thinking in Science and Science Education

> *I believe in intuition and inspiration [...] At times I feel certain I am right while not knowing the reason. When the [solar] eclipse of 1919 confirmed my intuition, I was not in the least surprised. In fact I would have been astonished had it turned out otherwise. Imagination is more important than knowledge. For knowledge is limited, whereas imagination embraces the entire world, stimulating progress giving birth to evolution. It is, strictly speaking, a real factor in scientific research.*
>
> Albert Einstein, in Cosmic Religion with Other Opinions and Aphorisms, p. 97

> *Knowledge is an island. The larger we make the island the longer becomes the shore where knowledge is lapped by mystery. It is the most common of all misconceptions about science that it is somehow inimical to mystery, that it grows at the expense of mystery [...] The extension of knowledge is an extension of mystery.*
>
> Chet Raymo, in Honey from Stone, p. 6

> *The cultivation of the imagination [...] should be the chief aim of education [...] we have a duty to educate the imagination above all else.*
>
> Mary Warnock, in Imagination, p. 9, p. 10

A student was once asked to answer the following question: how would you measure the height of a skyscraper using a barometer? The student replied as follows:

> Take a very long piece of string. Tie one end of the string to the barometer. Keeping hold of the other end, dangle the barometer off the roof of the skyscraper until it reaches the ground. Then measure the length of the string plus the length of the barometer and you get the height of the skyscraper.

The interviewing tutor did not accept the answer, and the candidate was rejected. But he appealed to the school authorities on the grounds that his answer, while perhaps unorthodox, was undeniably correct. So another professor was asked to arbitrate in the dispute. He asked the candidate to see him and gave him 5 min to reply to the same question in a way that demonstrated knowledge of physics. The young candidate was silent for the first 3 min. The professor warned him that time was

© Springer International Publishing Switzerland 2016
Y. Hadzigeorgiou, *Imaginative Science Education*,
DOI 10.1007/978-3-319-29526-8_1

running out. Then said the student, 'The problem is I've thought of several possible answers, but I can't decide which is the best'. At that time the professor warned the student that he had only 1 min left. Then the student began giving his answers to the problem:

> *You could take the barometer to the roof of the building and then drop it, using a stopwatch to measure the time the barometer took to reach the ground. If this t is time, and the acceleration due to gravity is g, then the height of the building would be gt/2.*
>
> *If there is sunshine, you could measure the length of the barometer, the length of its shadow, and the length of the skyscraper's shadow. Then it's just a matter of proportional arithmetic to work out the height of the skyscraper.*
>
> *Or you could walk up the stairs with the barometer and a pencil, marking off lengths of the barometer as you go, adding them up at the end. If you still want to be boring you could measure the air pressure on the roof and at ground level, convert millibars to meters and get the height of the skyscraper from that.*
>
> *And if you want to be highly scientific you could tie a piece of string to the barometer and make it swing like a pendulum, first on the roof and then on the ground. Then you could work out the acceleration due to gravity on the roof and on the ground from the period of the oscillation of the pendulum. Then from this difference you can calculate the height of the building.*
>
> *But in the end perhaps the best method would probably be to knock on the janitor's door and say, "Look here Mr. janitor, if you tell me how high this building is, I'll give you this lovely new barometer".* (Based on Calandra, 1968)

What does this incident, whether true of fictional, tell us about how we teach and assess science understanding in our classrooms? Certainly it tells us a number of things, but one of them—that most of us would agree with—is the neglected role of creative imagination in science education. The student was assessed for his understanding of the concepts and laws of physics and not for his creative/imaginative thinking. But can't these two be assessed together? If creative imagination is indeed central to science (as is argued in this chapter), should not science education reflect this kind of imagination? Should not science education encourage, foster, and assess it, as an integral part of the learning process?

ISE, by giving priority to the imagination in the context of teaching and learning school science, can be considered an approach that has the potential to encourage science learning in new ways. Of course, the view, that in the world of education most ideas are nothing but a refurbished version of old ones, contains an element of truth, given that the fundamental questions about teaching and learning, and education in general, have remained the same. However, the fact is that the answers to these questions have changed over time, and, most importantly, perennial problems and questions regarding the teaching-learning process could be approached, either through a new synthesis of old ideas or from the perspective of new research, which sheds new light on those old problems and questions. Imagination, an old idea in the world of education, can certainly be approached in the light of contemporary brain research, which has provided evidence for the importance of both hemispheres in the thinking process (Caine & Caine, 2001; Caine, Caine, McClintic, & Klimic, 2005; Semrud-Clikeman, 2012). It is therefore imperative that imagination, a central feature of the right hemisphere, be seriously considered when planning curriculum and instruction across all ages and all students.

No doubt one would feel more comfortable about discussing the role of imagination in school subjects, such as literature and the fine arts, rather than science. Yet it is a truism to claim that imagination is central to science. Indeed scientists, in their attempt to understand how the world works, visualize unobservable entities (e.g. atoms, electrons, lines of force) and phenomena that cannot be directly observed (e.g. electromagnetic induction, change in intermolecular distance), and also think of possible ways to explain phenomena by proposing various hypotheses and theories. In general, scientists play with ideas, with different possibilities, through thought experiments, analogies, and modelling.

Yet imagination is one of those concepts whose importance is not explicitly stated by scientists (i.e. in scientific reports and various research papers). There are, however, sources where one can find what scientists have said about the role of imagination in their work and also in other scientists' work. Van't Hoff, for example, in a letter to his father, wrote, 'The fact is the basis, the foundation. Imagination is the building material, the hypothesis the ground to be tested, and reality is the building' (Van't Hoff, 1967, p. 2). Maxwell, in admiring Faraday's exceptional imaginative thought, said 'Faraday, in his mind's eye, saw lines of force, traversing all space, where the mathematicians saw centres of force attracting at a distance. Faraday saw a medium where mathematicians saw nothing but distance' (cited in McAllister, 1996, p. 54), while Planck (1933, p. 215) did remark that 'Imaginative vision and faith in the ultimate success are indispensable. The pure rationalist has no place here' (meaning 'no place' in modern physics).

Thus, if the true image of science is to be presented to the students, imagination should become a ubiquitous element in the teaching-learning process. It is therefore crucial that the role of imagination be recognized and acknowledged. This chapter will discuss the role of imagination in the thinking process and its place in the nature of science, as well as its role in science education.

1.1 Imagination and Its Role in Thinking

No doubt imagination can prove people wrong. For example, it was the ancient people's imagination that helped shape their belief that the Earth was flat. But it was science itself that proved that the Earth was round. Indeed science can test what exceeds the boundaries or rationality and reality. Also the imagination can lead to fantasy and science fiction, which have nothing to do with objective reality. But again there is rationality that tests the constructions of the imagination (see Chap. 4, for the complementary role of the two modes of thinking, i.e. the paradigmatic and the narrative).

Notwithstanding the constructions of the imagination that fall beyond the boundaries of reality and rationality, the role of the imagination is crucial in the process of thinking. Indeed, it is the imagination that helps bring order and unity to all the elements of sense experience and intuition, while its role, through a combination of freedom and productive action, is central to creative thinking (McLean, 2009).

Even free will and consciousness can be associated with the imagination. As Bronowski (1978) argued, 'What we really mean by free will [...] is the visualizing of alternatives and making a choice between them' [...] 'the central problem of human consciousness depends on this ability to imagine' (p. 18). Recent philosophical work (Morton, 2013) seems to support Bronowski's argument.

But what is imagination and what specific role does it play in the thinking process? Starting with young children, thinking and imagination go hand in hand. Vygotsky's (1991, 2004) theory is based upon the idea that imagination is the internalization of children's play. For him imagination develops as children develop their speech through social interaction with adults, passing though through a stage of egocentric thought. In play, imagination enables children to change the meaning of an object from what it is to what they imagine it represents. It also allows them to combine various elements in general. Children can indeed form mental images of objects and events that do not exist in reality. These are products of their imagination, even though the material for these products is drawn from everyday reality (e.g. a soccer ball with eyes, a house on chicken legs, rain with water drops of various colours). Vygotsky stressed the fact that imagination is based upon prior knowledge and experiences, an idea that has implications for the development of creative thinking (see Chap. 5).

In adolescents, imagination and thinking (in concepts) begin to converge, and their combination becomes manifest in creative thinking first in adolescents and then in mature adults. In regard to the function of the imagination, he differentiated between children and adolescents and pointed out that 'imagination in adolescence is, from a developmental point of view, the successor of children's play' (Vygotsky, 1991, p. 77). However, what is important to keep in mind from the perspective of Vygotsky's (2004) theory of imagination is that imagination is a higher mental ability involving a consciously directed thought process and the basis of all creative activity, in both art and science. Unlike fantasy, which denotes an expression of personal needs and wishes, imagination denotes the generation of meaning (see in Chap. 4 the role of the narrative mode in generating meaning).

In line with Vygotsky's idea of imagination, as the basis of all creative activity, is Egan's (1990) conception of imagination as an ability that allows one to form mental images and also to think of the possible rather than just the actual. This mental ability to think in terms of possibilities is found in Gaut's (2003) analysis— 'playing' with different hypotheses is considered a central feature of the imagination—which points to imagination as the vehicle for creativity. Thus, although one can imagine without being creative, one cannot be creative without being imaginative. It is quite interesting to note that, in distinguishing between extraordinary, 'big C creativity' (BCC) and ordinary, everyday, 'little c creativity' (LCC), Craft (2001) identifies the latter with what she calls 'possibility thinking' (see Chap. 5 for a detailed discussion of the role of creativity in science and science education). She does point out the value of such thinking in everyday life, and, in so doing, she appraises the value of imagination in everyday life. It is therefore evident that whether creativity is viewed as the manifestation of imagination or as synonymous with it, the association between the two is indisputable. But what is the role of

1.1 Imagination and Its Role in Thinking

the imagination in everyday thinking? Dewey's ideas can help answer this question.

In his seminal work *How We Think*, Dewey (1998) provides an excellent justification of the role of imagination in thinking. For him thinking refers to the process of inferring what consequences would ensue from particular suggestions. As he points out, there is 'Imagination and observation in every mental enterprise' (p. 291). The data (facts) and the ideas (suggestions, possible solutions) form the two indispensable and correlative factors of all reflective activity, which are carried on by means, respectively, of observation and inference. The latter goes well beyond what is actually present and what is actually noted and 'proceeds by anticipation, supposition, conjecture, imagination' (Dewey, 1998, p. 104).

For Dewey (1998), considering various possibilities is central to human thinking: an 'idea after it is formed is tested by acting upon it, overtly if possibly, otherwise in imagination' (pp. 104–105). He goes on to say that the aim of the imagination is 'clear insight into the remote, the absent, the obscure' (p. 290), and its role is to envision realities and possibilities:

> *The proper function of imagination is vision of realities and possibilities that cannot be exhibited under existing conditions of sense perception* [...] *Imagination supplements and deepens observation; only when it turns into the fanciful does it become a substitute for observation and lose logical force.* (Dewey, 1998, p. 291)

It is for this reason that imagination should be considered an absolutely necessary element in the reconstruction of one's experience. It is indeed the imagination that enables one to go beyond one's concrete/lived experience and imagine alternative realities and possibilities. In the area of aesthetics in particular (i.e. when one admires a beautiful sunset, the starry sky, a work of art), reconstruction of experience, according to Dewey (1934), presupposes the workings of the imagination. In this context,

> [*Imagination is*] *a way of seeing and feeling things as they compose an integral whole. It is the large and generous blending of interests at the point where the mind comes in contact with the world. When old and familiar things are made new in experience, there is imagination.* (Dewey, 1934, p. 267)

Here Dewey is talking about the having of a holistic experience, which he called 'aesthetic'. In his theory of aesthetics, according to which aesthetic experiences (see Chap. 2) can be had even in everyday life—not only in ateliers, where art is created, and in museums, where art work is perceived and admired—imagination plays a crucial role.

But let us return to situations that require reflective thinking. Although for Dewey (1998), consciousness of an obstacle is the source of thinking, in the sense that this obstacle becomes a source of perplexity, confusion, or doubt, imaginative engagement is a prerequisite for thinking, in the sense that it initiates thinking when no obstacle seems to be present. Dewey's example of a person walking in unfamiliar territory, arriving at a fork, and thus faced with two different route options, with no prior knowledge of which route leads to which (or the desired) destination, is an example of a person who is confused and in doubt, and thus he or she starts to think.

But suppose that no obstacle was present while the person was walking in the woods. According to Dewey, no reflective thinking could be initiated. But also suppose that strange, unfamiliar, unusual, extraordinary things were present while the person was walking in the woods. In this case, also according to Dewey, thinking could be initiated in order for the person to understand what is going on around him or her. Thinking, in this case, was the result of the person's imaginative engagement with what seemed strange and unfamiliar. And this looks like the case of students in the classroom. Why?

Thinking, in the context of education, presupposes a degree of involvement with the object of study. Asking children or even inviting them to think, without some kind of motivation, is not something that takes place voluntarily, especially in the classroom or even outside of school (i.e. in the context of an assignment). It is for this reason that imaginative engagement with the object of study, namely, with science content ideas, is imperative, especially in the context of compulsory education. The problem of engaging students in science, as discussed in Chap. 2, is a complex one, but, as Dewey (1998) pointed out, 'thought must be reserved for the new, the precarious, the problematic' (p. 289). In short, it must be reserved for what captures the imagination. Egan (1997, 2003, 2005), in fact, has strongly criticized the view that teachers should start from the everyday and the familiar, since such an approach does not stimulate the imagination. The role, therefore, of the remote, the unfamiliar, and the mysterious needs to be seriously considered, especially in the context of early years education, since there is evidence that thinking does not have its origin in what we consider to be familiar and simply relevant to young children. This is discussed in detail in the last section of this chapter.

It is important though to point out that Piaget's research has provided evidence that the order of development of geometrical concepts in young children follows the reverse order to historical development. In other words, children master topological figures (i.e. forms without shape or size), while figures of the Euclidean geometry involving angles and sides are mastered later (Piaget, 1954, p. 409). Piaget's research also allowed him to answer Einstein's question (which was posed to Piaget during a conference on epistemology, where the two men met) about the order of development of the concepts of time and speed. According to Piaget (1970), the development of the concept of speed precedes the development of the concept of time. This, of course, means that speed is a fundamental concept in young children's thinking. Although unexplained, these research findings do suggest that in children's thinking there is an element that is closer to the unfamiliar and the remote than in adult thinking. Those of us familiar with the special theory of relativity, in which speed is considered fundamental and time is defined as a function of speed, may remember that this relationship between speed and time seemed and sounded somehow strange, given that in everyday life time is fundamental and speed is defined as a function of time. Yet this relationship is the natural outcome of children's thinking.

What needs to be noted here is that such findings by Piaget were based upon children's construction of logico-mathematical knowledge (i.e. knowledge of relationships among elements associated with children's actions on objects in order to change their position or form, as in the case in which a child constructs a relationship between the length of a pendulum and its period), but without reference to children's imagination, even though the imaginative element was indeed present. It is true that imaginative thinking is different from logico-mathematical thinking, even though the latter presupposes the ability of the imagination (e.g. to form mental images of objects and events, other than those physically present, in order to construct relationships among objects and among events). For example, in the case in which a child tries to understand why an empty plastic bottle, which was placed in the freezer for a while, looks squeezed—even though no one was in the freezer to squeeze it—she or he has first to form a mental image of the bottle, before it was placed in the freezer, and then put this mental image into a relationship with what she or he actually sees in the freezer. And in the case in which she or he is asked to predict the shape of a shadow, she or he forms a mental image of the shadow. Even though Piaget (1954, 1964, 1970) emphasized the crucial role of logico-mathematical knowledge for children's development and learning, he neglected the imaginative element inherent in the development of logico-mathematical knowledge.

It has been pointed out that exclusive emphasis upon logico-mathematical thinking results in the neglect of the 'development of those concepts out of which the more formal logico-mathematical concepts will grow' (Egan, 1988, p. 23). For example, although the formal concept of causality is indeed rare in very young children, an intuitive concept—the story concept, as Egan calls it—of causality is very common. And it is quite interesting to note that emphasis on children's logico-mathematical aptitudes—which in fact represent a limitation in their thinking since their thinking is perception driven (Piaget called that kind of thinking figurative)—has resulted in the neglect of the imagination, that is, a capacity that is very well developed in children. Egan's work has shown that a focus on children's imaginative lives can reveal intellectual activity that is dominated by abstract concepts (e.g. Egan, 1997, 1999, 2005, 2010). It is unfortunate, of course, that most empirical research into children's cognition has rather ignored children's imaginative powers (see Chap. 4 for the role of imagination in the narrative mode of thinking).

Egan (1999), contrary to Piaget's ideas that young children are concrete thinkers, has convincingly argued that children are abstract thinkers. However, Johnson's (1987) work on the role of imagination in the construction of image schemata, that is, general mental images which are created through children's early experiences (interaction) with the world (e.g. balance, compulsion, blockage, contact, attraction, equilibrium, in-out orientation, containment, trajectory), provides additional support for the role of imagination in thinking.[1] Even emotions like anger are said to be

---

[1] The fundamental idea in Johnson's (1987) book 'The body in the mind: The bodily basis of meaning, imagination and reason' is that almost all of our knowledge derives from bodily experiences through metaphorical projections into abstract domains. According to Johnson, bodily motion and forces give meaning both to our physical experiences at a preconceptual level and to many abstract concepts of our language through the use of metaphors (see Hadzigeorgiou, 2000a and Hadzigeorgiou et al., 2009 for empirical evidence of Johnson's claim). Johnson argues that schemata, as structures for organizing and understanding the world, 'fall between abstract propositional structures ... and particular concrete images' (p. 29). However, these schemata can also 'constrain our meaning and understanding' (p. 138).

represented through a schema involving a fluid within a container that bursts open. And our experience of symmetry is not in our perception of symmetrical objects but, instead, in our experience of bodily balance. Fuchs' (2015) work points to a perceptual gestalt called 'force of nature' which is structured metaphorically and narratively. This Force Dynamic Gestalt (FDG), which is characterized by three elements, namely, fluid substance, polarity of verticality, and force,[2] facilitates analogical reasoning, that is, an imaginative mental ability. For example, increases in temperature of the water in a pot are understood in the same way as increases in the level of the water in that pot, and electric current flowing in a wire (i.e. motion of electrons) is understood in the same way as water flow in a pipe. It also facilitates understanding of natural processes/phenomena (e.g. water flow due to pressure difference, heat due to temperature differences).

It is therefore quite evident that imagination plays a central role in thinking, regardless of the context in which the latter takes place. Certainly, in the context of education, where engagement (see Chap. 2) is an issue and, in fact, a problem to be dealt with, the stimulation of the imagination needs particular attention. This is the reason why new, unfamiliar, problematic situations, as was previously mentioned, should always feature in the curriculum. Of course, in Chap. 6, where the notion of wonder and examples from 'wonder-full' science are discussed, one can see that even familiar and ordinary objects, situations, and phenomena can stimulate thinking. However, even in these cases, in which objects and situations help evoke a sense of wonder, imagination is present. When one experiences a sense of wonder, for example, at and about an ordinary object or natural entity (e.g. a glass of water, a flower, a speck of sand) or an ordinary phenomenon (e.g. shadow formation, water condensation and boiling), the imagination is captured and stimulated by that ordinary object and phenomenon (e.g. through questioning, demonstration), which now, as a result of the experience of wonder, appears unfamiliar, unusual, and even extraordinary. Scientists become involved with their object of study and start to think about it because their imagination is stimulated by a sense of mystery and wonder, which brings me to the relationship between imagination and the nature of science.

1.2 Imagination and the Nature of Science

There is no doubt that imagination and science are related. Imagination is not needed only because scientists deal, by and large, with unobservable entities (e.g. atoms, electrons, photons), phenomena that cannot be directly observed (e.g. electromagnetic induction, change in intermolecular distances), and even imaginary constructions (e.g. lines of force) but also because they have to solve various

---

[2] The image schema of fluid substance corresponds to the concept of quantity, the verticality image schema corresponds to the concept of potential, and the image schema of force corresponds to the concept of energy (see Fuchs, 2015).

1.2 Imagination and the Nature of Science

problems, to design experiments, to interpret both observational and experimental data, and to propose theories that explain phenomena. Moreover, what should be noted is that science is not just logic and mathematics not just experiments and interpretation of results but also beauty and, as Richard Dawkins (1998) argues, 'poetry'. Inherent in this 'poetry of science' is imagination and wonder (see Chap. 6).

Imagination, as an ability that allows one to form mental images and also to think of the possible rather than just the actual (Egan, 1990) and 'to play with different hypotheses' and 'with different ways of making objects' (Gaut, 2003, p. 280), is not simply important but central to both art and science. In regard to science, in particular, the centrality of the imagination can be seen in the design of thought experiments. These are experiments set up in the 'laboratory of the mind' (Brown, 1991; see also Sorensen, 1992), and their role in the development of science has been decisive, since they were used to criticize and reject an established theory and also helped develop a new one (e.g. Galileo's balls were used to critique Aristotle's theory of motion and to develop the new mechanics) or simply to introduce an innovative idea (e.g. Heisenberg's microscope was used to introduce the uncertainty principle). The imaginative element inherent in all thought experiments is quite obvious, whether one considers thought experiments in classical mechanics (e.g. Galileo's balls, Newton's cannon) or in quantum mechanics and relativity (e.g. Heisenberg's microscope, Einstein's accelerating elevator). Popper (1959), in remarking upon Galileo's ingenuity, did point out that his experiments introduced the 'most ingenious arguments in the history of rational thought' (p. 442).

Perhaps it was Einstein's famous experiments in his (special and general) theory of relativity that were most imaginary and, in fact, were used to establish his 'strange' relativity ideas and also to popularize them (see Einstein, 1961). It was indeed his imaginative thinking that gave birth to the theory of relativity, and it must have been for this reason that Einstein considered imagination more important than knowledge, for he is reputed to have said that while knowledge points to what there is, imagination points to what there can be. And he also urged people who want to become scientists to take 30 min a day and think like nonscientists (Di Trocchio, 1997).

What Einstein has said about the value of the imagination and the value of thinking 'outside the box of science' may sound rhetorical, but it nevertheless contains an important truth and makes sense once one considers the shift from the standard, empirical view of science that has taken place over the last three or so decades. Even if one were unwilling to abandon the standard view of science, one would still have to acknowledge the role of the imagination in the development of a theory. 'The creation of a theory, whether inductively or deductively, presupposes inferences to facts or theories through imagination' (Hempel, 1966, p. 47). Even in scientific fields, where deduction is the primary form of thinking, imagination is still important: 'imagination and free invention play a similarly important role in those disciplines whose results are validated by deductive reasoning, i.e. mathematics' (p. 47).

However, once the standard, positivist, empirical view of science is abandoned, the view, which scientific ideas (i.e. hypotheses, theories) are not directly derived from observation data but are invented in order to account for them, becomes central

(Chalmers, 1990; Cobern, 2000; see also Hempel, 1945). This view, of course, has been shared by many scientists and philosophers (e.g. Einstein & Infeld, 1966; Popper, 1959). And, as Medawar (1984b) put it, central to the work of scientists is the 'ability to imagine what the truth might be'. Indeed the idea that a hypothesis comes first, as an imaginary construction, and that observational data are used to validate the hypothesis is well accepted by philosophers of science (Chalmers, 1999; Popper, 1972; see also Love, 2010, for the role of 'imaginary illustrations' in Darwin's theory of evolution).

Of course, from a postmodern perspective on science (compatible with the shift from a positivist, empirical view of science), the notion of truth loses its meaning (see also Chap. 4), since reality is considered a construction and not something that exists 'out there' waiting to be discovered. Apparently, Medawar's view can be understood from that perspective. But if one were to accept that there is indeed a reality 'out there', guesses (i.e. hypotheses, theories) could still be made, although one could never know what this reality might be. So even if the realism-vs-anti-realism debate cannot be settled, epistemologically one can still be a constructivist and a fallibilist, while ontologically one can continue to be a realist (see Chap. 4). From a realist perspective, for example, Einstein and Infeld (1966) wrote that 'Physical concepts are free creations of the human mind, and are not, however it may seem, uniquely determined by the external world' (p. 33). What needs to be noted though is that while the invention of theories, more often than not, requires extraordinary imaginative leaps, everyday scientific work (e.g. problem finding and problem-solving, hypothesis formation, visualization, modelling) also requires imaginative thinking. Thus the idea that science is a creative/imaginative endeavour seems and sounds very plausible. And this idea becomes quite indisputable, once one considers the way scientific knowledge develops and the so-called 'scientific method'.

Over the past three decades, work in the epistemology of science has questioned empirical inductivism, since it has been recognized that scientific concepts and theories are imaginary constructions rather than products of direct observation of physical reality. Such work has also led to a reconsideration of what could be called the 'scientific method' and the view of science as a rational activity (Duschl, 1994; Trefil, 2003). There is now consensus that there is no timeless and universal conception of science or scientific method; there is no ideology of science, which 'involves the use of the dubious concept of science and the equally dubious concept of truth that is often associated with it' and which is used to defend 'bodies of knowledge […] that have been acquired by means of 'the scientific method' and, therefore, must have merit' (Chalmers, 1990, p. 169).

Strong arguments against the so-called scientific method, that is, against a rational, step-by-step procedure, according to which scientists create knowledge, have been provided by Feyerabend (1993), whose celebrated motto has been 'anything goes'. He makes specific reference to scientific progress, as a process being full of accidents, conjectures, and a curious juxtaposition of events, all of which demonstrate the complexity of science as a human endeavour and the unpredictability of scientists' actions and decisions. But the word 'anything' means really 'anything', including even irrational factors. In actual fact, irrationality, according to

1.2 Imagination and the Nature of Science

Di Trocchio (1997), has played an important role in scientific progress, in the sense that what once sounded an irrational, even crazy, idea, it finally came to be accepted as a rational idea, which, in turn, became a standard scientific knowledge (e.g. Planck had advised Einstein not to try to include gravity in his theory of relativity; Poincare did believe that the transmission of radio waves on the surface of the Earth to a distance more than 300 km was impossible).

It should be noted that while scientific theories, as final products, appear—and they are indeed—orderly and rational, the conception and the development of the ideas that gave birth to those theories are quite complex and, more often than not, disordered. That is why Paul Feyerabend's (1993) motto makes real sense. Even though the claim that human beings are not rational beings is an exaggeration, perhaps a false claim, it is a fact that the source of their ideas, especially the innovative and revolutionary ones, is not their rationality.

It is interesting to note that Gell-Mann, a Nobel Laureate in physics, has said that 'Rationality is one the many factors governing human behaviour, and it is by no means always the dominant factor' (cited in Jenkins, 1996, p. 147). It is also worth noting that Michael Polanyi (1981, p. 93) talked about 'the vision of a hidden reality' that guides scientists and pointed out that:

> *Science is based on clues that have a bearing on reality. These clues are not fully specifiable; nor is the process of integration which connects fully definable; and the future manifestations of the reality indicated by this coherence are inexhaustible. These three indeterminacies defeat any attempt at a strict theory of scientific validity and offer space for the powers of the imagination and intuition.* (Polanyi, 1981, p. 97)

So what needs to be recognized is that the way scientists go from a list of observational data to knowledge and understanding is both remarkable and astonishing and that in this process of understanding, contrary to the positivist and empirical view of science, there is a place not only for reason but also for intuition, aesthetics, even serendipity, and, of course, imagination. Moreover, the shift from a positivist and empirical view of science has led to a conception of science, as an activity, in which the scientists' personal values, beliefs, and views determine the standards, which assess the adequacy of the proposed theories.

Apparently, such conception of science is in line with what science really is, that is, a very complex human activity. And this complexity may very well explain why an effort by philosophers of science to arrive at a satisfactory definition of science has not been fruitful (Ziman, 2000). Yet there has been a consensus that imagination is indeed an important element of the scientific process, which complements observation, reason, and experiment (Hadzigeorgiou, 2005a, 2005b; Holton, 1978; Hadzigeorgiou & Stefanich, 2001; McComas, 1998a). As McComas (1998b) has pointed out, 'close inspection will reveal that scientists approach and solve problems with imagination and creativity, prior knowledge and perseverance' (p. 58). This complementary, but at the same time central, role of the imaginative element in science has been stressed by Richard Feynman:

> *The test of all knowledge is experiment. Experiment is the sole judge of scientific "truth". But what is the source of knowledge? Where do the laws to be tested come from?*

Experiment, itself, helps to produce these laws, in the sense that it gives us hints. But also needed is imagination to create from these hints the great generalizations – to guess at the wonderful, simple, but very strange patterns beneath them all, and then to experiment to check again whether we have made the right guess. (Feynman, 1995, p. 2)

The history of science provides many examples that testify to the importance of imagination (Di Trocchio, 1997). Who would really doubt that Max Planck, the father of quantum physics, did not make a 'quantum' leap in order to shift his interest from radiation to the radiating atom itself? Indeed, he had remarked that the scientists 'must have a vivid intuitive imagination, for new ideas are not generated by deduction, but by an artistically creative imagination' (quoted in Shavinina & Seeratan, 2004, p. 90). Who would also doubt that it was Newton's imagination behind his view of the Moon as being attracted to the Earth with the same kind of force as is a stone or an apple? And who would not give Einstein credit for a great imaginary leap, when he and not Poincare, who had the same mathematical equations available to him, arrived at the theory of relativity (Di Trocchio, 1997)?

In line with these imaginative leaps, and, of course, in line with the shift from the positivist and empirical view of science, is Bruner's (1986) view that 'science proceeds by constructing worlds […] by inventing the facts against which the theory must be tested' (p. 14). Bruner's view does not simply make imagination the sine qua non of science, by noting its role in some indispensable mental abilities (i.e. reimagining problems, creating mental images, possibility thinking), but it stresses its central role in a kind of thinking that he calls 'narrative thinking' (see Chap. 4). This kind of thinking is completely divergent and complements 'logico-mathematical' (or 'paradigmatic' thinking (Bruner). There is evidence that narrative thinking is central to science (Hadzigeorgiou & Stefanich, 2001; Klassen, 2006a), and, as Nobel Prize winner Peter Medawar commented, scientists, in building exploratory structures, are in fact 'telling stories which are scrupulously tested to see if they are stories about real life' (Medawar, 1984a, 1984b, p. 133). Explanatory schemes (i.e. hypotheses, theories) are the result of both narrative and logico-mathematical thinking (see Chap. 4), since the latter checks and tests ideas generated by the former, no matter how imaginative they (ideas) may be, against reality and also in relation to what is already known about reality.

Perhaps the best example, at least from an instructional point of view, is the development of classical mechanics. Without mental jumps on the part of both Galileo and Newton classical mechanics would have not made any progress. It was indeed the thought experiment performed by Galileo, which helped him move beyond what he actually observed to the hypothetical, namely, from motion in everyday life to motion without friction or resistance.[3]

---

[3] The thought experiment he performed involved a ball going down an incline. He reasoned that if he placed another incline facing the first one, the ball will first accelerate, as it goes down the first incline and will then move up the second incline and finally stop to approximately the same height from which it was released on the first incline. This conclusion was reached because Galileo had mentally removed all friction. He then started to decrease the angle of the second incline until to the point that it became zero, that is, level with the ground. It was then apparent that, without any

1.2 Imagination and the Nature of Science

It would not be an exaggeration to say that without imagination there can be no science. Gerald Holton (1978)—one of the very few who have written about the scientific imagination—used interviews, notebooks, and personal communication in order to study the role that imagination played in the thinking of great scientists, such as Einstein, Millikan, Heisenberg, Fermi, and Oppenheimer. His work shows the implications of scientific imagination not only for the history and philosophy of science but also for education and for the place of science in our culture.

Holton (1996), in pointing out that 'logic, experimental skill, and mathematics are constant guides' but 'are by no means adequate to the task of scientific investigation' (Holton, p. 7), distinguishes between three types of imagination, that is, 'three closely related companions that are rarely acknowledged: the visual imagination, the metaphoric imagination, and the thematic imagination' (p. 78). Scientists make use of all three kinds of imagination. Faraday's lines of force, Tesla's electrical vortices, and Kekule's cyclic structure of benzene (which he daydreamed) are examples of the visual imagination, while the various analogies, like the one between sound and light waves (i.e. between their respective standing wave patterns) that were used by Thomas Young to explain the double-slit experiment, and the ball-and-stick models used in chemistry are examples of the metaphoric imagination. The thematic imagination, on the other hand, refers to general, broad, unifying ideas and presuppositions that determine a scientist's work, like simplicity, symmetry, stability, and unity.

It is interesting to note that Kind and Kind (2007), in reviewing the notion of scientific creativity, stress the fact that despite the difficulty and the challenge we face to build a reliable picture of science, as a human activity, a consensus about a set of ideas concerning the nature of science has been reached, and this set includes the idea of imagination (Kind & Kind, 2007, p. 14):

- Scientific theories are creative products (ideas) made by scientists.
- Many scientists work on the same problems and new ideas (theories, laws) emerge by common effort.
- Most science theories develop over a long period in small steps.
- Some scientists are highly creative and make substantial contributions in their fields, but they always build on other people's ideas.
- All scientists must use their imagination when contributing to the development of science.
- Scientific theories are created in many different ways. The processes are sometimes highly creative and/or highly logic, rational, and/or accidental.
- In science creativity and rationality always work together. Scientific creativity never works without rationality and strict empirical testing.

Apparently, this list of ideas gives an image of science as an imaginative/creative endeavour. This list, therefore, is an important one to consider not only when approaching scientific creativity (see Chap. 5) but also when planning for curriculum

---

frictional forces acting on the ball, as it was released from a point on the first incline, the ball would continue forever on the now level ground.

and teaching. An idea though that does not explicitly appear in the above list is the aesthetic dimension of science and scientific knowledge (see Girod, 2007a), for in talking about the nature of science, the role of imagination should also be sought in the area of aesthetics. The philosopher and historian Thomas Kuhn has stressed the importance of the aesthetic element in scientific revolutions: 'Aesthetic considerations can be decisive. Though they often attract only a few scientists to a new theory, it is upon those few that its ultimate triumph may depend' (Kuhn, 1970, p. 156).

The history of science can provide us once again with a number of examples. Paul Dirac remarked that 'It is more important to have beauty in one's equations than to have them fit the experiment' (Dirac, 1963, p. 47), while Maxwell said, 'I always regarded mathematics as a method of obtaining the best shapes and dimensions of things; and this meant not only the most useful and economical but chiefly the most harmonious and the most beautiful' (cited in McAllister, 1996, p. 42). It is also believed that De Broglie arrived at his famous formula, which connects wavelength and momentum, on the basis of symmetry, 'on the grounds of intellectual beauty' (Polanyi, 1958, p. 148). According to Shepard (1988), the thinking of both Einstein and Maxwell was 'rooted in an aesthetic appreciation of a more concretely visual kind of symmetry' (p. 157). As in Maxwell's equations, which express a symmetry between the electric and magnetic field, in the theories of relativity, there is a symmetrical relation between motion and rest, space and time, matter and energy, and gravitation and inertia (and of course between electric and magnetic fields after Maxwell's theory).

In talking, however, about the aesthetic dimension of science, the role of mystery should also be acknowledged (Hadzigeorgiou, 1999; Papacosta, 2008). In fact it has been argued that scientific inquiry as well as assertive rational judgement that is entirely based on empirical evidence is nonexistent (Ross, 1981):

> *There is no knowledge without mystery. Mystery is not the limit but its spirit* [...] *Were there no mysteries, there could be no query* [...] *knowledge and query depend only on primordial mysteries.* (Ross, 1981, pp. 128–129)

The idea that 'mystery generates wonder and wonder generates awe' (Goodenough, 1997, p. 13) is not just in line with Einstein's point that 'the fairest thing we can experience is the mysterious [...] which stands at the cradle of true art and science' (Einstein, 1949, p. 5); it is a strong argument for the catalytic role of the mysterious in scientific inquiry (see also Chap. 6, for the role of mystery as a source of wonder). This role refers specifically to the role of the emotional dimension of creative imagination, something that is not explicitly associated with ideas about the nature of science. Indeed, creative imagination can be the source of great satisfaction:

> *The employment of the creative imagination in the construction of scientific abstractions in the form of typologies, principles, models, hypotheses, and theories may be a source of great satisfaction and the fulfilment of a type of intellectual action that is signally contributory to living well.* (Phenix, 1982, p. 311)

It is therefore important that the role of the imagination in the creation of scientific knowledge be acknowledged. As the analysis so far suggests, imagination is a higher mental ability that should be explicitly linked with science and therefore with science education. This is discussed in the next section.

1.3 Imagination and Science Learning

Given the role of emotions in learning in general (e.g. Egan, 1999, 2005; Ormrod, 1999; see also Ortony et al., 1989) and in science learning in particular (Alsop, 2005; Matthews, 2005; Zembylas, 2002, 2005), and also the link between emotions and imagination (Morton, 2013)—all emotions require some form of imagination as Morton's analysis shows—the role of imagination in learning science should be given serious thought by science teachers and science educators. Indeed, if the link between emotions and imagination is closer than we preciously thought (Morton, 2013), then the link between imagination and learning should be much stronger than what we have hitherto considered (Hadzigeorgiou, 2002b).

It is true, of course, that in mainstream science education, imagination has been identified with the generation of analogies, with visualization, modelling, and inquiry science. And, as will be discussed in the chapters of this book, its (imagination) role in such approaches as teaching through storytelling, drama, and the experience of a sense of wonder is central. However, it is in thought experiments that the students' imagination is stretched to the utmost, even though one might distinguish between devising (i.e. setting up) and running a thought experiment and simply using one (designed by others) in order to understand a science idea or a natural phenomenon (Hadzigeorgiou & Fotinos, 2007). Nevertheless, thought experiments, in general, necessitate the use of the imagination and could be considered the strongest justification for the use of imaginative thinking in school science education.

In this section I will discuss how the imagination relates specifically to science learning, by focusing not only on the role of thought experiments but also on the role of emotions and also on the role of what students perceive as unfamiliar, strange, mysterious, and paradoxical. The reason why I choose to discuss these three ideas is that, with the exception of thought experiments, which, as was said, stretch the students' imagination to the utmost and therefore are directly linked to the idea of imagination, the role of the emotional element and the role of the unfamiliar, strange, paradoxical, and mysterious are central to students' engagement with school science. These elements, therefore, deserve particular attention in the context of early childhood, where the foundations of scientific knowledge are laid (Fleer, 2012, 2013; Hadzigeorgiou, 2001; Cabe Trundle & Sackes, 2015). The centrality of these two elements can be also seen in the fact that while the emotional element is present in all imaginative teaching/learning approaches that are discussed in the next chapters (i.e. in aesthetic experience, in storytelling, in the experience of wonder, in creative activities, in activities linking art and science), the element of the unfamiliar, the strange, the paradoxical, and the mysterious can be present in romantic understanding and also in some imaginative approaches such as storytelling.

1.3.1 The Role of the Emotional Imagination

In discussing the role of the imagination in the context of science education, the role of the emotional element, especially with young children, is of paramount importance. Recent studies provide evidence that even very young children (ages 4–6) can construct scientific understanding (see Cabe Trundle & Sackes, 2015). Notwithstanding the sociocultural element underpinning these studies, children's imaginative engagement has played a central role in this process of understanding (Egan, 1988, 1999; Hadzigeorgiou, 2001).

Fleer (2009), in studying the relationship between everyday and scientific concepts in the context of play in preschool settings, found that play-based contexts focused on concepts and not just on materials can encourage young children to think conceptually and thus encourage science learning. Fleer (2013) talks explicitly about the 'affective imagination', which is involved in children's most dominant activity, that is, play, especially if the latter is combined with story listening. In imaginative play, in particular, children create imaginary situations (in which they change the meaning of the objects they play with), while in story listening, they identify with characters and want to help them solve a problem and also feel an emotional tension and anticipation (see Chap. 2 where the notion of anticipation is linked with the notion of aesthetic experience).

According to Fleer (2013), listening to fairy tales and re-enacting them can be an effective teaching/learning strategy with young children. In flickering between real and imaginary worlds, children can construct scientific knowledge, by imagining objects and situations, like those involving the concepts of heat and light. In fairy tales, 'magical moments', as Siry and Kremer (2011) note (e.g. when children express such ideas as catching a rainbow or holding a rainbow), are crucial in this 'imaginative' process of knowledge construction. They have reported that children are able to demonstrate both sophisticated understanding of science and 'magical thinking' related to rainbows.

What is important to stress is that Fleer's (2009, 2013) and Siry and Kremer's (2011) studies have provided evidence that the affective imagination can help children approach abstract or non-direct observable science concepts and even develop scientific consciousness. Their findings concur with those from Andrée and Lager-Nyqvist's study (2013) with grade 6 students who participated in informal play during the teaching of the 'The Chemistry of Food' unit. This study provides evidence that 'students' spontaneous collective play offers opportunities for them to explore the epistemic values and norms of science and different ways of positioning in relation to science' (p. 1735).

The role of the emotional imagination is very evident in drama activities and especially in role playing, where both young children and older students actively and imaginatively explore science in a social setting. The results, as far as science learning is concerned, are encouraging (see Chap. 7, which discusses the role of art in science education), even though more studies are still needed to document the effect of role playing on learning outcomes.

However, in regard to science learning, what must be emphasized is that the emotional imagination is important for all students—at all levels of education—and even teachers. Indeed, if all emotions, as was said, require some form of imagination (Morton, 2013) and if emotions are a social construct and thus play a significant role in teachers' profession (Zembylas, 2002, 2005), then the role of emotional imagination in sociocultural settings should refer to all those involved. In considering the problem of women following careers in science and engineering and the fact that women do not lack the abilities to follow such careers (Ceci, Ginther, Kahn, & Williams, 2014; Wang, Eccles, & Kenny, 2013), the emotional imagination in the context of the science classroom can be the vehicle for attracting the next generation of scientists and engineers. It is indeed very important for the science education community to realize that 'Developing the affective domain could [...] help produce more scientists of both sexes' (Matthews, 2005, p. 185). And for this to happen, not only the students' but also the teachers' emotional state should be considered.

1.3.2 The Role of the Unfamiliar, the Strange, the Paradoxical, and the Mysterious

Having discussed in this chapter the blow that the epistemology of empirical inductivism has suffered, particular attention by science teachers and science educators should be paid to the idea that scientific concepts and theories are imaginary constructions rather than products of direct observation of physical reality (e.g. energy cannot be directly observed). Attention should also be paid to the complementarity of the two modes of cognitive functioning (see Chap. 4) and hence to the role of creative imagination in the generation of hypotheses. More specifically, the idea that a hypothesis comes first, as an imaginary construction, and that observational data are used to validate the hypothesis, as has already been discussed, needs to be seriously considered. Indeed, such idea helps science teachers and science educators understand what Ausubel, Novak, and Hanesian (1978) claim, namely, that 'when young children are discovering principles inductively they are really attempting to use empirical evidence to confirm their existing preconceptions' (p. 538). This is not only an argument against the epistemology of empirical inductivism (e.g. the idea that sugar dissolves faster in hot water is not something that children can directly observe) but also a warning for science teachers and educators, particularly those in early childhood education. Although sensory experience is extremely important for young children, attention should be paid lest such importance should result in an exclusive emphasis on sensory experience and thus have the opposite result, that is, 'the neglect of both logico-mathematical thinking and imagination, the two fundamental ingredients for the construction of scientific knowledge' (Hadzigeorgiou, 2001, p. 68).

On the other hand, exclusive emphasis on logico-mathematical thinking may also be ineffective in helping children learn science. Why? A focus on what children already know before formal instruction begins and on the various interrelationships of such knowledge (i.e. how the various ideas that children already have relate to one another) has been the starting point of what has been called the 'conceptual change' approach. Indeed, over the last three decades, a great importance has been placed on students' prior conceptions, their elicitation, and restructuring (Duit & Treagust, 2003; Limon, 2001; Schulz, 2009, 2014; Treagust & Duit, 2008; Weaver, 1998). Ausubel, Novak, and Hanesian's (1978) famous dictum, which the single most important factor influencing learning is what the learner already knows, has made a major contribution to the adoption of the idea that the teaching-learning process should start from students' prior conceptions, so that these prior conceptions, which are often misconceptions, are changed (see Driver, 1991; NRC, 2007).

Conceptual change, of course, is and should be an important goal of science education (Duit & Treagust, 2003; Treagust & Duit, 2008; Vosniadou, 2008). In actual fact, the traditional debate over whether science content should be considered more or less important than the skills and processes of science appears to have been settled with the conceptual change approach to the teaching and learning of science (see Driver, Asoko, Leach, Mortimer & Scott, 1994). Indeed, on the one hand, students can be active in exploring their environment, and, on the other hand, they can understand concepts, that is, the content of science. However, the conceptual change approach, because of its emphasis on the purely cognitive dimension of the process of learning, represents a narrow conception of what it means to be a learner. Apart from the neglect of the emotional dimension, the conceptual change approach, at least to date, has downplayed or totally ignored both aesthetic factors (Girod, 2007a, 2007b; Pugh & Girod, 2007) and the students' worldviews (Cobern, 1996, 2000). And this can be considered an important reason for the failure or rather the ineffectiveness of such an approach (Hadzigeorgiou & Konsolas, 2001; Hadzigeorgiou & Schulz, 2014). Moreover, according to what has been said in this chapter about human thinking, starting with what the students already know may very well be not the best way to engage students' thinking and hence learning. Egan (2003) convincingly argued that if our goal is student engagement, then the idea to start from what the students already know needs to be reconsidered.

Elliot Eisner, in criticizing curriculum-planning practices, argued that 'the unfamiliar may be much more stimulating than what is immediately at hand' (Eisner, in Forward to Egan, 1999). Eisner echoes what Dewey pointed about imaginative engagement and its importance in the teaching-learning process:

> Teachers who have heard that they should avoid matters foreign to pupils' experience are frequently surprised to find pupils wake up when something beyond their ken is introduced, while they remained apathetic in considering the familiar [...] Thought must be reserved for the new, the precarious, the problematic. (Dewey, 1998, p.289)

1.3 Imagination and Science Learning

Eisner also echoes Philip Phenix, who stressed the role of imaginative engagement in the process of meaning making:

> *Ordinary, prosaic, and customary considerations do not excite a vital personal engagement with ideas. One of the qualities of good teaching is the ability to impart a sense of the extraordinary and surprising so that learning becomes a continuous adventure. According to this principle, ordinary life situations and the solving of everyday problems should not be the basis for curriculum content. The life of meaning is far better served by using materials that tap the deepest levels of experience.* (Phenix, 1964, p. 12)

It is quite apparent that both Dewey (1998) and Phenix (1964) insist, as Egan (1999, 2003) does, upon using, in fact starting from, strange, remote, even new, and unfamiliar situations, in order to capture and stimulate the students' imagination and thus facilitate their personal involvement with a topic or subject. Here one may distinguish between familiar situations that produce disequilibration through discrepant events that challenge students to think, and new, unfamiliar, strange, and remote situations that are also challenging, by capturing their imagination. The distinction may seem to have only a theoretical and not a practical value, since what matters, in the end, is whether involvement with an idea, topic, etc. takes place, and thus learning ensues. Yet what is crucial to point out is that there is some evidence that supports not only what Dewey, Phenix, and Egan have said about the content of the curriculum but also the idea that spatial and temporal distance makes people more imaginative/creative (see Chap. 5).

What should be noted is that this approach does not contradict Ausubel's view, namely, that the starting point of the learning process should be the learner's preconceptions. Prior ideas will inevitably be used by the students to make sense of the new, the strange, and the remote, and these ideas, erroneous or not, will become explicit during the discussion about the topic in question. Ausubel's view presupposes a personal involvement with the topic of the lesson or in general with the ideas, which, however, is not always the case. Teachers should certainly consider these prior ideas and help their students organize and reorganize them, in a way that leads them to the learning of concepts. But they should bear in mind what Dewey said, namely, 'the old, the near the accustomed, is not that to which but that with which we attend; it does not furnish the material of a problem, but of its solution' (p. 290).

It is quite interesting to note that Lewis Thomas recommended not just the role of the mysterious to capture the student's imagination but its prominence in the science curriculum. His position is that all introductory courses in science, at all levels of education, should be revised, and, instead of starting from the fundamentals, we should start from the mysteries of cosmology.[4] And he argued that that 'part of the

---

[4] *It is the strangeness of nature that makes science engrossing. That ought to be the centre of science teaching. There are more than seven-times-seven types of ambiguity in science awaiting analysis [...] I suggest that the introductory course in science, at all levels from grade school through college be radically revised. Leave the fundamentals. The so-called basics, aside for a while, and concentrate the attention of the student on the things that are not known. You can possibly teach quantum mechanics without mathematics, to be sure, but you can describe the strangeness of the world opened up by quantum theory. Let it be known, early on, that there are deep mysteries and profound paradoxes, revealed in their distant outlines, by the quantum. Let it be*

Table 1.1 Topics that have the potential to capture the imagination

Topic	Concept(s)
The blackness of space or the invisibility	Scattering and reflection of light, the colour of objects
The hopping step of an astronaut on the Moon	Forces, motion, mechanical energy changes
Boiling eggs on the Moon	Heat, atmospheric pressure
Seeing an electron	The uncertainty principle
The transmission of Sun heat through an empty and freezing space	The electromagnetic wave
The mystery of universal attraction	Gravity, action at a distance vs property of space
Matter: its emptiness and its solidity	Wave-particle dualism, probability
Drinking coffee on an artificial satellite	The state of weightlessness
The fate of the Earth after the shrinking and total disappearance of the Sun	Interactions, particles, gravitons
Looking now and seeing the distant past (stars-ghosts)	The speed of light
Paying a visit on an anti-planet onboard a terrestrial spaceship	Matter, antimatter, the process of annihilation
Exploring the *Titanic*	Hydrostatic pressure

Source: Hadzigeorgiou (1999)

intellectual equipment of an educated person, however his or her time is to be spent, ought to be a feel for the queerness of nature' (Thomas, 1995, p. 66).

It should be pointed out that 'inquiry science' (Konicek-Morran, 2008, 2010, 2013; Martin, Sexton, Franklin, Gerlovich, & McElroy, 2008; Stefanich & Hadzigeorgiou, 2001) should start with a situation or a question that captures the students' imagination. Questions, in particular, have indeed been used to initiate student inquiry (e.g. How do plants feed themselves? What causes condensation and evaporation? How do airplanes fly?). However, it can be argued that the stranger, the more unfamiliar, and the more mysterious a situation is, the more likely it is to encourage student involvement. Indeed the topics in Table 1.1 encouraged involvement of boys and girls alike, even though some of those girls had been considered as 'outsiders' in the physics class (Hadzigeorgiou, 1999). There is evidence that both engagement with science in general and science content learning in particular can result from strange and mysterious objects and situations. Hadzigeorgiou (2010) studied the role of such objects and situations with preschool and lower and upper elementary school students. It became clear, through systematic observations, that students spent more time engaging in discussions about how an object worked or

---

known that these can be approached more closely, and puzzled over, once the language of mathematics has been sufficiently mastered [...] Teach at the outset, before any of the fundamentals, still imponderable puzzles of cosmology. Let it be known, by the youngest minds, that there are some things going on in the universe that lay beyond comprehension, and make it plain how little is known [...] The worst thing that has happened to science education is that the great fun has gone out of it (Thomas, 1995, pp 150–155).

how to explain a situation, that is, a physical or chemical phenomenon (e.g. a cylindrical can rolling uphill, a metal box crashed literally by the air, emptying water from one glass into another by using a tiny strip of towel), compared with the time spent on objects and situations, according to the mandated curriculum. These results reconfirm what Hadzigeorgiou (2001) had reported, namely, that emotions play a crucially important role especially in the context of early childhood science education.

A good example illustrating the pedagogical importance of initiating a lesson with strange and mysterious objects and situations, according to Hadzigeorgiou's (2010) study, is solar toy that was used in order to introduce upper elementary school students to the concept of energy. This toy became the focus of their attention. Students, working in groups of four, were busy discussing how strange and mysterious the toy was and how it could move, even though 'it did not use any batteries or wires'. During the students' discussions, one student said in his group that he had seen watches using sunlight. The other students in his group suggested that they tested this idea. So they placed the toy in the light and observed that it moved faster. They were really excited that they could explain how that toy moved. Although in the case of many students, the teacher had to assist them by providing suggestions and also clues through questioning, *all* students were engaged in discussions and possibility thinking and did come up with many ideas. (A group of students from another classroom did suggest that the toy had batteries that could be recharged by its own movement, while another group thought that the toy might use batteries that could last for several hours and then the toy cannot move at all.) In all cases, inquiry science, through the formation of hypotheses and experimentation (for testing these hypotheses), was encouraged, despite the fact that some hypotheses could not be tested in the context of that particular lesson (as they required a lot of time or literally the 'total destruction' of the toy, in order for students to 'look into what was inside it').

But why was that toy so engaging? I believe, along with Kieran Egan, Philip Phenix, and John Dewey, that it was the strangeness and mysteriousness of the toy that engaged the children's imagination and challenged them to think in order to come up with a plausible explanation. However, what should also be said is that children were somehow aware of their incomplete knowledge (i.e. they became aware that something can move, even without the use of batteries or electric wires that could help supply current electricity, so something else must be involved). In other words, a mysterious object (i.e. the solar toy) and a situation (i.e. its motion in the absence of a visible energy source) became the children's source of wonder (see Chap. 6 for the relationship between wonder and awareness).

A situation, such as the above, should be differentiated from a situation which is introduced through a question that asks students to find out or explain something. Even though such questions can be quite challenging for some, or even several, students, they are not mysterious in the sense that that solar toy was (or that situation involving the 'transference' of water from one glass into another using a tiny piece of towel). Konicek-Morran (2008, 2010), for example, recommends a number of 'mysterious situations' (e.g. Does the temperature of wood affects the heat of the fire? How does heat affect the strength of a magnet? What are the odds of a meteor hitting our house? How can a ball roll faster?). Notwithstanding the great importance of such questions in the context of student inquiry, the fact that they do not include or rather they do not make evident the elements of mystery, strangeness, and paradox (in the sense that the students do not find anything mysterious, strange, or paradoxical in the above questions) should be considered in the context of planning for engagement with science. Perhaps, if they were phrased differently, the elements of mystery, strangeness, and paradox might have been made more evident. This is the reason why we need more studies that document the effect of questioning (i.e. the type of questions teachers ask, how they are phrased) on student learning.

For example, Hadzigeorgiou (2010) found a difference, in regard to student involvement, between a question like 'What causes condensation?' or 'How can condensation be explained?' and the question 'Where do the water drops appearing on the outside of a cold and empty bottle come from?' (after the students themselves were able to observe the water drops forming on a very cold bottle). Even though the content of these questions is the same, it is in the second case that students feel a sense of mystery and experience a sense of wonder (while in the first case they are simply invited to think and explain). What must become clear here is that regardless of whether or not the first questions are challenging for students, it is the second question that better encourages student involvement with the actual object of study (i.e. condensation) and therefore with scientific inquiry. This distinction may be a delicate one—as is the distinction between curiosity and wonder—but nonetheless an important one in the context of teaching and curriculum design (Hadzigeorgiou, 2014). And this is the reason that they are discussed in Chap. 6, which provides an in-depth discussion of the notion of wonder in science and science education.

According to Egan's (1997) recapitulation theory (see Chap. 4), imaginative engagement through mysterious, strange, paradoxical, and 'wonder-full' situations can be very effective in producing considerable learning results (see Chap. 6). Such situations, if incorporated into the plot of a story, can also produce significant learning outcomes, as empirical evidence suggests (Hadzigeorgiou, Prevezanou, Kabouropoulou, & Konsolas, 2010; Hadzigeorgiou, Klassen, & Froese-Klassen, 2012). However, what should be stressed here is that such strategy, although effective, does not mean that familiar situations cannot excite thinking. In actual fact, there are cases in which familiar, ordinary objects or situations appear unusual and extraordinary through the experience of wonder (see Chap. 6).

It is quite interesting to note that Klassen and Froese-Klassen (2014b), in reviewing the literature on interest and its relationship to learning, identified a number of factors associated with situational interest. These include coherence (e.g. provided by a narrative or story, by a specific plot structure), character identification (e.g. when students identify with a character or hero/heroine in a story), physical activity with a cognitive aspect (e.g. when students participate in hands-on activities), and social element (e.g. when students participate in a collaborative setting). However, they also identified the following factors:

1.3 Imagination and Science Learning

- Obscurity
- Mysteriousness
- Suspense
- Anticipation and prediction
- Intensity
- Surprisingness
- Abnormality
- Challenging of beliefs

Given the emotional element involved in situational interest, and in light of what was discussed in this section of this chapter, these factors need to be seriously considered. Klassen and Froese-Klassen (2014b), in pointing out the role of interest as a cognitive tool, stress that 'its potential for re-focusing the design of educational environments cannot be ignored' (p. 137).

1.3.3 The Role of Thought Experiments

The literature on thought experiments in school science education is quite rich. There are a number of studies, especially with secondary school students, pointing to the potential of thought experiments as a teaching/learning tool. However, there are also a few studies with younger students. In most cases, the students are encouraged to design an experiment in their mind, to let the experiment 'run', and then to 'observe' the consequences (Klassen, 2006b). For example, according to Gilbert and Reiner (2004), 12- and 13-year-old students, working in small groups, were able to design and ran thought experiments. These researchers reported that students made progress towards scientific understanding by alternating between imaginary and physical models.

There is empirical evidence, according to which students use quite often thought experiments in order to solve conceptual problems (Kösem & Özdemir, 2014). Reiner and Gilbert (2000) also found that 'thought experimentation in which students construct imaginary situations are a frequently-used strategy for problem solving' (p. 502). This is something that the famous German physicist Ernst Mach had in mind, who, as a coeditor of a pedagogical journal, in which he posed thought-provoking questions, invited his readers to approach imaginatively various problem situations (see Klassen, 2006b).

However, the crucial role of a thought experiment as a teaching/learning tool lies in its potential to facilitate conceptual change (Helm, Gilbert, & Watts, 1985), which first necessitates an elicitation of students' tacit beliefs and intuitions. Indeed, according to Reiner and Burko (2003), 'thought experiments force a learner to access tacit intuitions, explicit and implicit knowledge, and logical derivation strategies, and integrate these into one working thought process' (p. 380). Matthews (1994) considers thought experiments effective teaching/learning tools, especially when students are first asked to predict the result of an experiment and then perform the actual experiment, in order to find out whether their predictions are confirmed.

It is interesting to note that Reiner (1998) found that students' thought experiments, unlike those of the scientists, develop as they (students) discuss their ideas in collaborative settings. In reviewing the literature (Blown & Bryce, 2013; Clement, 2008; Galili, 2009; Gilbert & Reiner, 2000, 2004; Gilbert, Reiner, & Nakhleh, 2008; Hammer, 1995; Helm et al., 1985; Matthews, 2015; Nikolic, 2012; Reiner, 1998; Reiner & Burko, 2003; Reiner & Gilbert, 2000), the pedagogical benefits of thought experiments can be summarized as follows:

- Development of the creative imagination
- Development of logical arguments and critical thinking
- Development of problem-solving skills
- Suggestion of modifications to real experiments
- Facilitation of conceptual clarity
- Facilitation of conceptual change

Imagining being inside an accelerating elevator, light beams passing near the Sun, situations that do not exist in everyday life (e.g. motion on frictionless surfaces, of celestial objects), situations that necessitate modelling (e.g. the motion of planets, various geological phenomena), or even situations that are difficult or even impossible to set up in real life or in the classroom (e.g. being inside an artificial satellite, riding a merry-go-round) can all be used profitably in thought experiments. There is, of course, empirical evidence that thought experiments, either as simplified versions or as variations of the original thought experiments, can be powerful teaching tools for the teaching of ideas from relativity and quantum mechanics, like the principle of equivalence and that of uncertainty (Velentzas et al., 2007; Velentzas & Halkia, 2011, 2013).

Given the fact that human beings rely on their narrative mode of thinking for the creation of meaning—which means that narratives and stories are much more engaging compared to logical arguments—thought experiments can be introduced through narratives and stories, despite the fact that they (experiments) are based upon logical arguments (see Chap. 4). Although inviting students to design their own thought experiments—a process that can be facilitated by the use of technology—may be a more creative activity, the power of storytelling, especially, if our goal is to help students understand certain ideas through thought experiments or the experiments themselves, is not to be ignored. However, even in this case, the students are helped to develop their imagination.

1.3.4 Teaching and Learning Possibilities

In general, as was pointed out in the introduction, the development of the imagination is a real challenge, since there is evidence that its development 'occurs before, outside of or perhaps in spite of such schooling – apparently through active but largely solitary interaction with physical objects of one's world' (Shepard, 1988,

p. 181). However, some possibilities for imaginative thinking in the context of school science education appear to exist. Based on Egan's (1988, 1887, 1999) work on imaginative engagement, Hadzigeorgiou (1999) proposed the 'start-from-the-unfamiliar-and-remote' strategy, which can be incorporated in a number of teaching-learning models. The 5-E (engage-explore-explain-extend-evaluate) model (Bybee & Landes, 1990), for example, can incorporate this strategy in the 'engage' phase. What needs to be stressed though is that, in order for students to become involved with their object of study, their imagination needs to be captured. Engaging students in science certainly poses a great challenge for science teachers, but the needs to borne in mind is that what teachers consider interesting and engaging situations, in order to introduce students to the object of study per se, are not necessarily interesting and thus engaging for students (see Chap. 2). The following steps represent a constructivist model and show how to implement a strategy for bringing curiosity, mystery, and wonder into the classroom by starting from remote, unfamiliar, and strange situations and topics (Hadzigeorgiou):

- Introduce a challenging topic in a variety of ways: questions, narratives, stories, and video watching.
- Provide time and encourage students to think, first on their own and then in groups, about the challenging topic and to come up with own explanations and predictions. Elicit and record the students' idea about the topic.
- Encourage further discussions through comparing and contrasting the students' ideas.
- Help students to test their ideas in the laboratory or elsewhere.
- Discuss and clarify erroneous (or not) ideas about the topic and related topics.
- Introduce the concept that explains the topic.
- Encourage students to compare their own ideas with the new concept that explains the topic.
- Guide the students to apply the new concept in other familiar and unfamiliar contexts.

This model is based upon the view that students, especially those of the upper elementary and early secondary level, can become aware of the mysterious 'texture' inherent in all science concepts and that such awareness requires imaginative thinking and therefore constant challenge: How about reliving New Year's Eve or Valentine's Day by just crossing the International Date Line? What about an imaginary trip to Everest, a trip to the bottom of the Atlantic to visit the *Titanic*, a trip to outer space to explore the surface of a star, or even a trip back in time to the beginning of the universe itself? It is these topics that are really challenging for young students, and they could be used to introduce all kinds of concepts (see Table 1.1). These are certainly topics that appeal to young students' imagination and provide them with opportunities for ongoing discussions. In fact, they can secure the 'romantic' engagement with scientific ideas, which, as Whitehead (1957) had pointed out, is a necessary step before students begin to study science in detail at a later stage in their education. There are, of course, many strange and mysterious situations and phenomena, taking place in familiar contexts (e.g. two friends standing at the opposite sides of a lake can hear each other in the morning but not in the afternoon, clocks running fast or slow, depending on their place on the planet) that can also be used to engage the students' imagination (see Chap. 6 how such situations can be used in the teaching/learning process).

What must be stressed, however, is that starting from the remote and the unfamiliar is the first step towards moving beyond the purely rational element that has dominated science education over the past years and towards helping students use their imagination, but it is certainly not the only one in the process of understanding. As Stinner and Williams (1993, p. 100) have pointed out, 'Understanding science requires non-mechanical approaches that appeal to the imagination and involve such procedures as analogy, limiting case analysis, thought experiments, and, especially, on-going discussion'. Indeed, the role of discussion—stimulated by questions and demonstrations—is central to the development of students' scientific knowledge and understanding (see Chap. 6 for the roles of language and discussion in the teaching/learning process).

Marzano (2007) also proposed a four-phase model, which is based on capturing and developing the students' imagination. He believes that such model has the potential to engage both the affective and the cognitive domain and therefore be considered an effective teaching/learning model:

Phase 1—imagine: The students are invited to imagine a situation and more specifically to form a mental image of it (e.g. the Earth and the Moon) and 'feel' the effect of their interaction (e.g. how the Moon exerts a force on the water of the ocean that is closest to it) and visualize it (e.g. see in their mind's eye the bulge on the ocean's surface).

Phase 2—explore: The teacher invites students to broaden their initial mental image (e.g. by including the Sun) and also encourages them to explore such an image, by mentally manipulating it (e.g. how the relative position of the three planetary objects has an effect on the tidal cycle).

Phase 3—describe: The students are asked to explain the results of their explorations (e.g. why tides are high when all three planetary objects are perfectly aligned).

Phase 4—confirm: Working collaboratively and seeking out information from a number of sources, the teacher and the students seek to confirm their findings and explanations.

Even though one might argue that pictorial representations (i.e. drawings, photographs) can be more effective in helping students think about tides—see Matthewson (1999), Gilbert et al. (2008), and Statham (2014) for the role of visualization in teaching and learning science—Marzano's (2007) approach is 'imaginative', as it taps the students' imagination, by inviting them first to imagine a situation or a phenomenon and then to explore and explain it.

In Chap. 5, where the notions of creativity and scientific creativity are discussed, one can see a number of possibilities for students to employ specifically their creative imagination. Here, however, is a list of possibilities that can be offered to those

1.3 Imagination and Science Learning

who are interested in capturing and developing the imagination in general (Hadzigeorgiou & Fotinos, 2007):

- *Presenting ideas which conflict with everyday common sense* (e.g. the straight-line motion of a spaceship at thousand miles per second but in the absence of an external force; the equivalence of rest and straight-line motion; the 'emptiness' of solid matter; the increase of mass of an object with the increase of its speed)
- *Presenting ideas through mysteries and paradoxes* (e.g. the mystery of universal attraction; the twin paradox in the special theory of relativity; the transmission of electromagnetic radiation through a freezing and empty space; radiation that penetrates matter and makes it visible)
- *Presenting ideas through the extremes of reality and human experience* (e.g. the lowest and highest temperatures in the lab, on Earth, and in the universe; the smallest and the biggest molecule; the fastest particle; the longest long jump; the strongest wind; the fastest athlete; the tallest tree; the deepest roots; the greatest shadow on Earth)
- *Having students keep daily journals where they record and write about their everyday experiences—their personal stories—which can illustrate a science idea* (e.g. the reverse thrust they experienced while riding the bus; the spectacular colours of a sunset; the immense biodiversity, and specifically of plant life, witnessed during a field trip; the breathtaking twisting somersault of a gymnast)
- *Using questions that challenge students to find connections among apparently unconnected facts and ideas* (e.g. between sound and heat; between an apple, the Moon, and a distant planet; between a thief, the police, and the speed of light; between Newton's laws, a nurse, and a soccer player; between light, electrons, and a surgeon; between a glass of wine, the age of the universe, and the evolution of stars)
- *Investigating topics from everyday life that call for a creative approach to inquiry* (e.g. investigating possible factors that might have an effect on the illumination of a room; the construction of a flashlight from simple materials; ways to produce electricity for the house in a case of emergency; ways to heat water in the absence of metallic containers; calculating the density of a proton, of a black hole)
- *Investigating topics and problems that might confront humankind in the future* (e.g. investigating alternative sources of energy; the possible effects of new technologies on the production of electricity; ways to protect the planet from various kinds of dangers, like asteroids, magnetic storms)
- *Investigating imaginary situations* (e.g. the fate of the Earth after the total disappearance of the Sun; the fate of the Earth if it was near a black hole; life on Earth without atmosphere; life on the ocean floor; life without gravity; what would happen if the Earth's magnetic field ceased to exist; if water did not exist in three states, that is, if no water would evaporate because of the Sun's heat)
- *Inviting students to imagine that they are parts of physical reality and participate in physical and chemical processes, while listening to a narrative* (e.g. they are free electrons in a wire, and they 'feel' their motion and the resistance to this

motion, as they move in the wire; they are water molecules during the transformation of ice into water and then into vapour)
- *Evoking in students a sense of wonder about ordinary and familiar objects and phenomena* (e.g. a glass of water, a piece of ice, a tree leaf, the water cycle, the transformation of motion into heat, the burning of a candle)
- *Presenting the great ideas of science through storytelling, based on real events from the history of science* (e.g. the idea of the nature electricity through the Galvani-Volta conflict; the idea of heat energy through the historical events that led to the abandonment of the idea of caloric; the idea of the atom; the discovery of the X-rays; the magnetic effects of the electric current)
- *Presenting mysterious situations/problems through short stories and narratives* (e.g. a group of friends try to explain what actually takes place in a mystery house, in whose basement they discovered potatoes, wires, and lamps; a little girl tries to explain the disappearance of some liquid)
- *Encouraging students to create their own analogies to understand phenomena and ideas* (e.g. the phenomenon of resonance, that of interference, the ideas of nuclear fission, fusion, and chemical bonding)
- *Using thought experiments to understand phenomena and ideas* (e.g. the idea of inertia; motion on a frictionless surface; the motion of a cannon ball fired above the surface of the Earth with different speeds; the motion inside an accelerating elevator and inside a back hole)
- *Approaching the teaching and learning of science through the arts* (e.g. using photography and making a collage to present the results of a study of a topic such as the effect of modern technology on everyday life; using sculpture and technologies to construct scientific models; using drawing to represent a phenomenon, such as photosynthesis or the water cycle)
- *Using science fiction* (e.g. speculating about possible applications which are based on established principles, such as how one might utilize nuclear waste in order to produce usable forms of energy, but also speculating on innovations which are based on hypothetical principles, like space travel, transporters)
- *Providing opportunities for learning science through experiences in museums* (e.g. practical experiences involving a sense of magic, such as clapping the hands and turning on the lights on a Christmas tree; following the historical development of ideas through various exhibits, such as the development of the ideas of heat and light, and their applications, like thermometers and photography)

Some points, of course, need to be made here. Some strategies appeal more to the imagination than others. For example, speculating on new science ideas, constructing a scientific model, or investigating a problem that humankind might face in the near or distant future appear to be more appropriate than having pupils discuss some of their own experiences in connection with their physics course. However, even in this case, both the narrative element and the attempt on the students' part to try to identify from an everyday experience the application of a scientific idea (e.g. the law of conservation of energy or momentum) can help to develop their creative imagination (Hadzigeorgiou & Fotinos, 2007).

Another point refers to the presentation of the ideas. It would be naïve on anyone's part to believe, for example, that just the presentation of an idea in a way that it conflicts with everyday experience is in and of itself sufficient to develop the imagination. Although conflict with everyday experience or even conflict with accepted beliefs is crucial in capturing the students' imagination, it is the discussion that will follow and the opportunities for students to think that will lead to an understanding. The way questions are posed, the opportunities given to pupils to respond, and the discussion that ensues are all crucial for imaginative thinking. Even an exciting experience at a science museum will remain simply an ordinary experience (see Chap. 2 for the difference between an ordinary experience and an aesthetic experience) if it is not followed by a discussion. The historical development of ideas, as these are presented through the various exhibits, should be followed by a discussion of the possibilities these ideas can open for humankind. For example, the students should discuss the various possibilities of applying the idea of radio waves. If the possibility of sending messages to the other side of the Atlantic or to the other side of the world had captured Marconi's imagination and the imagination of people who had become aware of the power of the idea of radio wave, then the pupils should also share in and extend that discussion.

Perhaps the most challenging idea in regard to capturing and developing the imagination is the speculation about new ideas. How can these ideas be taken seriously? What if these new ideas are indeed 'crazy' ideas? Two arguments can be advanced here. First, the history of the evolution of ideas in physics has provided evidence that ideas, which first appeared, not simply revolutionary, but very irrational indeed in order to be taken seriously, did prove in the end to be 'great ideas'. For example, Planck had advised Einstein not to try to include gravity in his theory of relativity because that was an almost impossible task. He also told Einstein that even if such an attempt was to be successful, no one would pay any attention. Poincare did believe that the transmission of radio waves on the surface of the Earth to a distance more than 300 km was something impossible, and Lord Rutherford initially thought that the energy which could be derived from the splitting of the atom was an absurd idea (Di Trocchio, 1997). And second, the hypothetical principles can be introduced in such a way that they don't contradict accepted principles that have been directly tested. These principles must satisfy what might be called the 'negative impossibility' criterion: nobody should be able to prove that it is impossible at present (Schmidt, 1980).

In talking, of course, about 'crazy ideas', a few words should be said about the role of science fiction in science education (see Allday, 2003; Brake & Thornton, 2003; Taylor, 2003). No doubt, some of science fiction ideas do present problems in regard to the possibility of their application. For example, 'transporters'—which function by scanning the human body and recording all of the information regarding its atoms and molecules and then by converting the mass of the human body into energy, which is stored as energy in a beam that is in turn focused on an arrival point, where the energy is reconverted into mass and thus the human body is recreated—pose a serious problem as regards the initial kinetic energy of the human body (see Allday). Nevertheless, I personally see four reasons—beyond the opportunities for visualization, which helps students develop their imagination—why this particular genre can be important in the context of school science.

First, science fiction can provide excellent opportunities for discussing metaphysical issues (e.g. human consciousness, personal identity, the origins of the universe, the nature of space and time), which, in turn, can help encourage involvement with science (see Hadzigeorgiou, 2005b). Second, science fiction can provide opportunities for discussing scientific principles that could explain science fiction ideas (e.g. space travel, transporters, artificial gravity). Third, it can provide opportunities for the identification of misconceptions behind science fiction ideas (e.g. about gravity, Newton's laws). And fourth, it can provide opportunities for raising awareness of the relationship between science, culture, and society. As Brake and Thornton (2003, p. 34) point out, 'science fiction can be used to help demystify science, highlight its social and cultural context, and act as a bridge to public consciousness'. However, as far as imaginative/creative thinking is concerned, the discussion of new possibilities opened up by science fiction ideas deserves particular attention. Discussing such new possibilities in the context of accepted scientific principles can be really rewarding and is an excellent way to help students develop their imagination (Hadzigeorgiou & Forinos, 2007). Stutler (2011), drawing on Dabrowski's theory of overexcitabilities (which posits intellectual, emotional, and imaginational linkages as the basis for highly creative intelligence), recommends science fiction for the case of gifted students. Because it is emotionally engaging and intellectually intriguing, science fiction is an excellent tool to tap students' creative imagination and also help initiate self-directed research and even scientific inquiry.

Lastly, it should be pointed out that imaginary situations, apart from their potential to develop the students' imagination and, of course, to initiate student inquiry, they help with the integration of a number of science content ideas and even help with the integration of various disciplines, especially those of science, history, art, and language. For example, an imaginary situation like 'life on Earth without an atmosphere' can help students investigate a number of concepts or ideas, such as vacuum, gravity, atmospheric pressure, sound, and such phenomena as motion of an object in the absence of atmosphere, hearing, candle burning, and lightning and their interconnections. Thus students become aware that all phenomena are linked in one way or another, which, in turn, can help evoke a sense of wonder (see Chap. 6 for an in-depth discussion of the notion of wonder). Moreover, this imaginary situation can help upper elementary and high school students become aware of the connections between history and science, if famous experiments from the history of science are presented through storytelling (e.g. Magdeburg's experiment, Torricelli's experiment) and also between art and science (e.g. when students represent science ideas and phenomena through art, like those of sound and motion in the absence of atmosphere, the effect of vacuum on human life) and between science and language arts (e.g. when students write an essay describing how their daily life would be affected if there was no atmosphere, when they write about the applications of the concept of vacuum in everyday life and create a PowerPoint presentation for their class).

1.4 Concluding Comments: The Need for Engaging the Imagination in Science Education

It has been pointed out that despite the blows that empiricist and logical positivist philosophies suffered over the last three decades (Duschl, 1994, Duschl Schweingruber, & Shouse, 2007), it may be still difficult for both students and science teachers to completely abandon such philosophies. It is very common for them to be engaged in laboratory work involving the investigation of the relationships among various variables and, sometimes, the confirmation—through an experiment—of an idea (e.g. a law, a principle) (see Monk & Dillon, 2000). However, if a true constructivist philosophy were to be considered as the foundation for science education, and especially physics education, then students should be given opportunities to propose hypotheses and to test them, to be involved in modelling, in problem-solving, and in finding diverse connections among ideas and generally opportunities for divergent thinking. From such a perspective, imagination, evidently, becomes a crucially important factor to be considered.

There is empirical evidence to support the view that the people who have opportunities to operate in imagined worlds become more creative (Lehrer, 2012; Mellou, 1995). And although the evidence that imaginative skills in science education are transferable to other areas is not convincing, there is good reason to believe that 'imagination offers the promise of making scientific creativity more concrete and helping to identify a potential starting point for further research' (Kind & Kind, 2007, p. 25). The work done by the *Imaginative Education Research Group* (IERG), directed by Kieran Egan at Simon Fraser University, is promising, at least as far as the role of imagination in learning is concerned. More research is certainly needed, especially in regard to creativity (Kind & Kind), but, nonetheless, such research needs to be based upon a theoretical framework, which gives primacy to imagination.

The chapters that follow focus on imaginative approaches to science education, which (approaches) have begun to receive attention by some science educators. The empirical evidence so far, although limited one might say, is quite encouraging as far as their (approaches) contribution to emotional involvement and learning outcomes is concerned. The following chapter discusses specifically the problem of engaging students in science and in a way complements the ideas discussed in the section on imagination and science learning in this chapter. Moreover, it helps one understand that it is not only important to consider the effect of students' experience on their learning—that is what most intervention studies explore—but also the effect of learning on the quality of students' experience. The notion of 'aesthetic experience' discussed in the next chapter helps one understand what it really means for a student to have a truly educative experience.

Chapter 2
Engagement and Aesthetic Experience in Science Education

> *Experience is the result, the sign, and the reward of that interaction of organism and environment, which, if it is carried to the full, is a transformation of interaction into participation and communication [...] Experience in this vital sense is defined by those situations and episodes that we spontaneously refer to as 'real experience'; those things of which we say in recalling them 'that was an experience'.*
>
> John Dewey, in Art as Experience, p. 22, p. 37

> *Contact with science by non-scientists and deeper involvement in science by scientists can only be truly beneficial and fulfilling to the individual and society when science itself is given deeper foundation in self.*
>
> Klaus Witz, in Science with Values and Values for Science Education, Journal of Curriculum Studies, vol. 28, p. 597

> *The only acceptable point of view appears to be one that recognizes both sides of reality [...] the physical and the psychic [...] it would be most satisfactory if physics and psyche [...] could be seen as complementary aspects of the same reality.*
>
> Wolfgang Pauli, in dialogue with Carl Jung, in Mindell's Quantum Mind: The Edge Between Physics and Psychology, p. 19

> *I was sitting by the ocean one late afternoon, watching the waves rolling in and feeling the rhythm of my breathing, when I suddenly became aware of my whole environment as being engaged in a gigantic cosmic dance. Being a physicist I knew that sand, rocks, water, and air around me were made of vibrating molecules and atoms, and that these consisted of particles, which interacted with one another by creating and destroying other particles. I knew also that the earth's atmosphere was continually bombarded by showers of 'cosmic rays', particles of high energy undergoing multiple collisions as they penetrated the air. All this was familiar to me from my research in high-energy particles, but until that moment I had only experienced it through graphs, diagrams, and mathematical theories. As I sat on that beach my former experience came to life; I 'saw' cascades of energy coming from outer space, I 'saw' the atoms of the elements and those of my body participating in this cosmic dance of energy, I felt its rhythm and 'heard' the sound.* (Capra, 1977, p. xii)

What Capra (1977) describes here is a 'wonder-full' and, in fact, an awe-filled experience. Some might even talk of a 'mystical' experience. Apparently, this particular experience is quite unique and, as such, rather unrealistic to be expected of students—let alone young children—in traditional classroom settings. However, the above passage also describes an experience that points to two crucially important factors, with implications for school science education, namely, (a) the perception of physical reality by, and its relationship with, the self and thus to the role of phenomenology in school science education and (b) the process of engaging students in science and the students' transformation of outlook, as a result of their experience. This chapter will deal with the latter.

Indeed, it has been recognized that engagement with science is central to science education, albeit a complex and challenging problem to be tackled by science teachers and curriculum designers (Hadzigeorgiou & Stivaktakis, 2008; Klassen & Froese-Klassen, 2014b; see also Duschl et al., 2007). The complexity of the process of engagement can also be seen in the fact that research findings on students' engagement cannot be 'neatly' categorized (i.e. social/emotional development, cognitive development, physical behavioural development). However, the main features, and, in fact, the benefits, of a truly engaging experience can be summarized as follows (see Fleer & March, 2009):

- Student learning extends beyond the classroom.
- Students transfer learning to different times and contexts.
- Students demonstrate long-term satisfaction with the learning process.
- Students reflect on their learning (e.g. in informal discussions, in journal entries).
- Students demonstrate long-term satisfaction with the learning process.

Of course, it would be a truism to state that, unless children and older students alike become involved with the learning process, 'significant' learning cannot take place. For this is quite self-evident. What is not self-evident though is that there may be involvement with the learning process, and yet learning may not take place. How can this be? The answer is that students may become involved with the teaching/learning process but not with their object of study per se (i.e. science content knowledge).

Cognitive conflict is a strategy that inevitably results in involvement with the object of study. And this strategy, because it can facilitate conceptual change, has been dominant in science education. However, over the last decade, there has been an increasing interest in alternative ways of understanding school science that have been called 'aesthetic' (Girod, 2007a; Girod, Rau, & Schepige, 2003; Wickmann, 2006), 'transformative' (Pugh, 2004, 2011; Pugh & Girod, 2007), and 'romantic' (Hadzigeorgiou, 2005b; Hadzigeorgiou, Klassen, & Froese-Klassen, 2012). While the first two kinds are based on Dewey's (1934) notion of 'aesthetic experience', the third derives directly from Egan's (1997) recapitulation theory and more specifically from a particular kind of understanding that is called 'romantic' (see Chap. 3). Despite the differences among these alternative ways of understanding, they share one common element: the facilitation of deeper involvement with science.

Are these approaches cures or even treatments for students' 'anorexia learnosa'? The evidence from those really few studies (see last section in this chapter) that report on the effect of aesthetic, transformative, and romantic understanding on science learning is certainly positive and quite encouraging (e.g. Girod et al., 2003; Girod, Twyman, & Wojcikiewicz, 2010; Hadzigeorgiou, 2012; Hadzigeorgiou, Klassen, & Froese-Klassen, 2012; Pugh, 2002, 2004; Pugh, Linnenbrink-Garcia, Koskey, Stewart, & Manzey, 2010a; Wickmann, 2006). However, given the complex nature of the learning process, the question how to get students involved with science content knowledge still remains quite pressing and challenging. The idea of 'aesthetic experience' deserves particular attention when it comes to encouraging involvement with science content ideas, which, of course, can be seen as a prerequisite for an aesthetic/transformative experience (Hadzigeorgiou, 2012; Hadzigeorgiou & Garganourakis, 2010).

This chapter will attempt to shed some light on the problem of student involvement with science and will then discuss the notion of aesthetic experience and (the need for) aesthetic approaches in science education. However, a discussion of the role of cognitive disequilibrium and cognitive conflict in the process of engagement, as well as the limitations inherent in the notion of cognitive conflict, are imperative.

2.1 The Problem of Engaging Students in Science

The problem of motivation in connection with learning is well known (Brophy, 1987, 1999; Bruner, 1966; Franken, 2001; Raffini, 1993). However, what has not been recognized or adequately addressed in the past is students' motivation with regard to the content of learning (Pugh, 2004). More specifically, in the context of school science education, what is at issue is not just the problem of students' motivation in general or the problem of considering the affective component of learning science but their motivation in connection with the object of study itself, for there is a distinction to be made between participation in a learning activity and involvement with the object of study itself (Hadzigeorgiou, 1999; Hadzigeorgiou & Stivaktakis, 2008). Pugh also distinguishes between peripheral things (e.g. humour, interaction with peers, flashy demonstrations) and engagement with science content.

It goes without saying that during participation in an activity, the object of study can be disconnected from the emotions of the student, which can arise mainly from participating in the activity. For example, a student may be interested in investigating the socio-scientific issue of genetically modified food or the topic of magnets in a cooperative setting and enjoy them, but his/her emotions may arise from the social context of those activities and not from the objects of study themselves. In contrast, when there is involvement with the object of study, emotions are not disconnected from it (object of study).

Although emotions are not the same as the motivation to learn, in the sense that they (emotions) do not necessarily have a goal orientation associated with them, emotions are nevertheless crucial for initiating involvement with the object of study itself and therefore discourage what Dewey (1934, 1966) had called the 'spectator theory of knowledge'. Certainly, contemporary learning models are not based on such a theory, but on constructivist theories, since they encourage students' active participation in learning activities. Yet 'most theories of constructivism remain within the dualistic framework which Dewey opposed' (Dahlin, 2001, p.456). This dualistic framework, in fact, is encouraged by science teachers, in the case in which their main interest is how to 'sugar-coat' school science, as Pugh (2004, p. 194) put it. It is quite apparent that 'sugar-coating' can motivate students, but it is debatable whether it can lead students to become involved with school science, as their object of study (Hadzigeorgiou, 2005a; Pugh & Girod, 2007).

The problem, therefore, for science educators and science teachers is how to move away from that dualistic framework by encouraging students' involvement with their object of study. What should become abundantly clear, however, is that the process of involvement is a complex one.[1] Although interest may be identified as a crucial factor that can encourage involvement (see Klassen & Froese-Klassen, 2014b), there are such factors as personal identity, purpose, and awareness of the significance of the object of study that can determine, to a large extent, student involvement (Hadzigeorgiou, 2005b). Here I will discuss three factors/problems, which have been documented empirically, and the questions raised in regard to the first one, that is, personal identity.

The first factor concerns the role of personal identity, which is central to the process of involvement. According to the literature, the construction of personal identity takes place through social relationships (Apple, 1999; Gallas, 1997), through lived experiences of practice (Wenger, 1998), and through narratives (Sfard & Prusak, 2005; see also Brown, 2004). Contrary to the prevalent modernist notion of identity as something predefined and fixed, identity is considered 'in terms of the possibilities, in terms of freedom, and the power to choose' (Greene, 2000, p. 295; see also Greene, 1988, 1991). The implications of this idea for school science education are important: the science classroom becomes a place where students and teachers negotiate ways of being, knowing, and acting; and science learning is considered not just a matter of acquiring or constructing knowledge but also a matter of deciding what kind of persons students are and what they aspire to be (Brickhouse, 2001; Cleaves, 2005; Hadzigeorgiou & Garganourakis, 2010; Kozoll, & Osborne, 2004; Reveles, Cordova, & Kelly, 2004). These implications, of course, raise important questions in regard to science teaching and learning, as I will discuss below. But first let me discuss the other two factors related to the process of student involvement with science, namely, that students' worldviews and the object of study itself.

In regard to the students' worldview (which is related to the idea of personal identity), one has to recognize the possibility of a clash between their worldviews and school science. This can happen if a student's metaphysical frame is in conflict

---

[1] Researchers in psychology use the term 'engagement' to refer to the intensity of one's involvement with an idea, a task, and an activity. The complexity of the process of engagement can be seen in the fact that it has behavioural e.g. persistence), emotional (e.g. interest), and cognitive (e.g. learning outcomes) components. It has been found that 'engagement is associated with positive academic outcomes' (Fredricks, Blumenfeld, & Paris, 2004, p. 87).

either with the teacher's metaphysical frame or with that of school science. For example, a student may experience a sense of beauty, which though derives from his/her own metaphysical frame and not from that of school science (i.e. when the latter is presented in a way that its aesthetic dimension is absent). Another student, who enjoys reading mysteries or feels inspired by a sense of wonder in reading science fiction and other relevant magazines, may very well find that a teacher's approach to teaching science is rooted in mechanistic, reductionist, and 'scientistic' view of science. In considering the issue of the incompatibility between the teacher's personalized worldview and that of a student (see Cobern, 1991, 1996, 2000), one can see how involvement with science, as students' object of study, becomes problematic. This issue of incompatibility, of course, becomes extremely crucial in addressing the problem of involvement with science in the case of students from other (non-Western) cultures (Hadzigeorgiou, 2005b).

As far as the third factor is concerned (which may also relate to personal identity and worldview), the object of study determines to a large extent whether students will engage in science. For example, some students may become involved with the study of evolution and the greenhouse effect but not with the study of chemical bonding and the laws of thermodynamics. Or, some students may become involved with physics and earth science but not with chemistry (Hadzigeorgiou, 2005b; see also Cobern, 1996).

However, there are two other factors that may also account for the problematic nature of student engagement in science. The first, which may or may not relate to personal identity, is the fact that involvement itself is something personal and is therefore determined by how meaningful the object of study is (i.e. how closer the connection between the object of study and self is). A female student, for example, who is helped to connect the law of universal attraction to her own menstrual cycle, may very well become more involved with the science lesson, in comparison with a male student who is presumably interested in sport and is helped to understand Newton's laws through the motion of a soccer ball. As for the second factor, prior experience may also play a role in the process of involvement. A student, for example, who has already burned her mouth by biting at a slice of pizza that did not appear at first to be too hot or who has experienced the shaking of the earth and had the feeling of impending death, as a result of an earthquake, will more likely find more meaningful the concepts of heat capacity and simple harmonic motion, respectively, than students who did not have such prior experiences. Although imaginative engagement with ideas can and does take place through unfamiliar, remote, and strange situations and events, as was discussed in the previous chapter, the possibility of the creation of existential meaning, through the connection of science content knowledge with one's self and personal life, and therefore with something that is already familiar, is also something to consider (Hadzigeorgiou, 1997, 2005b).

Of all factors, however, personal identity seems to be the most critical factor contributing to, or rather determining, involvement and learning outcomes. The issue of personal identity, however, begs two questions: (a) Are identities only the product of significant narratives? (b) Are students to be viewed only as products of social interaction? For Sfard and Prusak (2005) 'identities are products of discursive diffusion' (p. 18). So, in this sense, narratives become the most important determinant of students' involvement with the object of study and hence a determinant of their success in school.

On the other hand, the fundamental premise in Gallas' (1997) work is that identity 'is a continuing piece of work, constructed in relation to the other, in conversation with the other, and, in the best of all possible worlds, in communion with the other' (p. 140). Cummins (2003) supports the same view: student-teacher interactions communicate to students' messages concerning their identity—who they are in their teacher's eyes and who they are capable of becoming. According to this view, empowering students becomes central to the learning process. In fact, this is the perspective espoused by Apple (1999): the construction of identity should be considered in terms of social relationships that lead to empowerment. In Wenger's (1998) work, the central idea is that identities are constructed 'in lived experience of engagement in practice' (p. 151). For him the role of language, although important, cannot account for the construction of identities. Given that a key element of his work is the notion of full participation in a community of practice (in order for people to create meaning), the construction of identities is viewed in relation to these communities.

What is not acknowledged explicitly by Wenger (1998) and also by Gallas (1997), Cummins (2003), and Apple (1999) is the role of the actual object of study in the process of identity construction. The awareness of the significance of the object of study (see Chap. 6), which can also help students to change their outlook on the world, may also be considered an important factor in the process of identity building (Hadzigeorgiou, 2012; Wong, Pugh, & Dewey Ideas Group at Michigan State University, 2001).

It deserves to be noted that the emotional significance of scientific ideas can be linked to the process of identity building. Why? If an awareness of the significance of an idea can result in a change of students' outlook on the world, then a change in the way students perceive themselves is also a possibility. This change of perception can facilitate the envisioning of possible identities, an idea that is also noted by Wong et al. (2001). In linking the process of identity building to the subject matter of science, they argue as follows:

> *Powerful subject matter ideas create anticipations about a way of being in the world, and that way of the world includes not only an image of the world, but also an image of the student in that world.... As many biographies and auto-biographies reveal, an important part of scientists' identity is that ideas hold a special sway on them. They are motivated to continue to be the kind of people who experience the beauty and drama of science ideas.* (Wong et al., 2001, p. 335)

Although there is a difference between scientists and students (e.g. difference in motives, conceptual frameworks), change of perception can take place even in the case of students, both at the elementary and high school levels, provided that teachers use an 'aesthetic pedagogy' (Pugh, 2011; Pugh & Girod, 2007; see last section in this chapter), or they try to evoke a sense of wonder (Hadzigeorgiou, 2012; see Chap. 6 for such a change of perception).

In considering though the notion of identity building, another question needs to be raised: is science learning a possibility only for those who envision, for some reason, a career in science and engineering? According to Head (1997), those who make the choice to study science are those who resolve to be scientists from an early age. This 'resolution', of course, is supported by the view that personal identity and understanding are intricately linked (Gallagher, 1992, pp. 156–168; Godon, 2004): what one can understand is directly related to 'the totality of one's self-understanding and it is in that context that its significance gets decided' (Godon, p. 590). This is an important message, since it points to the close relationship between identity, understanding, and learning. However, Marcia's (1988) work on identity statuses (see also Kroger, 1996)—which are defined in terms of the dimensions of exploration and commitment—makes one consider not only the statuses of 'identity achieved' (commitment to an identity has been made) and 'identity foreclosed' (adoption of parental values) but also those of 'identity diffused' (no commitment to any life options has been made) and 'moratorium' (failure to see the importance of choosing one option over another). Is involvement with science an 'impossibility' for those students who have a diffused identity or for those who are in a moratorium? This question is imperative and points to the exploration of ways to encourage involvement with science, even when a commitment to becoming a scientist has not been made.

Two approaches, as research has shown, can be effective: the cognitive conflict approach, which can help evoke a sense of wonder (see Chap. 6), and the so-called aesthetic approach, which can provide students with opportunities for aesthetic experiences. Both approaches can encourage involvement with science, particularly with science content knowledge. The fact that there is some empirical evidence to support the view that an aesthetic experience, through students' imaginative engagement with science content knowledge, can lead to identity options, and more specifically to the possibility of considering a career in science and technology (Hadzigeorgiou & Garganourakis, 2010), should certainly be considered. But what is really an 'aesthetic experience'? Before this notion is discussed, the role of cognitive disequilibrium and cognitive conflict in the process of engagement, as was said, deserves particular attention.

2.2 Engagement Through Cognitive Disequilibrium

This idea of cognitive disequilibration is central to the Piagetian epistemology. In Piaget's (1977) most recent work, it is disequilibrium that produces the driving force of thinking and hence cognitive development. This disequilibrium occurs when there are contradictions between one's expectations about an event or about the behaviour of an object and the actual results of one's actions. For example, a child might believe, and therefore expect, that an orange will fall faster than a peanut, if they are both left to fall simultaneously from the same height to the ground. A simple demonstration provides a first-hand contradiction of this, since both the orange and the peanut reach the ground simultaneously. It is this disturbance of a prior knowledge system and 'the dissatisfaction with existing conceptions', as Posner, Strike, Hewson, and Gertzog (1982, p. 214) had pointed out, that will lead to the construction of new knowledge. According to Glasersfeld (1989, p. 128):

> *The learning theory that emerges from Piaget's work can be summarized by saying that cognitive change and learning take place when a scheme, instead of producing the expected result, leads to perturbation, and perturbation, in turn, leads to accommodation that establishes a new equilibrium.*

Over the last three decades, this idea of disequilibrium has proved very fruitful in the area of science education, especially in physics education, since it was found to be a good strategy to challenge students' misconceptions (Duit & Treagust, 2003; Treagust & Duit, 2008; Hadzigeorgiou, 1999; Hadzigeorgiou & Schulz, 2014). Situations causing confusion through perturbations of existing knowledge enable students to become aware of obstacles to be overcome in order for these situations to be understood. Discrepant events appear to be a good starting point in this strategy. These events can be introduced through (Hadzigeorgiou, 1999):

- Concrete (hands-on) activities (e.g. pouring water into a glass tumbler and seeing a coin at the bottom of the tumbler disappears, dropping objects of different weight from the same height to the floor and observing that they hit the floor simultaneously)
- Computer simulation (e.g. observing that an object can move in a straight line with constant speed even when no net force is acting on it)
- Questions that make students aware of explicit verbal discrepancies (e.g. if warm air rises, how come it is cold on a mountain top? If stars fall, why is the sky not empty after some time?)

Popper's (1972) argument that 'the vital first step towards understanding a theory is to understand the problem situation in which it arises' (p. 182) is in line with the idea of 'obstacle overcoming' and provides food for thought for science teachers and science educators. According to Popper, it is not just the historical context that is crucial for understanding a science concept or idea, but the problem, the obstacle, that the scientists of the past faced and which they had to overcome in order to explain certain phenomena.

In the context of science education, engagement with science can also be identified with students' awareness of obstacles or problems. Indeed, according to Dewey (1998), as was discussed in the previous chapter, consciousness of an obstacle is the source of reflective thought, and it is this consciousness that makes one feel perplexed, confused, or in doubt. Thus, in light of what Dewey said, a student, for example, trying to calculate the value of the specific heat capacity of a substance, or the magnitude of a force required to accelerate an airplane, experiences neither confusion nor doubt. In other words, the calculation of the specific heat capacity or the magnitude of the accelerating force cannot be sources of reflective thinking if there is not an awareness on the student's part of an obstacle to be overcome, unless, of course, such an obstacle refers to the formula or the procedure to be followed, in order for the calculations to be successfully made. On the other hand, a student trying to understand how a piece of pizza that did not feel so hot did burn his/her mouth, or how a soldier, according to a newspaper article, was electrocuted by touching a metal pole, may very well be perplexed and confused, since she or he cannot understand how these situations work. Thus cognitive disequilibrium is central to the process of involvement with science.

But although cognitive disequilibrium, as the initiator and the driving force of thinking, is of paramount importance, it is not the only 'key factor' involved in the teaching and learning of science. It is true, as was pointed out above, that instructional models developed over the last three decades placed a great emphasis on students' prior ideas, their elicitation, and restructuring (Driver & Oldham, 1986; Driver, Asoko, Leach, Mortimer, & Scott, 1994; see also Schulz, 2009). However, two major problems appear to emerge from this emphasis on the notions of disequilibration/cognitive conflict and restructuring/conceptual change: one practical and one theoretical. The practical problem is about whether science learning can be based exclusively on perplexity and confusion. If everything that is to be learned were presented in a confusion-and-perplexity fashion, then many topics could very well be left out.

Using discrepant events, as a starting point of instruction, seems to be a good way to facilitate student involvement. Indeed the challenge of fundamental beliefs (e.g. heavier bodies fall faster than lighter ones, cold flows from cold to hot bodies, there is a net force in the direction of a body's straight line motion) produces confusion and perplexity, but what about topics like light interference patterns, crystal formation, the uncertainty principle, black holes, electric current and resistors, and bonding and chemical reactions? Certainly the practical problem seems to be resolved once we introduce all topics through discrepant events. But, until we develop curricula based solely on disequilibration, we cannot accept the view that only cognitive conflict leading to conceptual change will lead to science learning (Hadzigeorgiou, 1999; see also Limon, 2001, Pugh & Girod, 2007).

In regard to the theoretical problem, this refers to the fact that the conceptual change approach per se has been considered too narrow. Why? This approach has been based mainly on three assumptions: First, 'learning is a rational activity', and therefore we should 'focus attention on what learning is, not what learning depends on' (Posner et al., 1982, p. 212). Second, 'ontogenetic change in an individual's learning is analogous to the nature of change in scientific paradigms' (Pintrich, Marx & Boyle, 1993, p. 169). And third, 'all concepts regardless of their origin and source are evaluated by the standards of science' (Cobern, 1996, p. 582).

These assumptions, of course, have been criticized (see Hadzigeorgiou, 2005b; Hadzigeorgiou & Konsolas, 2001; Limon, 2001; Schulz, 2009, 2014). In fact, Strike and Posner (1992), the two of the four science educators who proposed the conceptual change model (which was based on the idea that the primary goal of school science education was to help students change their conceptual ecology, by replacing a set of concepts with another set, incompatible with the first) in acknowledging the role of intuition, emotions, motives, and goals in the process of conceptual change, point out that 'the idea of a conceptual ecology [...] needs to be larger than the epistemological factors suggested by the history and philosophy of science'

(p. 162). Apparently, motivational and generally affective factors, which were downplayed by constructivist/conceptual change models (see Treagust & Duit, 2008), need to be seriously considered by science teachers and science educators, if they are to encourage involvement and facilitate learning.

And yet, over the past two decades, science education has been dominated by such notions as conceptual change and science inquiry, in line with a constructivist perspective on teaching and learning science. Pugh and Girod (2007) call these notions 'dominant paradigms' that have shaped the current landscape of science education and propose art and aesthetics as potential paradigms to join the ranks of the dominant ones. The idea of aesthetics in science education may seem at odds with the other dominant ideas of conceptual change and inquiry, but it is not. In fact it is complementary, but, at the same time, central to both inquiry and conceptual change (see Chaps. 6 and 7).

> *Science education is still largely focused around standards, conceptual change, and inquiry. These are some of the dominant paradigms that shape current science education pedagogy. What if art and aesthetics were to join the ranks of these dominant paradigms?* (Pugh & Girod, 2007, p. 10)

What must be recognized, with regard to the process of cognitive conflict per se, is that it does make students intellectually curious about what is really going on in the situation that produces conflict. But what also should be recognized is that intellectual engagement may very well be fostered through an 'aesthetic awareness', that is, through an awareness of the aesthetic dimension of science (i.e. awareness of the mysterious and also of the beauty of science). Bachelard's idea, that 'all knowledge is the answer to a question' (cited in Gil-Perez & Carrascossa-Alis, 1994, p. 307), is certainly a valid one. But what kind of questions are we talking about? Wonderment questions, as empirical evidence suggests, do foster thinking (Chin, 2007; Chin, Brown, & Bruce, 2002; Hadzigeorgiou, 2012), but their source is not only discrepant events that produce cognitive disequilibrium, thus making students aware that their knowledge is incomplete or erroneous. Their source can be also found in mysteries and paradoxes and in the experience of wonder (see Chap. 6). In other words, Bachelard's point is well taken, provided the question to be answered does not have its source only in cognitive disequilibration but also in the experience of mystery and wonder, for it should be noted that wonderment questions can be raised through one's aesthetic appreciation of nature and its phenomena. Such an appreciation has been crucial in scientific progress (see section on imagination and the nature of science in Chap. 1). In the case, for example, of some students who wondered about how Newton thought about the law of action and reaction and who expressed an admiration for Newton's genius, one can clearly see that a wonderment question can have an aesthetic dimension too (see Chap. 6).

Indeed, it could be argued that everything around human beings can be taken to be a mystery and a source of wonder that capture their imagination and makes them think (Hadzigeorgiou, 1999). Although growth in scientific knowledge is due in large part to the identification and subsequent solution of a problem, as Popper (1972) argued, science, as a form of inquiry, was developed not because all problems produced disequilibration and confusion (i.e. cognitive conflict) but rather because human beings have an inborn bewildered curiosity and a sense of wonder, which is evoked by a sense of mystery (Hadzigeorgiou & Schulz, 2014). Thus, the desire for explanations and understanding cannot be seen as the sole result of cognitive conflict. And it is for this reason that the possibility of providing students with opportunities for aesthetic experiences is quite crucial.

2.3 The Notion of Aesthetic Experience

The notion of experience is quite common, not only in everyday language but also in the language of education. We talk, for example, about experiential learning, which is distinguished from a learning experience in general, as well as about students' experiential repertoire and educative experiences. And we all know that although students have many experiences, in and out school, these are not necessarily educative. More often than not, students' experiences cannot even be called 'learning experiences'. Given that an 'experience' is the unique outcome of a student's interaction with his/her environment—therefore the notion of experience should not be confused with that of activity—participation in a planned activity does not necessarily result in learning. The reason is that a student's experience depends upon what Dewey (1938) called the 'internal' and the 'external' conditions. The former refers to a student's interest, motivation, and prior knowledge, while the latter refers to the total social and physical set-up of the classroom and also to what the teacher does and says. It is evident that an interaction of these two conditions can have diverse outcomes for the students, despite the fact that they have participated in the same activity. However, even if an experience can result in some kind of learning, there is still a question concerning the quality of that experience.

It is for this very reason that Dewey (1934) proposed the term 'aesthetic experience'. According to him, there is a difference between an ordinary experience and an aesthetic experience. The latter is a holistic and fulfilled experience, in the sense that feelings, thoughts, and actions form a unified whole.[2] The role of the imagination in such a holistic experience is also acknowledged by Dewey: 'Esthetic experience is imaginative […] all conscious experience has of necessity some degree of imaginative quality (p. 272).

What must be pointed out is that while the senses play an important role in an aesthetic experience (e.g. in the case of an artistic creation, in the case of observing a natural phenomenon), their role is not to be restricted solely to certain sense qualities, such as colour, odour, or texture, as is the case with traditional empiricism (i.e. Hume's empiricism). For Dewey (1934), it is the meaning attached to these sense qualities and their integration into a unified whole that determines whether or not an

---

[2] According to Dewey (1934), the origins of an aesthetic experience are both artistic/creative experiences (e.g. those of artists) and everyday, commonplace experience. Therefore the students' experience in school or outside of school can also be aesthetic experiences (see also Alexander, 1987, for an analysis of Dewey's idea of aesthetic experience and a better understanding of his aesthetic philosophy).

experience will be an aesthetic experience. In other words, unless one creates meaning from what one abstracts from current experience, on the basis of one's values and past experiences, one cannot have an aesthetic experience.

Scholarly work on aesthetic experience points to such descriptions of it as 'vital experience' (Aguirre 2004, p. 259), 'transformative experience' (Pugh, 2011, p. 107), 'compelling and dramatic' experience (Girod et al., 2003, p. 578), and 'powerful' experience (Wong, 2002). However, a distinctive feature of an aesthetic experience is that it is 'consummatory', in the sense that there is a completion and not a cessation of the experience. It is this consummation, at the end of the aesthetic experience, which brings a kind of satisfaction and fulfilment. The following excerpt from Dewey's *Art as Experience* is illustrative of what an aesthetic experience is:

> *Oftentimes, however, the experience had is inchoate. Things are experienced but not in such a way that they are composed into an experience. There is distraction and dispersion; what we observe and what we think, what we desire and what we get, are at odds with each other [...] In contrast with such experience, we have an experience when the material experienced runs its course to fulfilment. Then and only then is it integrated within and demarcated in the general stream of experience from experiences. A piece of work is finished in away that is satisfactory; a problem receives its solution; a game is played through; a situation, whether that of eating a meal, playing a game of chess, carrying on a conversation. Writing a book, or taking part in a political campaign, is so rounded out that its close is a consummation and not a cessation. Such an experience is a whole and carries with it its own individualizing quality and self-sufficiency. It is an experience.* (Dewey, 1934, pp. 36–37)

The intense feeling of 'consummation' is related to that of 'anticipation'. As Dewey (1934) argued, consummation 'does not wait in consciousness for the whole undertaking to be finished. It is anticipated throughout and is recurrently savoured with special intensity' (p. 55). Thus, an aesthetic experience is dramatic. And like drama, 'it involves a build-up and resolution of anticipation that gives the experience its completeness and uniqueness' (Pugh & Girod, 2007, p.11). In education, apparently, it is desirable for students to have experiences that lead to consummation, not cessation. Although an aesthetic experience cannot be a pre-planned experience, some activities are more likely to provide opportunities for anticipation and consummation in the context of science education. Storytelling, for example, has the potential to create anticipation through the plot of a story[3] (see Chap. 4). Certain

---

[3] An excellent way to develop in students a sense of anticipation is through storytelling (see Chap. 4). We all know that we anticipate a good story, and in the case of school science education, an introduction is quite crucial. The following introduction played a catalytic role in creating anticipation for the Tesla story (Hadzigeorgiou, Klassen, & Froese-Klassen, 2012): *The following Saturday we are going to listen to a story about something that we take for granted, but which literally transformed our lives. In actual fact ever since, our world has not been the same. This 'something' is the electric current, which makes almost all of our home appliances work, which is used in industry and, generally, which changed completely the world about a century ago. No one would dare to imagine what our life and the world at large would be like without the use of electric current. Moreover, many of us are unaware of the fact that useable electric current was an idea of a man; a product of his legendary, extraordinary, even heroic qualities; and the victor of a long controversy and battle. This battle, known as the 'War of the Currents', was finally won by that man, in 1895, when, for the first time, he was successful in transmitting electrical power from Niagara Falls to Buffalo City, about 34 km away. That feat was*

2.3 The Notion of Aesthetic Experience

immersion activities—especially those connecting science and art—can also provide students with opportunities to experience anticipation (see Chaps. 5, 6, and 7).

In the literature one can find a few approaches to the notion of 'aesthetic experience' (e.g. Girod et al., 2003; Jakobson & Wickman, 2008; Wickmann, 2006). Girod et al. have identified certain behaviours associated with the having of an aesthetic experience as follows:

- Trying to learn more about what has been taught in the classroom
- Thinking about it outside of the classroom
- Talking to other people about what has been taught
- Trying to see examples of the taught ideas in everyday life
- Seeing the world differently

For Girod et al. (2003), the students who demonstrated those behaviours, after a teaching intervention, developed an 'aesthetic understanding'. The idea of seeing the world differently is also captured in what Pugh (2002, 2004, 2011) calls 'transformative experience',[4] whose qualities though are the same as those of 'aesthetic understanding'. Indeed, the conception of transformative experience 'as a learning episode in which a student acts on the subject matter by using it in everyday experience to more fully perceive some aspect of the world and finds meaning in doing so' (Pugh, 2011, p. 111) is based upon Dewey's (1934) theory of experience and more specifically upon the idea of engagement with content knowledge, which results in an expansion of one's perception. And this expansion of perception, as a result of learning, is the first step towards seeing the world differently.

Even though a number of philosophers and educators have defended significant learning as a transformation of one's outlook (see Hadzigeorgiou & Schulz, 2014), and therefore such a learning outcome should not be linked exclusively to Dewey's aesthetic theory (which is the basis for the aesthetic teaching/learning approaches to school science education), seeing the world differently can be facilitated by aesthetic approaches. A review of the studies on aesthetic approaches to science education (see later in this chapter) does show that the notions of 'aesthetic understanding', 'aesthetic experience', and 'transformative experience' have identical or similar characteristics. The main difference between aesthetic approaches can be found in

---

something unthinkable, even unimaginable, at that time and that's why it was covered by the world press and that man was praised as a hero worldwide. That man was Nikola Tesla, who, according to Life magazine's special issue (September 1997), is considered among the 100 most famous people of the millennium. Although current electricity was not a new idea at the time of Tesla, the transmission of current electricity over such a distance, and at that time, was a heroic achievement, which is directly associated with Tesla's extraordinary mind and especially his passion for, and even obsession with, a particular kind of current that is called alternating current (AC) (see Appendix A for the Tesla story itself).

[4] Pugh (2011) explains why we need a concept like 'transformative experience': 'We do have concepts like 'meaningful' or 'authentic' learning, but such constructs are vague and do not specifically target the *consequences of school learning on everyday experience*. In addition, we have constructs that represent discrete components of this phenomenon, including the constructs of transfer, conceptual change, task value, and individual interest. I propose transformative experience as a composite at the individual level that integrates these discrete components and represents the phenomenon of school learning transforming everyday experience' (p.107).

the way one goes about to find out whether an experience is aesthetic or transformative.

Wickmann (2006), for example, focused on, and studied, the way students express themselves during science activities (rather than on learning outcomes after an intervention). In pointing out that 'learning science from the primary school to the tertiary level is in necessary and inseparable ways dependent on aesthetic experience' (p.145), he found that an aesthetic experience can play four different roles in the process of learning science;

- It helps students learn how to act and work in the science class.
- It helps students to connect their everyday/informal experiences with the classroom science experiences.
- It helps students use aesthetic language effectively in order to convey experiences over time by the use of aesthetic words.
- It is integral to the facts and logic/reason of science.

An important notion for Wickmann (2006) is that of 'aesthetic judgements', which he defines as:

> *[…] utterances or expressions that either deal with feelings, or emotions related to experiences of pleasure or displeasure, or deal with the qualities of things, events, or actions that cannot be defined as qualities of the objects themselves, but rather are evaluations of taste – for example, about what is beautiful or ugly.* (Wickmann, 2006, p. 9)

By focusing on the aesthetic judgements and the language used, as students react and communicate their feelings and emotions about the experiences of phenomena, Wickmann (2006) studied an aesthetic experience from both a positive and negative perspective. He used such 'binary opposites' (see Egan, 1997, 2005) like beautiful/ugly and pleasure/displeasure in order to analyse students' feelings, since the latter influence children's approach to learning science. It is apparent that Wickmann acknowledges the fact that engagement in science has two dimensions, an emotional and a cognitive, and that the two are inextricably linked (Hadzigeorgiou & Stivaktakis, 2008).

Jakobson and Wickman (2008), in line with Wickmann's (2006) approach, studied elementary school children's aesthetic experiences, that is, their experiences related to what they found beautiful/ugly and also pleasing/displeasing when learning science. In studying what young children liked and disliked, they found that children's judgements are not trivial utterances but of great significance in the situation in which they are uttered. These judgements are important for learning not only science ideas but also the normative aspects of doing science and also important for engagement with science. Indeed, given that the idea of beauty and what counts as beautiful are rooted in one's worldview, aesthetic judgements and more specifically judgements regarding personal tastes (e.g. beauty/ugliness, liking/disliking, nice/disgusting) which are involved in many science activities (e.g. an arrangement of some things during an experiment, dissecting an earthworm) determine the extent to which one becomes involved with science (Hadzigeorgiou, 2005b). In short, aesthetic judgements can be seen as a tool to study interest in

2.3 The Notion of Aesthetic Experience

science, which (interest in science), according to Klassen and Froese-Klassen (2014b), deserves more attention by mainstream science education. As Jakobson and Wickman pointed out 'we should look closer at the possibility that interest in school science does not come as a lightning flash, but is slowly learnt in the numerous aesthetic encounters of science class'. (pp. 62–62)

Milne (2010), on the other hand, attempted to classify aesthetic experiences on the basis of their potential to generate a sense of awe and wonder and therefore interest in science. His list includes the following types of experiences:

- *Spiritual*: Experiences from both a religious and secular perspective
    Indicators: (a) Statements or feelings expressed when looking at stars and appreciating nature. (b) Direct reference to God or a creator
- *Utilitarian*: Experiences motivated through need or problem solving
    Indicator: Expressions or feelings communicated when faced with and overcoming challenges associated with problem solving
- *Fashion/marketing*: Experiences in the context of fashion/marketing
    Indicator: Expressions or feelings expressed when affected by such context
- *Value/respect*: Experiences associated with an appreciation of the power of nature or the power of position
    Indicators: Expressions or feelings expressed when confronted with (a) the awesome power of nature (e.g. as in tsunamis, earthquakes) and (b) the awesome influence of people in a certain position (e.g. as in the case of the Pope or a great athlete)
- *Beauty*: Experiences resulting from an appreciation of natural form and structure of nature
    Indicator: Expression or feelings expressed as one responds to interactions or close encounters with natural entities (e.g. stars, flowers, rocks)
- *Mathematical*: Experiences resulting from an appreciation of the natural patterns of nature both in form as for beauty and abstraction for number
    Indicators: Expression or feelings expressed when experiencing the beauty of form and patterns and time associated with nature/exploring and appreciating the structure and rules associated with working with very small or very large numbers
- *Personal enjoyment or pleasure*: Personal experiences that result in an interest over time
    Indicator: Attachment and/or reaction to both pleasurable and non-pleasurable experiences
- *Curiosity*: Personal experiences that result in pure curiosity
    Indicator: Expression of curiosity in order to understand the world

From the aforementioned approaches, it becomes apparent that an aesthetic experience requires more than what the constructivist approach to teaching and learning science prescribes. This is extremely crucial to stress, given the argument that 'the aesthetic dimension of knowledge is more likely to be appreciated, if knowledge is seen as something that is constructed and not as something that we merely discover or find' (Eisner (1985, p. 8). But the appreciation of the aesthetic dimension of scientific knowledge—for example, when one perceives beauty in scientific knowledge (see Chaps. 5 and 7)—is more realistic in the context of professional science rather than school science education. In light of this difference between scientists and school students, one should be careful not to conflate the aesthetic dimension of scientific knowledge with the having of an aesthetic experience, even though the former can sometimes lead to an aesthetic experience in the context of school science. In any case, what should be clear at this point is that an aesthetic experience is not an ordinary educational experience, and a constructivist approach to teaching and learning science does not guarantee the having of an aesthetic experience. In short, it is not easy for a student to have an aesthetic experience, no matter how 'constructivist' such an experience may have been.

For Dahlin (2001), who recommends attention to 'the aesthetic dimension of knowledge formation' (p. 130), aesthetic is 'a point of view which cultivates a careful and exact attention to all the qualities inherent in sense experience' or 'an approach to natural phenomena' which (approach) 'would not merely be to appreciate their beauty, but also understand them' (p.130). Capra's (1977) experience, as described at the beginning of this chapter, is based on 'exact attention to all the qualities inherent in sense experience', and therefore it can be called an aesthetic experience. Thus, an aesthetic experience results from a holistic experience, and this requires a phenomenological, not simply constructivist, approach to teaching and learning science (see Hadzigeorgiou & Schulz, 2014; Østergaard, Dahlin, & Hugo, 2008). The reason is that a constructivist approach does not necessarily require holistic experiences (in which emotions, imagination, reason, and action are united, and there is involvement with the object of study per se). Indeed, an important distinction between a constructivist and a phenomenological approach to learning should be pointed out, whereas the former is based upon the purely cognitive aspect of the construction of knowledge, 'phenomenology has a stronger emphasis on the precognitive phase, including the roles of sensing and feeling as different from purely conceptual cognition' (Østergaard et al., 2008, p. 98).

I believe that, in the context of science education, it is preferable to say that an aesthetic experience presupposes a phenomenological/aesthetic approach to science, that is, a holistic approach, which, on the one hand, places primacy upon 'lived experience' and not simply upon sense experience (e.g. listening, observing) and, on the other, considers the aesthetic dimension of scientific knowledge (Hadzigeorgiou & Schulz, 2014). In other words, such an approach includes sense experience, emotions, somatic reason, and also the appreciation of the beauty of science (especially the role that aesthetic perception plays in the construction of knowledge). Thus, it provides students with opportunities for 'careful and exact attention *to all the qualities* inherent in sense experience' (Dahlin, 2001, p. 454, my emphasis).

By the same token, an aesthetic experience in science is not necessarily a 'wow experience'. The 'wow factor' (Feasey, 2005) in school science is certainly important, but it does not necessarily lead to an aesthetic experience, unless the characteristics of an aesthetic experience are present. Put simply, in an aesthetic experience, it is not only the emotional but also the cognitive dimension of involvement that needs to be present (see Chap. 6).

Perhaps the most important educational outcome associated with an aesthetic experience is the change of one's outlook on the world. This means that one sees the world differently as a result of an aesthetic experience. Such change or transformation as a result of learning goes beyond an 'artistic expansion of perception' (i.e. seeing details and nuances) even though the latter can often lead to the former. When Feynman said that the world looks beautiful after one learns science, he made reference to one's ability to see the world differently.

It is interesting to note that the need for an experience, which makes students see the world differently and also apply science in their everyday life and beyond their classroom and school, had been pointed out by both Barnes (1987), who stressed the importance of 'action knowledge', and R. S. Peters (1966), who argued for the development of 'cognitive perspective'.[5] Barnes differentiated between 'action knowledge' and 'school knowledge'—the former representing knowledge that is applied in one's everyday life and makes one see objects and phenomena differently—while Peters thought that knowledge and understanding should not be inert, but part of one's everyday life, and should allow one to see the place of knowledge 'in a coherent pattern of life' (Peters, 1966, p. 45). Thus, while there is no mention of aesthetics, or aesthetic experience, in Barnes' and Peters' arguments about school learning and education, their views do express a transformative pedagogy (i.e. learning that can be applied to everyday contexts and learning that produces a change in perception).

The following section provides empirical evidence for aesthetic experiences in real classroom settings. Although such evidence is still limited, it is nonetheless encouraging, as it shows that, for some students, the having of an aesthetic experience is indeed a possibility.

2.4 What the Research Shows

The discussion thus far begs one question: how realistic is the having of an aesthetic experience in classroom settings? Notwithstanding the methodological difficulties inherent in the process of conceptualizing and evaluating an aesthetic experience, due to the conceptual complexity of the latter (see Pugh, 2011), the evidence so far, albeit limited, is quite encouraging. Intervention studies with both elementary and

---

[5] It is unfortunate that the term 'cognitive perspective', which Peters (1966) used as one of the criteria by which to judge the educated person, might lead one to take it as representing a limited view of knowledge. But if cognitive perspective is about the ability of the learner to see the place of knowledge 'in a coherent pattern of life', then cognitive perspective is closely related to emotions, aesthetics, and ethical conduct. Therefore it is a holistic notion. Scheffler (1996), in fact, has pointed out that 'the notion of cognitive perspective is related to the idea of wholeness' (p. 84).

junior high school students in the USA demonstrate that teaching for aesthetic/ transformative experiences has a positive effect on science learning.

Girod and Wong (2002), in a case study of three grade 4 students learning about rocks, found that one of the students did have an aesthetic experience, given that she created a 'rock book', which, in fact, was not part of any class assignment. She reported during an interview that she thought about rocks differently than she did before the lesson and that whenever she had nothing to do, she looked at a rock and then try to tell its story (i.e. what its name is, where it came from, where it formed). During this interview she also said that she could imagine herself being a geologist, which is evidence that her learning experience made her consider an identity option.

Pugh (2004) found similar results in a case study of two grade 7 students, who were learning about Newton's laws of motion. One of the students became involved with the lesson and then began to associate everyday phenomena of motion with Newton's laws of motion. An example illustrating the fact that Newton's laws became part of his life is when he was observing his young niece running across the recently mopped floor and not being able to stop till the door acted on her and stopped her. This example and others that this student cited show the influence that the learning experience had on him. His learning experience was transformative in the sense that he began to look at everyday phenomena involving force and motion differently (i.e. through a scientific lens) and also he began to look at himself differently (i.e. he described himself as a science person).

Experimental studies by Girod et al. (2003, 2010), and Pugh (2002) also produced evidence that teaching for aesthetic understanding can have a significant impact on students. In regard to the study by Girod et al. (2003), the sample was two 4th grade classes (ages 9 and 10 and 28 students in each class). One of the classes was taught for conceptual understanding (control group), while the other class for aesthetic understanding (treatment group), through discourse emphasizing engagement with science content knowledge. This science content was taught over a period of 10 weeks and concerned geological topics such as fossils, minerals, rocks, weathering and erosion, as well as rocks, and volcanology. The assessment of aesthetic understanding was made possible through a survey that included a story (vignette) about a student who learned about friction and found it to be powerful and important, and a questionnaire. First students in both classes were read this vignette and then asked to respond to questions, which investigated the degree to which they have had experiences similar to the experience of the student described in the story. Each question related to some element of aesthetic understanding, such as the perceived transformation of person and world, learning that brings unification or coherence to science ideas and phenomena, and the dramatic nature of learning in this way. Girod et al. (2003) found a statistically significant difference between the groups, with students in the experimental group reporting greater levels of engagement. The results show that 'for the most part, students in the experimental classroom experienced clear, and in some cases, profound progress toward the three conditions of aesthetic understanding' (p. 583).

As regards the intervention study by Girod et al. (2010), two 5th grade classes were used over a whole school year (9 months)—the experimental group was taught

2.4 What the Research Shows

for aesthetic understanding and the control group was taught in a traditional manner, emphasizing the cognitive aspects of the learning process. The science content that was covered involved three units on weather, erosion, and matter. The researchers looked at levels of interest in science, efficacy beliefs (about themselves as learners) and identity beliefs (about themselves as 'science-type' people), levels of conceptual understanding, retention of conceptual understanding 1 month after instruction, and also the possibility of acting differently in the world as a result of new learning. The differences between the groups were small, but nonetheless statistically significant (even though the differences between the two classes, control class and treatment class, judged from the scores on pretests for each of the factors (interest, efficacy beliefs, and identity beliefs) were very small). Girod et al. found that the experimental group displayed greater learning at both the follow-up and the posttest assessment, but with a greater effect occurring at the follow-up assessment. The main differences between the students of the two groups in regard to conceptual understanding—both immediately after instruction and in the delayed (after 1 month) posttest condition—were that students who were taught for aesthetic/transformative experience appeared to view the world differently and continued to investigate the world using ideas learned in class. On the other hand, the students from the control class were more focused on science terminology and sought to learn and apply new knowledge not because they found it interesting but because 'it made them feel smart'.

Similar positive results have also been reported by Pugh (2002), who conducted an intervention study with two high school biology classes. Both classes were taught a unit on animal adaptation and natural selection. One class (the experimental group) was taught for aesthetic/transformative experiences. In the other class (the control group), the inquiry model was used. The results showed a statistically significant difference between the two groups, with a greater percentage of students in the experimental group displaying engagement in transformative experiences. For example, one student said that learning about animals stimulated an interest to learn more about them, while another student reported that he applied the ideas of adaptation and evolution many times in his everyday life and whenever he saw an animal he wondered if it related to him in some way. And he was aware that the concept of adaptation made him try to understand more about it. Pugh reports that the students from the experimental group displayed greater learning immediately after intervention but not at the posttest compared to students from the control group.

Pugh et al. (2010a), on the other hand, in their study with high school students, found that most students in their sample did not show any evidence of deep engagement with science, which is a prerequisite for an aesthetic/transformative experience. These students exhibited, as Pugh et al. say, low levels of transformative experience. However—and this is important to stress—'those students who strongly identified with science and who endorsed a mastery goal orientation were more likely to report engagement in higher levels of transformative experience' (p. 1). Pugh et al. also found that high levels of engagement in transformative experience were associated (a) with conceptual change regarding natural selection but not inheritance at both the follow-up and the delayed posttests and (b) transfer at the follow-up assessment. Pugh et al. proposed that students with a mastery orientation would be more likely to undergo transformative experiences than those with a performance orientation.

Even though the aforementioned studies did not assess conceptual change, it is very likely that those students who had an aesthetic/transformative experience changed their conceptual framework too. However, it is also very likely that students (from the experimental or the control group) who experienced conceptual change did not have an aesthetic experience. Pugh, Linnenbrink-Garcia, Koskey, Stewart, and Manzey (2010b) compared the effects of implementing the TTES (Teaching for Transformative Experience in Science) model in combination with a conceptual change model with the effects of implementing the conceptual change model alone. The teacher created three treatment groups. One was the control group (2 classes, 40 students), while the other two groups (2 classes, 42 students and 2 classes, 44 students) would receive two different instructional treatments aiming at conceptual change and transformative experience, respectively. The results obtained through a comparative analysis of the three instructional conditions show that the combined TTES and conceptual change models fostered student engagement and learning, compared with both the control instructional condition and the conceptual change condition. The TTES model was found to foster transformative engagement and support transfer of knowledge and conceptual change (when combined with the conceptual change model). Pugh et al. (2010b) report that students who were taught for transformative experience displayed greater transformative engagement in class but did not report a higher level of transformative experiences. In the case of regular-level students, the combination of TTES and conceptual change models fostered greater change in basic knowledge, while the TTES model helped them display more enduring transfer. In the case of honours-level students, the combination of TTES and conceptual change models helped with inducing greater conceptual change, but not greater transfer.

In the context of Swedish schools, Jakobson and Wickman (2008) selected a variety of schools (e.g. a big city school, a school in a rural area, and three medium-sized town schools) in order to study elementary school children's (ages 6–10) aesthetic judgements during hands-on inquiry activities in physics, chemistry, biology, and earth science. They recorded children's talk in the context of nine units (e.g. changes/the transformation of matter, solids and liquids, electric circuits, mixing and separating, buds, shadows, and soil). The results showed that children experienced not only anticipation but also consummation (i.e. feelings experienced when a science activity does not simply end but instead is brought to fulfilment) and that through aesthetic judgements, the children talked about their own place in their science class and whether they belonged there.

However, what should be pointed out here is that imaginative engagement with science content knowledge can also lead to an aesthetic/transformative experience. According to empirical evidence from three studies in Greece with junior and senior high school students (Hadzigeorgiou, 2012; Hadzigeorgiou et al., 2012; Hadzigeorgiou & Garganourakis, 2010), the development of romantic understanding (see Chap. 3) and the experience of a sense of wonder (see Chap. 6) fostered an aesthetic/transformative experience. Such a claim can be based on several findings, such as (a) students spending time outside the school to search for information regarding ideas and phenomena that they found mysterious, amazing, and astonishing; (b) students spending time to make journal entries during their free time; (c) their perception regarding science ideas, natural entities, and phenomena as well as their view of science as a school subject that changed as a result of learning science; and (d) the possibility to consider a career in a science-related field (e.g. electrical engineering). Even though these findings are based upon data that came from observation and students' journal entries, they are nonetheless indicators of an aesthetic/transformative experience. Even if such a claim is considered too bold, there is evidence for deeper involvement with science, which, in turn, helped some students change their perception of the natural world and of science as a school subject (see Chap. 6).

The studies that were cited in this section provide evidence that an aesthetic experience is not simply an abstract concept or an ideal situation, perhaps necessary as a standard for judging the quality of students' everyday educational experiences. This is, of course, important, but the message from the aforementioned studies is that aesthetic experience is indeed a possibility, at least for some students, and that the having of an aesthetic experience determines interest and involvement with science. What must be stressed, therefore, is that even though such an experience represents the ideal educational experience, and its evaluation is not quite straightforward,[6] neither the possibility on students' part to actually have such an experience nor the effort on teachers' part to foster it should be downplayed or dismissed.

2.5 Implications for Science Education

The notion of aesthetic experience has important implications on school science education. The first, and perhaps the most important one, I believe, is students' emotional involvement with science, particularly its ideas. The reason is that such an emotional involvement can help bring science outside the school classroom (Hadzigeorgiou, 2005a; Hadzigeorgiou & Stivaktakis, 2008; Pugh, 2004). As Pugh and Girod (2007) pointed out, 'As science educators, we often obsess over misconceptions but fail to ask whether students ever apply their 'correct' conceptions outside of school and use them to have aesthetic experiences in the world' (p. 10).

---

[6] If a student reports that she or he had a transformative experience (i.e. by saying that the lesson made her or him find a science topic interesting or intriguing and that she or he experienced change in the way she or he perceives reality and science or by saying yes, when she or he is asked, to all the questions regarding the criteria whereby one judges a transformative experience), it cannot by itself guarantee transformative experience. Multiple data are certainly needed, and inferences must be carefully drawn.

But in order for students to become emotionally involved with science and apply their conceptions outside of school, teacher awareness of the centrality of ideas in the process of meaning making is imperative. It has been pointed out that

> One of the major failures of our present educational efforts is that so many schools leavers have little sense of what ideas are or how to use them and control them [...] The development and manipulation of ideas are central to rationality, to romantic understanding, and to the Western intellectual tradition. (Egan, 1990, p. 225)

Focusing on ideas as potential sources of wonder (see Chap. 6), as tools that help bring out the drama inherent in their conception and historical development, especially if presented through storytelling (see Chap. 4; see also Appendix A) or through dramatization (see Chap. 7), should be seen as the number one priority in curriculum planning. Ideas, such as electric current, force and motion, photosynthesis, light reflection, chemical valence, can be seen as curriculum items to be learned or as ideas that can have an impact upon one's life. Even though 'learning meaningfully' such ideas can be—as it has been—based upon their various interrelationships, as in a concept map (Novak & Gowin, 1984), one becomes aware of their emotional significance (and hence existential meaning) if one becomes aware of their relationship with human life, especially one's personal life (Hadzigeorgiou, 1997, 2005b; Hadzigeorgiou & Schulz, 2014).

Of course, the most important and practical, at the same time, implication of the notion of aesthetic experience regards the question about how a teacher can foster such an experience. What steps she or he can follow? What she or he can focus on in order to increase the possibilities for an aesthetic experience? Pugh and Girod's (2007) aesthetic pedagogy, based upon Dewey's notion of aesthetic experience, is worth considering in regard to what a teacher can do in her/his classroom. They recommend the following six guidelines:

- Crafting ideas out of concepts
- Restoring concepts to the experience in which they had their origin and significance
- Fostering anticipation and a vital, personal experiencing
- Using metaphors and 're-seeing' to expand perception
- Modelling a passion for the content
- Enculturating students into ways of valuing and experiencing science idea

These guidelines can indeed help one 'bring out' the emotional significance of science ideas and thus encourage deeper engagement with them. The task of crafting ideas, for example, which refers to the process whereby the curriculum content is transformed into a list of beautiful and powerful ideas, becomes 'something that is relished' (Girod et al., 2003, p. 579) and not just a list of concepts to be learned (e.g. the idea that we can use the same laws—contrary to what Aristotle and Mediaeval scholars believed—to study the motion of a water drop, a soccer ball, a dancer, an asteroid, and a distant planet). And because science ideas have their origins in human experience and are tied both to those who created and developed them and to the specific events that facilitated or even hindered their creation and development, restoring concepts to the original experience can help bring out the drama that has been involved in the development of scientific knowledge.

Take, for example, the idea of a heliocentric solar system. Long ago this was a frightening, provocative, even terrifying idea – one that forced students to think about the world and their place in it, very differently. Today, however, the notion of a heliocentrism is taken for granted as something always known or understood. Heliocentrism has lost its artistic power to shape our understanding in profound ways. The first step in teaching for aesthetic understanding is to recapture or reanimate existing content into the artful and compelling ideas they are (or were at one time). (Girod et al., 2003, p. 579)

Anticipation also is quite important as students are prepared to feel the suspense, the excitement, even the drama inherent in science ideas. A project or an artistic/creative activity, in which students will participate, a story that will be told in the classroom, and even a brief introduction that imaginatively outlines the unit or the lesson can all help foster anticipation. As for the ability to see differently something as a result of learning, teachers can cultivate an artistic kind of perception—Girod et al. (2003) talk about 're-seeing' as 'an attempt to focus our perception on the nuance and detail of the world' (Girod et al., pp. 579–580). They can also evoke a sense of wonder at and about the ideas of the lesson, by using metaphoric/poetic language, by pointing out to students' unexpected relationships among phenomena and among ideas, and by revealing the mystery associated with science ideas and phenomena (see Chap. 6).

As regards the modelling of passion, it is imperative that teachers first feel the fun, the excitement, and the wonder of science before they attempt to make their students experience such feelings. They must show their students how the ideas of science have an impact on their own (teachers') life. Thus passion for science and its ideas can motivate students to experience what their teacher has experienced.

It is apparent from the aforementioned guidelines that an aesthetic approach should not be conflated with a constructivist approach, as the former is more phenomenological.[7] On the other hand, constructivism's main focus has been on the construction of knowledge per se, not on the quality of students' experiences, which may also change the way they view learning and the world, and themselves.[8]

---

[7] It should be noted that even though Dewey's aesthetic philosophy (Dewey, 1934) is not explicitly phenomenological, its potential to make a contribution to the literature concerning the phenomenological approach to learning science needs to be acknowledged (Hadzigeorgiou & Schulz, 2014). One could talk, for example, about a 'phenomenological/aesthetic' approach to science as a holistic approach, which, on the one hand, places primacy upon 'lived experience' and not simply upon sense experience (e.g. listening, observing) and, on the other, considers the qualities of an aesthetic experience. In other words, such an approach includes sense experience, emotions, somatic reason, and also the appreciation of the beauty of science and its ideas (especially the role that aesthetic perception plays in the construction of knowledge). Thus, it provides students with opportunities for 'careful and exact attention *to all the qualities* inherent in sense experience' (Dahlin, 2001, p. 454, my emphasis).

[8] Both phenomenological and constructivist approaches are based on the notion of subjectivity, even though in the case of social constructivism, interactions between the individual and the social environment are central to developing knowledge and understanding. However, while in both constructivist and phenomenological learning, subjectivity plays a central role, in phenomenological

Wong's (2002) view that constructivist approaches with their emphasis on logico-mathematical thinking, self-regulated learning, and rational reflection may very well be an obstacle to the having of transformative experience, as such notions presuppose a dualism between students and their object of study. Wong's (2002) notion of 'opposite of control' provides food for thought, as an emphasis on conscious control in regard to learning (e.g. emphasis on metacognition) cannot foster an aesthetic/transformative experience: '[…] an excess of conscious control and self-awareness is more likely to obstruct rather than facilitate the having of transformative experiences' (Wong, 2002, p. 204).

Another important implication is to help students develop the ability to see what they normally do not see, that is, the ability to observe details and nuances, which, in turn, can help them view everyday objects and phenomena in new light. Hadzigeorgiou and Schulz (2014) recommend that science education reclaim the value of sense experience, which, according to the constructivist perspective, was considered the source of students' misconceptions and therefore an obstacle to learning. But if a change in students' outlook is considered an important instructional goal, then a 'new kind of empiricism', which gives primacy to sense experience and which aims at enabling students to closely observe nature and natural phenomena, is imperative. This kind of empiricism is akin to Goethe's 'delicate empiricism', which refers to empathetic and prolonged observation and which is necessary for gaining first-hand knowledge of the thing in itself, be it a natural entity, such as a flower, or a natural phenomenon, such as a flash of lightning (Hadzigeorgiou & Schulz, 2014). Of course, the development of such observation skills necessitates a restructuring and retooling of science teacher education programmes.

Approaches that have the potential to foster aesthetic experiences deserve particular attention and can be called 'aesthetic approaches'. For practical purposes, notwithstanding the obstacles and/or difficulties that need to be overcome in the context of compulsory education, there are some indicators that can help a teacher, if not to tell whether or not an experience is aesthetic and to judge the possibilities that his/her students will have an aesthetic experience through those approaches. These are:

- Curiosity and wonder about how the world works

---

learning there is always intentionality—thought and consciousness are always involved and cannot be separated from the object of study. This last characteristic of phenomenological learning is not necessarily characteristic of constructivist learning; the purpose of learning, for example, is either ignored or downplayed by constructivist learning (Hadzigeorgiou, 2005b, see also van Manen, M, *Researching Lived Experience*, Albany, NY: SUNY Press, 1990). The most important questions that are associated with the constructivist perspective are (a) *ontological* (what do we know?), (b) *epistemological* (how do we know?), (c) *communicative* (how can we communicate what we know?), and (d) *technical* (what can we do with our knowledge?) (see Osborne, 1996). Without a question of purpose, however, intentionality is discouraged, as is the case with many constructivist approaches (Hadzigeorgiou, 2005b). Moreover, a question regarding how our knowledge changes the way we see science and the world, in general, is not asked by those who approach science teaching and learning from a constructivist perspective (Hadzigeorgiou & Schulz, 2014).

- Interest in science in general or in a particular area or topic
- Enthusiasm and excitement about science ideas and phenomena
- Becoming absorbed in a science activity, in and outside the classroom
- Pursuing further reading and carrying out investigations in out-of-school settings
- Change of perspective (i.e. seeing the world differently) after learning science
- Inspiration

The idea of inspiration is an important one to consider, not simply because it has both a cognitive and an emotional dimension, but also because inspiration can lead students to some kind of action. This action may refer to further reading and further thinking, an experiment that students might do at home, or even the development of a special interest in science or a particular topic. In short, it is inspiration, rather than conceptual understanding, that has the potential for taking science learning outside the school classroom. And this is why the notion of aesthetic experience—which increases the possibilities for inspiration—should be seriously considered by mainstream science education. It would be too narrow a way to define teaching effectiveness in terms of goals and objectives to be attained in the classroom, even if these goals and objectives referred to conceptual change (Hadzigeorgiou, 2002b, 2005a). Moreover, the notion of inspiration can help us reconsider what it really means to be a good teacher.[9]

2.6 Concluding Comments: The Need for Aesthetic Approaches to Teaching/Learning Science

The recent increasing attention to the notion of aesthetic/transformative experience in the context of school science teaching and learning can be justified on the grounds that such an experience shows 'the intimate connections between learning science and interest in science' (Wickmann, 2006, p. 145). In the context of some proposals that put an emphasis on utility and citizen science and also on sociopolitical conceptions of science (see Roth & Desaultes, 2002; Roth & Lee, 2004), the idea that science learning can be an aesthetic experience is worth considering. Richard Dawkins (1998, p. 10) has been quite explicit on this:

> *Far from science not being useful, my worry is that it is so useful as to overshadow and distract from its inspirational and cultural value. Usually even its sternest critics concede the usefulness of science, while completely missing the wonder.*

Elsewhere I have argued that there is a difference between students' involvement with the events of instruction and their engagement with the scientific ideas

---

[9] William Arthur Ward wrote: 'the mediocre teacher tells. The good teacher explains. The superior teacher demonstrates. The great teacher inspires' (cited in The Royal Bank Letter, 1989, p. 2). How many of those in the teaching profession would have liked to instil a zest for learning in their students? And how many would have liked to 'inject life into inert symbols' (Egan, 1990, p. 252)?

(Hadzigeorgiou, 1999, 2005b). Although the former could be considered a prerequisite for science learning, it is the latter we should aim at. It is for this reason that special attention should be paid to the notion of 'inspiration' as was previously discussed.

While it is true that an aesthetic experience, like any experience, cannot be predicted, in the sense that any experience is the unique outcome of one's interaction with the environment, and more specifically the interaction between one's internal conditions (i.e. prior beliefs, psychological state, interest, motivation) and the external conditions (i.e. physical and social set-up), as Dewey (1938) pointed out, curriculum developers and science teachers can increase the possibilities for students to have such an experience. The chapters that follow provide food for thought to all those interested in how to encourage and foster aesthetic experiences. Based on empirical evidence, as the respective chapters discuss, activities like storytelling, which encourage narrative thinking and 'romantic understanding', and also activities that refer to 'creative science', 'artistic science', and 'wonder-full science' have indeed the potential to encourage aesthetic experiences, through anticipation, inspiration, and deeper involvement with the object of study. This deeper involvement is crucially important since science can be 'truly beneficial and fulfilling to the individual and society when science itself is given deeper foundation in self' (Witz, 1996, p. 599).

The question how realistic it is for students to have an aesthetic experience in the context of compulsory education may very well translate to how realistic it is to have teachers who can implement an aesthetic pedagogy, like the one proposed by Girod et al. (2003) or by Pugh and Girod (2007) or the romantic approach to school science proposed by Hadzigeorgiou and Schulz (2014). In other words, we need teachers with passion—how many studies have indeed been conducted in order to investigate the role of teachers' passion for knowledge and understanding in their students' learning process? And, of course, we need teachers with a good grasp of the science content of their area. However, with an appropriate retooling of science teacher education programmes, future teachers can acquire the skills to implement and evaluate an aesthetic pedagogy.

Chapter 3
Teaching for Romantic Understanding

> *Poets say science takes away from the beauty of the stars – mere globs of gas atoms. Nothing is 'mere'. I too can see the stars on a desert night, and feel them. But do I see less or more? The vastness of the heavens stretches my imagination – stuck on this carousel my little eye can catch 1 million year old light. A vast pattern – of which I am a part. [...] It does not do harm to the universe to know a little about it. For far more marvelous is the truth than any artists of the past imagined it. Why do the poets of the present not speak of it? What men are poets who can speak of Jupiter if he were a man, but if he is an immense spinning sphere of methane and ammonia must be silent?*
>
> Richard Feynman, in The Relation of Physics to Other Sciences, The Feynman Lectures on Physics, Vol. 1, Lecture 3, p. 3

> *We can enlarge [the significance of a styrofoam cup] by considering it as a part of the heroic journey that is the human struggle to shape the world more closely to our desires, to find release from the constant toil, sickness, and pain that have been the lot of most people most of the time [...] The knowledge of chemical and physical processes that have gone into its design and making is prodigious. And we have learned the environmental costs entailed in applying this knowledge to create this convenience, and we are as a society recognizing that we must satisfy this particular desire in other ways that do not threaten our harmony with the natural world. One can flash such thoughts through the mind in less than half a second [...] They are associations that come with romantic image of the broken cup.*
>
> Kieran Egan, in Imagination in Teaching and Learning, pp. 76–77.

The notion of a 'romantic understanding' of science, in the context of mainstream science education, would have seemed naïve, if not absurd, even 20 years ago, since the focus was on conceptual understanding (Schulz, 2009). Even nowadays, a romantic understanding of science may well raise a number of arguments on metaphysical and epistemological grounds. While it is not unusual for someone to describe a sunset romantically, or to experience it in a poet's work, it is uncommon to read about a physicist's romantic understanding of a sunset. While both the poet's and the physicist's descriptions have aesthetic dimensions, the latter's understanding of it differs markedly, not only in terms of the language used but also in terms of the actual process used for understanding (Hirst, 1972). The physicist's approach will be to look for connections among facts, phenomena, concepts, and ideas, to search for patterns, and to apply known laws that may explain the observations made.

Generally, a concept or an idea, in order to be understood, needs to be placed in relationship with other concepts or ideas (e.g. as in a concept map) and applied in a variety of contexts (AAAS, 1990; Klassen, 2006a; Mintzes, Wandersee, & Novak, 1997; Resnick, 1983a, 1983b). Additionally, Popper (1972) points out that in order to understand a concept, its historical origin and evolution have to be understood. Such an approach, however, does not negate the fact that science itself has an aesthetic dimension, allowing for a physicist's aesthetic experience and understanding of a phenomenon or an idea, as well (see Girod, 2007a, 2007b; Hadzigeorgiou, 2005a, 2005b; Root-Bernstein, 2002; Tauber, 1996; Wong, Pugh, & Dewey Ideas Group at Michigan State University, 2001). It is, therefore, reasonable to consider the idea of a romantic understanding of science and its implications for science curriculum and teaching.

Whitehead (1957), in talking about the rhythm of education, identified three stages—romance, precision, and generalization—and maintained that any subject of study first needs to be approached in a romantic way before it is studied in some depth, with precision, and students are required to apply their ideas and make generalizations. If the idea of romance refers to the initial stage of engagement with science, then the term 'romantic engagement' might be preferable to 'romantic understanding'. In this sense, romantic engagement could be considered the impetus for learning science, whereas understanding would result from exploring the subject matter in greater depth, inevitably leading to stages of precision and generalization.

Whitehead's (1957) three-stage model acknowledges the role of motivation in understanding, and other research findings support its role in learning (Franken, 2001; Ormrod, 1999; Pintrich & Schunk, 1996), although it is not particularized to the students' understandings of the world at specific ages. Kieran Egan's (1997) recapitulation theory, however, addresses this latter issue. According to Egan, people make sense of the world in five distinctive ways, which he calls forms or kinds of understanding. He sees education conceived as a process during which students recapitulate, that is, repeat in the same order, kinds of understandings—somatic, mythic, romantic, philosophic, and ironic—in the way that these have appeared in our cultural history. Each kind is best developed at particular ages when students learn to use an array of cognitive tools that are characteristic of each kind of understanding.

These kinds of understanding, not to be confused with the Piagetian stages of cognitive development occurring at certain ages under some psychological impulse, are produced when students acquire certain sets of cultural tools that, through their use, are converted into cognitive tools. With the exception of somatic understanding (knowing through the physical senses), all the other kinds of understanding are language dependent; they are our 'language engagement' with the world, and the connection between 'cultural development in the past and educational development in the present' (Egan, 1997, p. 27) are those kinds of understanding and the cognitive tool characteristic of each.

It is the purpose of this chapter to examine the concept of 'romantic understanding' from the perspective of Egan's (1997) theory, by outlining its features, its potential role in science education, and also the issues, which are raised in regard to its evaluation. A teaching framework, based on these features, is outlined in Chap. 4, which focuses exclusively on storytelling, as a teaching/learning tool and its specific roles in science education. One such role is the development of romantic understanding. But first the notion of romantic science and its main features are worth exploring, given the similarities between these features and some of the characteristics of romantic understanding.

3.1 The Notion of Romantic Science

Romanticism, as a movement that emerged in Germany and spread to Europe in the late eighteenth and early nineteenth centuries (i.e. roughly between 1770 and 1830), was based on a new conception of nature, which, in sharp contrast to the mechanistic/Newtonian conception that prevailed in England and Continental Europe at the time, was 'organic' (Berlin, 2001; Bortoft, 1996, Heringman, 2003; Kloncher, 2013; Richards, 2002). Due to the changes that the romantic movement brought to literature and philosophy, it would not be an exaggeration to say that the romantic movement initiated a cognitive revolution (Abrams, 1971). In fact, some scholars even talk of a 'Second Scientific Revolution'[1] due to the 'romantic' conception of science and the fruits of that science (Cunningham & Jardine, 1990, p. xix; see also Watson, 2010).

The romantic movement has been viewed as a movement meant to counter the rationalism that had dominated the arts, philosophy, and music (Beiser, 1992; Cunningham & Jardine, 1990; Poggi & Bossi, 1994). It would be misleading, however, to think of it simply as a *Weltanschauung* that placed primacy upon sensibility or simply upon the unification of reason and feelings. The romantics gave great importance not only to social and political education—since it was through education that human beings *became* human and a citizen (Beiser, 2003)—but also to science, neither of which is well known or typically associated with Romanticism. It was 'romantic science', in fact, as a movement that grew in reaction to the eighteenth-century Enlightenment rationalism, with its allied mechanistic philosophy (based on objectivity and determinism), that succeeded in actually transforming

---

[1] While the First Scientific Revolution was associated with the names of Newton, Locke, and Descartes, and also with the founding of the Royal Academy in London and the Academie de Sciences in Paris, the Second Scientific Revolution referred to the sudden series of breakthroughs in chemistry and astronomy. And it was the poet Coleridge who first talked about it during his philosophical lectures in 1819 (Heringman, 2003).

the latter by emphasizing imaginative/creative thinking and excitement about scientific work and discoveries (Holmes, 2009). However, this transformation was even deeper than that, since, according to Berlin (2001), there took place a transformation in the lives and thought of the Western world.

The romantics had accepted the view of the German poet and playwright Friedrich Schiller that 'the arts, rather than religion and philosophy, should be the leading power behind the moral education of mankind' (Beiser, 2003, p. 141; Schiller, 1795). Schiller is considered one of the most important figures of the romantic movement, as he influenced with his thinking other romantics, especially Goethe. It was indeed Schiller who first put forward the idea that 'nature inspires us with a sort of love and respectful emotion' and that 'the object that inspires us with this feeling must be really nature, or something we take for nature' (Schiller, 2006, p. 1). And he also wrote that 'our culture should lead us along the path of reason and freedom back to nature' (Schiller, 1993, p. 180). However, he argued that 'the satisfaction that nature causes us to feel is not a satisfaction of the aesthetical taste, but a satisfaction of the moral sense' (Schiller, 2006, p. 1). Thus, Schiller could be considered the romantic[2] who laid the foundation upon which romantic science, as a science that enriched natural philosophy with both aesthetic and moral elements, was developed.

As a movement, romantic science spread across Europe and quickly came to influence scientists not only in Germany, where the 'romantic' conception of science originally took hold and flourished, but in many other countries, such as England (with prominent scientists such as Davy, Owen, Darwin), France (Lavoisier, Lamarck), and Denmark (Ørsted), among others (Cunningham & Jardine, 1990; Poggi & Bossi, 1994; Cunningham & Jardine, 1990). Moreover, it should be acknowledged that during the period of the romantic movement, many scientific accomplishments of romantic science took place. Although these accomplishments cannot be considered necessarily fruits of 'romantic science' per se, the influence of the latter needs to be acknowledged (Cunningham & Jardine, 1990; Holmes, 2009; Poggi & Bossi, 1994).[3]

It should be noted though that the meanings of the terms 'Romanticism' and 'romantic science' have been debated over the years, and scholars have admitted the

---

[2] Other important figures of the movement in Germany include Schelling, Novalis, and the Schlegel brothers. In England, important Romantics were Wordsworth, Keats, and Coleridge. However, the intellectual architect of the Romantic Movement was Karl Friedrich Schlegel, who coined the term *romantisch* to refer to imaginative literature, which was to be contrasted with other kinds of literature, especially that of Homer and of the Greek poets and dramatists (Richards, 2002).

[3] For example, Sir Humphrey Davy's discovery of the laughing gas (nitrous oxide), Volta's invention of the first battery, and infrared radiation (all three in the year 1800!), as well as the theory of evolution, the placebo effect, the discovery of the carbon cycle, and the awareness of the immensity of the universe, were all fruits of the romantic age (Holmes, 2009). Included is Karl Burdach, a seminal figure in the history of physiology and neuroanatomy, who first coined the words 'morphology' and 'biology' (also in 1800), the latter understood initially only as the science of the study of morphology, physiology, and psychology of humans (Meyer, 1970; Poggi & Bossi, 1994). It should also be recognized that some contemporary notions as 'the mad scientist' and the 'diabolical genius' have their origins in the romantic period (Holmes, 2009).

3.1 The Notion of Romantic Science

difficulties and the problems inherent in any attempt to define them (Berlin, 2001; Heringman, 2003; Kloncher, 2013). The reason behind such difficulties and problems is that any attempted definition fails to capture 'the rich diversity of a movement that involved extraordinary individuals, their luxury of their art, and the prodigality of their ideas' (Richards, 2002, pp. 6–7). In this sense, it cannot be said that the romantic movement was characterized by a specific group of individuals (who were philosophers, painters, artists, poets, theologians, and scientists) that displayed a unanimity of ideas. However, what can be said with some certainty is that what best describes the romantic movement is the intellectual and personal interactions of those individuals who appropriated the name 'romantic' (e.g. the Schlegel brothers, especially Friedrich, Schiller, Novalis, Schelling, Schleiermacher, and Goethe). Indeed, according to Richards' study of the life and work of those individuals, it was their personal and intellectual lives that gave birth to, and shaped, the romantic movement and its ideas.[4] Richards, in attempting to describe the romantic personality, writes that 'it is Goethe in Rome lying with his arms about his sleeping mistress and composing a poem, while counting out its measures softly on the vertebrae of her back' (p. 14).

Notwithstanding the important role that the personal/erotic lives of the romantics played in the romantic movement, what really characterized those individuals was their attempt to understand nature through philosophy, poetry, and science (Abrams, 1971; Berlin, 2001; Richards, 2002). Therefore, it would be misleading to link the romantic movement to some kind of 'emotional mysticism'. In addition, the unity that the romantics sought both in their thinking and nature does show that their primary concern was an understanding of nature and in such a way that human beings are resituated in the universe (see Abrams, 1971, for a discussion of the roles of romantic poetry and imagination in human beings' effort for personal salvation through their communion with nature and how this effort led to a change in worldview).

It should be stressed that the mechanistic/Newtonian conception of nature did not make human beings 'feel at home' in nature, and this is the reason why an 'organic' conception of nature was proposed by the romantics (Abrams, 1971; Bortoft, 1996; Friedman & Norman, 2006). Such an 'organic' conception was based upon the metaphysical position of monism, that is, the idea that matter and spirit are considered two features of the same reality. This metaphysical position had one implication: nature was both a product and a producer, both a cause and an effect.[5] Moreover, monism led the romantics to adopt the idea that God and nature are one and that nature and the self reflect each other.

No doubt, Kant's analysis of the similarities between scientific and aesthetic judgement, as well as between aesthetic and moral judgement, provided the romantics with an aesthetic and moral grounding of their conception of nature (Abrams,

---

[4] For example, Goethe was inspired and influenced by Schiller, while Schelling both inspired and was inspired by Goethe. Humboldt also inspired and influenced Darwin.

[5] Schelling's view was that all natural creatures were products that are purposive but without a purpose (Richards, 2002, p. 162).

1971; Bortoft, 1996; Richards, 1992, 2006), which allowed them to approach nature holistically. This holistic approach to the study of nature was central to the 'romantic' conception of science. A good illustration of this holistic approach is Darwin's study of nature. Indeed, Darwin, according to Richards' (1987, 2002) interpretation of the romantic influence on Darwin's conception of nature, was a revolutionary, in the sense that his conception of nature had both aesthetic and moral elements. In addition, Darwin, as Richards (2002) points out, did not invoke God—like British natural philosophers of his time did—but 'rooted archetypal structures back in nature' (p. 10), and his conception of nature 'exemplified archetypal patterns beneath the wild frenzy of their variations', which 'gradually changed not under the aegis of Paley's God' but 'through the power of a creative nature' (p. 553).

This inclusion of aesthetic and moral elements in the study of nature made romantic science quite different from natural philosophy (*Naturphilosophie*), which simply and specifically focused upon the organic side of nature and its relationship to the human mind (Bortoft, 1996; Poggi & Bossi, 1994). Thus, while romantic scientists accepted and adopted the metaphysical and epistemological positions of natural science, they enriched the latter with aesthetic and moral considerations (Abrams, 1971; Richards, 2002). However, the holistic approach of German natural philosophy (*Naturphilosophie*) and thus its contribution to romantic science should be recognized.[6]

With the rise of positivism[7] around 1840 with Auguste Comte, romantic science declined. However, positivism itself did undergo a crisis in the 1880s due to several reasons. First was the expansion and increasing specialization of the domains of each science, and second the new philosophical questions concerning the relationship between scientific knowledge and man's knowledge of himself and his action (Poggi & Bossi, 1994, p. xiv).[8]

---

[6] It is quite interesting to note the influence exerted by the Naturphilosophie on scientific thinking of the time. More specifically, it can be seen in the idea of a fundamental unity between the inorganic and the organic (living) worlds. Thus, the concept of energy was approached from a holistic perspective (i.e. from both the perspective of biology and that of physics). The 'Berlin school of physiologists' also approached physiological phenomena on a physical basis. This holistic approach cannot be found in Great Britain, where the British scientists approach the concept of energy through engines, both thermal and electrical, without considering 'life' as a factor to be included in their thinking (Frontali 2014).

[7] Positivism, as a movement, emerged with as a reaction or rather as a response to the inability of philosophy to solve its own philosophical problems (i.e. it rejected philosophical speculation as a way to obtain knowledge of the world). It entirely rejected metaphysics and accepted the view that only empirical science can provide data, which then be used to account for true knowledge of the world. However, its own fundamental tenet, that is, the rejection of speculation, cannot be evaluated through empirical research (see Pickering's *Auguste Comte: An Intellectual Biography*. Cambridge, England: Cambridge University Press, 1993).

[8] In the 1920s, positivism did reemerge under the name of logical positivism and was linked to the Vienna Circle. This form of positivism retained the value of empirical research, but logic was also added, in order to account for valid knowledge claims (see Friedman's *Reconsidering Logical Positivism*. Cambridge: Cambridge University Press, 1996).

3.1 The Notion of Romantic Science

Of course, nowadays, the term romantic science may sound like an oxymoron, even a paradox, given that science developed (although not exclusively) because of its emphasis on rationalism, deductive thinking, reductionism, and the mathematization (modelling) of nature. But to the romantics, those aspects were all too 'cold' an approach, having, as they did, one ultimate goal: to control nature and not to be in harmonious coexistence with her (Poggi & Bossi, 1994).

Yet, the 'romantic' conception of science contributed to some strong anti-science undercurrents in the criticisms of several writers during the romantic period (Poggi & Bossi, 1994; Richards, 2002). And more recently, strong criticisms have also been levelled at romantic science. For example, Harvard biologist Edward Wilson has charged the romantics with overprivileging emotions and especially imagination and even fantasy over scientific reason (Wilson, 1998). Snelders (1990) also believes that Schelling's science did not consider experimentation an important factor in science, and hence empirical study cannot be the foundation of natural science. It is true that historians of science have considered the romantic movement in the natural sciences 'an aberration from the path of healthy scientific progress'. However, 'there is now widespread recognition of the importance of particular romantic contributions to the natural sciences' (Cunningham & Jardine, 1990, p. xix; see especially Holmes, 2009). The view by Heringman (2003) is also worth considering: most modern scientific disciplines derived their energy from eighteenth-century natural history.

Today, given the keen interest in romantic science, which has developed over the past three decades (e.g. Hadzigeorgiou & Schulz, 2014; Heringman, 2003; Kloncher, 2013), a look at its main ideas is quite timely. On the other hand, romantic ideas, which played a central role in progressive education, like the self, freedom, imagination, creativity, and joy, can also be revisited and seriously considered in an attempt to formulate an alternative conception of childhood education (Dahlin, 2013). These ideas, according to Dahlin (2013, p. 36), help us to view Romanticism as 'neither anti-modern nor pre-modern, but alternative-modern'. In fact, Dahlin invites the education community to seriously consider romantic ideas. As he points out, the romantics were as committed to progress and development through science and technology as their rational Enlightenment counterparts. The difference, however, was that their vision of science and technology, and also of progress and development, was based on other assumptions, principles, and ideals. These assumptions, principles, and ideals, which constituted the romantic worldview, presuppose 'a certain de-conditioning of our perceptual faculties', and it should be remembered that such 'de-conditioning or de-habituation of perception was a major concern of the Romantics' (Dahlin, 2013, p. 36).

But what were the main features/ideas of romantic science? According to the literature (Beiser, 2003; Berlin, 2001; Holmes, 2009; Poggi & Bossi, 1994) these features/ideas were:

- A revolt against 'Newtonianism' and the mechanistic conception of nature
- The humanistic dimension of science
- The aesthetic dimension of scientific knowledge

- The unity between human beings and nature
- The importance of facts derived from first-hand experiences with nature
- The desire for exploring new worlds, new ideas
- The belief in an infinite and mysterious nature
- The trend for making new discoveries accessible to the public
- The spread of science to society (the popularization of science)
- The image of the romantic male hero/scientist, as a solitary genius, in search for adventure, exploration, and truth questing for truth

The question, of course, is this: which of these ideas have value for contemporary science education? With the exception of the image of the romantic male hero/scientist (which has long been discarded, even though the idea of 'heroic/transcendental quality', as will be discussed in the next section, has an important educational value), the aforementioned features/ideas point to six ideas/possibilities that deserve particular attention by mainstream science education, although their value has been recently recognized and acknowledged by some science educators (see Hadzigeorgiou & Schulz, 2014):

- The emotional sensitivity towards nature
- The centrality of sense experience
- The importance of 'holistic experience'
- The importance of the notions of mystery and wonder
- The power of science to transform people's outlook on the natural world
- The importance of the relationship between science and philosophy

These ideas are interrelated, in the sense that they all derive from the metaphysical position of monism (which grounded the philosophical foundation of romantic science) and hence to that of unity among phenomena, ideas, and between epistemic and epistemic object. However, they are also directly linked to two 'romantic' ideas, namely, the complementarity of the aesthetic and scientific conceptions of nature and the importance of aesthetic perception in the study of natural phenomena (Hadzigeorgiou & Schulz, 2014).

In the light of what was discussed in Chap. 2, it is evident that, with the exception of two ideas, namely, those referring to the relationship between science and philosophy and to the importance of sense experience, all ideas, and especially the ideas of mystery and wonder and the emotional sensitivity towards nature, point to a particular kind of understanding science that could be called 'romantic'. Central to this kind of understanding is a romantic engagement with anything that is supposed to be studied and understood. In other words, a 'romantic' quest for knowledge and understanding is an emotional kind of understanding, which involves exploration, excitement, amazement, mystery, and wonder. These characteristics, as discussed in the next section, are akin to what Kieran Egan has called 'romantic understanding', even though the latter has been further elaborated in Egan's educational theory (see Egan, 1997, 2005, 2010).

3.2 The Notion of Romantic Understanding

Romantic understanding, as the term 'romantic' suggests, is a kind of understanding that places primacy upon the idea of romance and therefore primacy upon the emotional dimension of understanding. However, as a kind of understanding that children begin to develop when they move from orality to literacy, romantic understanding is, on the one hand, a kind of understanding associated with literacy and, on the other hand, a transitional kind between mythic understanding[9] and conceptual understanding (Egan, 1988, 1997).

It has been observed that with the onset of literacy, children between the ages of 8 and 15 start to develop a more realistic kind of understanding. More specifically, as children move from orality to literacy, they move beyond magic and fantasy towards a logical and realistic conception of the world. At this stage, children are attracted, for example, to literary characters who do heroic, but possible, things. Although partly illogical or irrational, these elements represent rational thinking that still retains some aspects of the mythic understanding, whose main characteristics are the use of binary opposites and the recognition of mystery, storytelling, mental imagery, and humour (Egan, 1997).

The emotional dimension of romantic understanding, along with the notion of literacy, can help provide a tentative conception of romantic understanding as the 'motivating insight that emerges through the combined engagement of the emotions and intellect in response to a specialized text'[10] (Hadzigeorgiou, Klassen, & Froese-Klassen, 2012, p. 1113) However, one has to acknowledge that this definition, while it captures the meaning of romantic understanding, does not include any specific characteristics of it. Egan (1990, 1997) has conceived romantic understanding as a particular way of making sense of the world and of experience through an association and even identification with heroes and heroic qualities, through the experience of a sense of wonder, through the contesting of conventions and conventional ideas, and through focusing on the extremes and limits of reality and experience. Central to romantic understanding, according to Egan (1997), is the creation of human meaning through the realization of the humanistic dimension of all knowledge. It is interesting to note Egan's (1990) own definition of romantic understanding:

> *It is a way of making sense of the world and of experience that highlights certain features and suppresses others. It serves as a kind of mental lens that brings particularly into focus .... a sense of reality and nature as vividly present to the senses and rich in meaning; the extremes of reality and its most exotic, strange, and mysterious features; a sense of the self as located within the head, distinct from the natural world and from social roles, and director of the imagination and an ambiguous, partly rebellious, desire to transcend everyday reality but also to recognize its bounds.* (Egan, 1992, p. 36)

---

[9] Mythic understanding is associated with orality and is developed by children of the age range 2–7, who rely upon oral language to interact with and understand the world.

[10] The 'specialized text' refers to a scripted story and this definition was used by Hadzigeorgiou, Klassen, et al. (2012) who used the Nikola Tesla story in order to encourage students develop romantic understanding of the ideas relating to AC electricity (see footnote 12 in this chapter).

Table 3.1 Understanding trees and electricity conceptually and romantically

Conceptual understanding	Romantic understanding
Developing classification schemes for trees and tree leaves	Exploring the highest and shortest tress and those with the deepest roots
Explaining the utility of trees	Experiencing wonder at and about a tiny leaf, responsible for producing life-supporting oxygen
Learning about the chemical equation of photosynthesis	Listening to a story about the heroic adventures of scientists who studied rainforests and exotic trees
Explaining how a simple circuit (i.e. a battery, light bulb, and wires) works	Feeling a sense of wonder about electricity and how it transformed the world
Explaining the relationship between the length and the type of wires (material) and the light produced by the bulb	Wondering about how simple an idea a circuit is and how many uses in everyday life it can have
Develop a theory about the difference in the light produced by using same length wires of different materials	Searching information on extreme values of lightning hitting the planet
	Listening to a story about the Galvani-Volta dispute and also about Tesla's work and his struggles to defend his ideas

Table 3.1 can help one develop a sense of romantic understanding, in the context of science education, even though the examples provided here do not include all the characteristics of romantic understanding.

In the light of these characteristics, romantic understanding can be seen as a transitional kind of understanding between mythic and conceptual (or philosophic, according to Egan's terminology) understanding. This is an important point to stress, given that neither Donald's (1991) distinction between mythic and rational thinking nor Bruner's (1986) distinction between narrative and paradigmatic (or logico-mathematical) thinking (see Chap. 4) can explain a transitional stage from mythic or narrative understanding to conceptual understanding. Moreover, the specific characteristics of romantic understanding make it a quite distinctive kind of understanding, which is not to be confused or conflated with narrative understanding in general (see Chap. 4 for the notion of 'narrative thinking'). While it is true that both mythic and romantic understanding are narrative in their nature (in the sense that children rely on the narrative mode of thinking), they nonetheless represent two distinct ways of making sense of the world. It is perhaps for this reason that Egan has delineated specific characteristics for romantic understanding and has also called these specific characteristics 'cognitive tools' (Egan, 1997, 2005, 2010). These cognitive tools are:

- The humanization of meaning
- Association with heroes and heroic qualities
- Confrontation of the limits of reality and extremes of experience
- A sense of wonder
- Contesting of conventions and conventional ideas[11]

---

[11] Egan (1997) uses the term 'revolt against convention'.

It deserves to be noted that the notion of 'cognitive tool' makes the distinction between narrative and romantic understanding quite apparent. While narrative understanding is a natural kind of understanding, in the sense that, under minimal contextual constraint, it comes spontaneously into being (see Chap. 4), romantic understanding, in order to be developed, requires, and in fact presupposes, the use of an array of specific cognitive tools, as already mentioned, beyond those that characterize narrative understanding (i.e. similes, paradoxes). Moreover, in comparing the aforementioned romantic tools with those of philosophic (or conceptual) understanding, which include searching for truth and probing for generality (i.e. general laws and theories, looking for patterns and connections among ideas), it becomes apparent that fostering a romantic understanding of school subjects like science and mathematics represents a real challenge for both curriculum designers and teachers.

It is also interesting to note that in the context of Egan's (1997) theory, the idea of 'cognitive tool' is broader than in the Vygotskian context, which focuses largely on oral language. Egan (2005) includes a range of cognitive tools that accompany both oral and written language. It may sound strange to call the characteristics of romantic understanding, or of any kind of understanding, 'cognitive tools', but it should be pointed out that these cognitive tools are aids to thinking, that is, mental devices that simply enlarge students' ability to think and understand: the larger the number of tools that a student uses, the better his/her understanding (i.e. mythic, romantic, philosophic). One could also use the term 'learning tool' since such a tool can facilitate learning. It is beyond the scope of this paper to provide a conceptual analysis of those two terms, but it should be pointed out that the idea of 'tool' is a useful one in approaching the problem of fostering various kinds of understanding.

The question, however, is how romantic understanding can be developed in the context of science and science education, both of which require logico-mathematical thinking, that is, the building of relationships and connections among things and ideas, problem-solving, hypothesis formation and testing, and the learning of laws and theories, which are tools that Egan (1997) associates with a higher level of understanding, which he calls 'philosophic'. To answer the question, then, necessitates establishing the requirement of a romantic understanding as an aspect complementary to logico-mathematical understanding. But before I discuss this (see romantic understanding in 'Science Education' section), three practical questions concerning (a) how romantic understanding is different from conceptual understanding, (b) whether romantic understanding can lead to conceptual understanding, and (c) a possible distinction between teaching for romantic understanding and the development of romantic understanding are imperative.

In regard to the first question, romantic understanding is different from conceptual understanding for four reasons. First, it is different in that it is 'human' understanding. Although a truism, this human dimension needs to be stressed, given the dehumanizing of knowledge that often takes place in the world of education. Romantic understanding gives primacy to the idea that knowledge is a human construction, and therefore such knowledge cannot be even considered outside of the context of its construction. This, of course, has a special implication for science education, since students are helped to see that knowledge is not as objective as it is often presented in textbooks and generally through the hidden curriculum but the outcome of a human endeavour, which includes successes and also failures. Thus students are helped to see that knowledge is always embedded in a human context and in some cases even identify famous scientists, for example, Einstein himself, who had difficulty and conceptual problems in accepting the ideas of quantum physics (Hadzigeorgiou, 2005a).

Second, romantic understanding represents, as has already been mentioned, a different way of making sense of the world and human experience through an attraction to their most exotic, strange, and mysterious features and the desire to transcend reality (Egan, 1990).

Third, romantic understanding makes use of the students' imaginative powers. It is a way of understanding the world through what Bruner (1986) called the narrative mode of thinking (see Chap. 4). However, as is discussed later, romantic understanding is not to be conflated with, let alone considered identical to narrative understanding.

And fourth, romantic understanding has an aesthetic dimension. This aesthetic dimension is more likely to be appreciated, if, as Eisner (1985) has argued, knowledge is seen as something that is constructed and not as something that we merely discover or simply find in books. And the inclusion of this aesthetic dimension in the curriculum can lead to 'a fuller, more complex understanding of what makes schools and classrooms tick' (Eisner, 1991, p. 8). Although from a constructivist perspective on teaching and learning knowledge is considered as a construction, it is debatable whether students' constructions have this aesthetic dimension. This becomes more easily understood if the aesthetic dimension is also associated with what Dewey (1934) called 'consummation' (see in Chap. 2 the notion of aesthetic experience).

Now these differences along with the sense of mystery and especially the sense of wonder that accompany romantic understanding can help one see that conceptual understanding presupposes romantic understanding (see Chap. 5). Indeed, all great scientists who propound new theories and change our perspective on the world make use of the narrative mode of thinking (see Chap. 4) and experience a sense of mystery and wonder (see Chap. 6). These scientists approach and understand reality 'romantically'. Did Einstein not approach the world of atoms and of planets and outer space romantically since he commented that 'The most beautiful experience that we can have is the mysterious' (Einstein, 1949 p.5)? Of course, the comparison between scientists and students is unrealistic for a number of reasons, but romantic understanding is a possibility even in the context of compulsory education, provided one is aware that teaching for romantic understanding does not necessarily mean that students will develop romantic understanding. This issue is explored in the last section of this chapter. However, what must be emphasized here is that Hadzigeorgiou, Kabouropoulou, and Fokialis (2012) have provided empirical evidence (see Appendix D) demonstrating that there is indeed an overlap between these two kinds of understanding and that a romantic teaching framework can indeed foster conceptual understanding. This is the reason why Klassen and Froese-Klassen (2014b) point out that the notion of romantic understanding deserves particular attention by science teachers and science educators.

3.3 Romantic Understanding in Science Education

The question as to whether romantic understanding can be developed in the context of science education can be answered both generally and specifically in the affirmative. The general answer considers the view that there are two modes of thinking, that is, the paradigmatic and the narrative. Bruner (1985, 1986, 1990) has proposed and documented narrative thinking as complementary to paradigmatic (or logico-mathematical) thinking (see Chap. 4). Romantic understanding, as a narrative kind of understanding, can be considered complementary to conceptual or logico-mathematical understanding, and, in this sense, it can and should be developed in the context of science education. Its significance lies in its potential, on the one hand, to inspire students and, on the other, to help them become aware of the emotional significance of scientific ideas (Hadzigeorgiou, 2005a). Moreover, romantic understanding has the potential to lead to an aesthetic experience (see Chap. 2).

On the other hand, the concept of romantic understanding, itself, provides the specific answer to the question of whether romantic understanding can be developed in the context of science education. Based on Egan's (1990, 1997) general conception of romantic understanding, a romantic understanding of science can be defined as a narrative kind of understanding, which enables students to become aware of the human context of the science content that they are supposed to learn, by associating, at the same time, such content with heroic human qualities, with the extremes of reality and experience, and with a contesting of conventional ideas and also by experiencing a sense of wonder. This definition of romantic understanding,[12] while different from that of conceptual understanding, nevertheless relates to the content of science in that science is full of extremes, can evoke a sense of wonder, and provides ample opportunities for associating the concepts with people and even things that have heroic qualities. Moreover, scientific content can be associated with the

---

[12] A definition by Klassen and Froese-Klassen (2014b), based on Egan's (1997) original conception, as 'the ability to grasp the meaning of the features of subject matter in a manner that tends to be idealistic in expectation and glamorously imaginary, possibly even exotic, and involving the potential for heroic achievement' (p. 140), is also helpful, but attention should be paid to its last part, that is, 'potential for heroic achievement', for it might be interpreted as an attempt on the teachers' part to encourage students to achieve something 'heroic' (even though this can be a consequence of the development of romantic understanding). According to Egan's (1997, 2005) conception, the heroic element refers to the possibility for students to identify with transcendental/heroic values, which are embodied by people (e.g. in the case in which romantic understanding is developed through storytelling, it is the hero or the heroine or any character in the story who does heroic things, and students identify with the values associated with such things (see 'Associating with Heroic Elements and Qualities' in this section).

contesting of convention if such content is associated with scientists who struggled against conventional and prevailing ideas and beliefs (Hadzigeorgiou, 2005a).

It should be noted that the aforementioned romantic elements can be directly or indirectly linked to the target concept. For example, in the case in which a teacher aims to teach students about electromagnetic waves, extreme values (i.e. lowest and highest frequencies), perhaps with their connection to human life, are directly linked to the target concept, while in the case in which she or he aims to teach about projectiles, the longest long jump represents an extreme that is indirectly linked to projectile motion. The same, of course, holds for all romantic elements. There has to be a connection, a link between the target concept and the romantic element.

3.4 Fostering Romantic Understanding

In light of the idea that the characteristic elements of romantic understanding can be considered cognitive tools, the problem of approaching the development of romantic understanding can be translated into the specific problem of how one can effectively use these tools (or, in other words, of how to experience the romantic elements). The question, however, is how many of these tools should be used in order for students to develop romantic understanding? Apparently, the design of any teaching framework, if it is to be effective (i.e. increase the possibilities for students to develop romantic understanding), should be based on the use of as many tools as possible. Thus, one may very well say that even though romantic engagement with science can be encouraged if teachers focus on at least some of the romantic elements/tools, the development of romantic understanding does presuppose that all the elements be considered (since the elements are all intertwined in a new conception of reality, that is, a romantic conception of reality).

Given the narrative nature, as was discussed, of romantic understanding, narrative or storytelling appears to be the best basis for a romantic teaching framework, since it can include them in its plot (see Chap. 4 for such a teaching framework). Egan (1992), in considering narrative as a means for the development of the imagination, has pointed out that storytelling can be more effective in enlarging the imagination if all of the romantic elements (which will become students' cognitive tools) are incorporated in the plot of a story. Such a view about effective storytelling indirectly answers the question regarding the number of tools that students should use in order for them to develop romantic understanding. What follows is a short description of these cognitive tools.

3.4.1 Humanization of Meaning

This tool is crucially important in enabling students to see knowledge through human emotions. Through the use of this tool, students grasp the deeper human meaning of what they are supposed to learn. To put this tool to use, students need to become aware of the human context of the knowledge and content to be learned. Egan (1997) contends that since all knowledge is, actually, human knowledge, it cannot be separated from the hopes, fears, ambitions, and struggles of those who created it; in other words, '[to] bring knowledge to life in students' minds, we must introduce it to students in the context of the human hopes, fears and passions in which it finds its fullest meaning' (Egan, 2005, p. xii). Humanization of meaning is facilitated through the use of the other cognitive tools. For example, a student, in associating with Galileo's heroic quality (i.e. the pursuit of scientific truth despite the possibility of his being sentenced to death), inevitably becomes aware of the human context of scientific knowledge. So although humanization of meaning is a distinctive tool, other tools may very well facilitate its development so long as those tools help one become aware of the human context, which created knowledge.

Examples are Curie's difficult living conditions in Paris; Tesla's ambition to go to the USA and work on his idea of AC electricity, as well as his humanitarian purpose to send free energy to all parts of the planet (which brought him fame but also made him work as a ditch digger at some point in his life, in order to make a living); and the controversy between Galvani and Volta with regard to the nature of electricity (ending with Galvani's humiliating defeat, even though his ideas received posthumous recognition).

3.4.2 Associating with Heroic Elements and Qualities

Students use this significant learning tool in their attempt to gain confidence in dealing with the real world. By associating things or people with heroic qualities, students gain confidence that they, too, can face and deal with the real world. This may be exemplified in creativity, ingenuity, imaginativeness, tenacity, and other heroic qualities in people or even in objects and phenomena (Egan, 1997). Of course, it might be argued that creativity or ingenuity are not heroic qualities. On the other hand, there may be a question concerning the confidence young students can gain by associating with a heroic quality. How can that be? How can a boy whose hero is a famous soccer player or a girl whose heroine is a famous singer gain confidence and cope with the real world?

Certainly being creative, courageous, or tenacious is not heroic. But it should be pointed out that heroic qualities are heroic in the sense that such qualities are crucial when it comes to coping with a threatening world (see also empirical evidence of this in the empirical study described in the last section of the present paper). When children move from orality to literacy and hence from a magical to a more realistic world (see Egan 1997), they come face to face with a world that provides many challenges and, at the same time—especially teenagers—feel threatened by that world, preoccupied with such issues as success in school and in life, getting along with this or that person, being likable by the opposite sex and the peers. Obviously such issues develop a sense of insecurity, and as Egan (1997) points out, by transcending reality, that is, by associating, even identifying with important human qualities or values, young people gain confidence that they themselves possess such, or some of those, human qualities. Of course, a youngster might associate more easily with creativity than with courage and self-determination when the latter may put one's own life at risk or even lead one (like Galileo) to lose one's life. Yet even in this case, in which the quality is really transcendental and might even be called superhuman, there may be an inspiration derived from that quality.

The strange thing here is that young people tend to associate with those qualities that they themselves think they need, in order to successfully cope with the real world. So they themselves choose the human qualities they will associate and identify with. And because all those human or even superhuman values are embodied by certain people, like a pop star, a soccer player, or even a scientist (i.e. Curie, Einstein), these people, that is, young people's heroes or heroines, are freely chosen. In other words, a hero is a hero for a youngster simply because that hero was chosen to be a hero by that youngster. In this sense, it does not make sense to raise any argument about who can become a hero or even about whether some human qualities are, and some others not, heroic. So from the moment a youngster picks his or her hero or heroine because they embody, for example, creativity or courage, those values become heroic. In other words, although soccer players or singers and even scientists do not represent heroes or heroines, in the real sense of the world, they become heroes and heroines for the young people because of the values that they embody, in which values are crucial for enabling those youngsters to cope with the world.

Examples are Galileo's ingenuity to abandon philosophic speculation about *why* objects move and, instead, focus on *how* they move by using mathematics to describe their motion, his ingenuity to imagine motion in the absence of frictional forces, Tesla's superhuman mental abilities that allowed him to continue his work (e.g. despite the loss of all his notes on innovative ideas and experiments, due to the fire that destroyed his laboratory in New York, Tesla could recover all of them due to his remarkable memory and his power of visualization), the amazing journey of a plant seed, the tireless journey of a water drop during the water cycle, and the tenacity of a wild flower growing on the surface of a rock.

3.4.3 The Extremes of Reality and Human Experience

This vital cognitive tool of romantic understanding enables students to gain security and confidence in dealing with reality by establishing the limits of the environment and its most outstanding features. The preoccupation with extremes and limits, as recorded and described in *The Guinness Book of Records*, explains young people's attempt to understand proper scale and norms and, thereby, what is possible or impossible for them to achieve. As Egan (1997) points out, if reality was infinite, then we would have a problem finding security and also creating meaning. By establishing the limits of any new environment, we feel secure in that that environment also creates meaning. Knowing, for example, about the tallest and the shortest person on the planet, we are in a position to develop a sense of the meaning of things.

Focusing on the extremes and limits is crucial for imaginative engagement and makes us reconsider the idea that teaching should always start from what is already known and familiar (Egan, 1997; see also Hadzigeorgiou 1999). In actual fact, most youngsters at about the age of ten, even younger sometimes, are more interested in the unfamiliar, the distant, and the exotic. This idea had been also clearly pointed out by Dewey (1998, pp. 288–289) in his book *How We Think*, which makes one wonder why the principle to start the teaching-learning process from what is already familiar has been so prevalent in the education system.

Examples are the highest tree, the shortest time interval ever recorded, the slowest animal on the planet, the smallest and the biggest animal living in a forest, the greatest shadow on Earth, the biggest sea waves, the shortest and longest wavelength of electromagnetic radiation, the strongest earthquake ever recorded, the most powerful flash of lightning ever recorded, the lowest and highest temperatures on the planet and the universe, the longest long jump, the highest and lowest frequency relating to man-made and natural forms of waves (e.g. radar waves, microwaves, waves that bats and sea creatures use), the highest electric current that can pass safely through the human body, the hardest/strongest (and softest) metal, the lightest gas, the heaviest molecule, the lightest and the heaviest element in the universe, the heaviest metal, the most abundant (and the scarcest) element in the universe, and the hardest substance on/in the human body.

3.4.4 A Sense of Wonder

A sense of wonder is conceived as astonishment mingled with bewildered curiosity, admiration, and the awareness that one's knowledge is incomplete or erroneous or that some extraordinary phenomenon exists. Wonder is an essential cognitive tool since it encourages involvement with the object of study (Hadzigeorgiou, 2005b, 2014) and discourages what Dewey (1966) called 'the spectator theory of knowledge', a view that is still prevalent in constructivist teaching models (Dahlin, 2001). Given that wonder has both an emotional and a cognitive dimension, its use as a learning tool is essential (Hadzigeorgiou, 2001, 2012). Because of its capacity to capture attention and enlarge the significance of almost anything we focus on—thus making something ordinary to appear extraordinary—wonder is the key tool for the initial exploration of physical reality. Egan (1997) points out that without initial wonder, it is difficult to understand how a systematic theoretical inquiry can take place (see Chap. 6 which focuses specifically on the role of wonder in the teaching and learning of science).

Examples are the fact that light is invisible (we see only what light 'hits') and that regardless of how fast a light source moves, the speed of light remains always constant; the fact that it is always cool on the top of a mountain (although we know that warm air always rises); that fact that in wave swingers in amusement parks, both empty and loaded chairs form the same angle to the vertical (i.e. their motion is independent of their mass); the mysterious properties of water; the beauty of water crystals; and the interconnection of natural phenomena (e.g. the connection among the breakfast an athlete eats, his/her muscular strength, and solar energy).

3.4.5 Contesting of Conventional Ideas and All Kinds of Conventions

This characteristic element becomes evident in students as they enter adolescence. Students, while they begin to explore roles they will take in the adult world, simultaneously resist those roles, and they seek greater freedom and independence from the rules and conventions that shape their everyday life. In the context of science teaching and learning, students question the required learning, prescribed by someone whose authority they question. They may also question the status quo of science. Contesting of conventions manifests itself in many other ways, such as hairstyles and disobeying rules, and can be seen as a learning tool that complements the other tools (Egan, 2005). This specific characteristic can become a cognitive tool when any kind of content knowledge is presented through a human struggle against a conventional idea. Perhaps science is a subject which offers many opportunities for students to become aware of such a struggle, but all subjects can be presented through the barriers of convention that people faced as creators of new knowledge. In science, of course, there have been struggles in the sense of conceptual problems to accept an innovative idea (e.g. Einstein's problem to accept the ideas of quantum mechanics, the academic establishment to immediately accept Einstein's general theory of relativity), but this is different from contesting a conventional idea.

Examples are as follows: Galileo went against the scientific status quo by contesting Aristotelian ideas about motion, Robert Boyles experienced a conceptual struggle to accept the idea of vacuum, and Tesla literally struggled against the idea of DC electricity, that is, the standard form of electricity at his time, before his idea of AC electricity was finally accepted.

3.5 Evaluating Romantic Understanding

It is evident that, in order for students to develop romantic understanding, a teaching framework based on the use of the aforementioned tools is imperative. However, what should be also borne in mind is that such a teaching framework cannot guarantee the development of romantic understanding. Indeed one should not necessarily expect that designing a teaching framework based upon these features naturally leads to the development of romantic understanding. Can we say that, because our teaching is 'romantic' and aims at communicating a romantic understanding, students develop a romantic understanding? This is an important question that points to a differentiation between 'romantic teaching' and 'romantic understanding', no matter how delicate—and perhaps elusive—such a differentiation is. There is indeed a crucial question to be asked by any science teacher: how can one really know that the students' understanding is indeed 'romantic'?

Teaching for romantic understanding, or teaching romantically, is about presenting the humanized, contextualized content so as to make students aware that such content is associated with heroic qualities, the extremes of reality and human experience, the contesting of conventional ideas, and a sense of wonder. Romantic understanding, on the other hand, is developed when students *actually* become aware of the human context of knowledge and of the aforementioned associations and, at the same time, experience a sense of wonder in their attempt to learn content knowledge. In other words, it is one thing to encourage romantic understanding and another to claim that students actually develop such an understanding. What one may very well say is that romantic teaching will most likely result in romantic engagement with content knowledge (as this can be assessed, informally and formally, through the time the students take to become involved in discussions about content knowledge within and outside the school classroom, the comments they make, the questions they ask, their behaviour), which, in turn, may result in an understanding that is not necessarily romantic.

On the other hand, there may be an argument that the distinction between romantic teaching and romantic understanding, although theoretically valid, is not practical; therefore, it may be immaterial to pursue it further. Given that all learning, and hence all understanding, is a possibility (i.e. it is well known that there is not a cause-and-effect relationship between teaching and learning), the development and use of a romantic teaching framework is the best we can do in order to encourage understanding. And we can hope that this understanding is romantic. We do our best to encourage romantic understanding, because students of a particular age range understand the world romantically. So if we teach romantically, and students actually develop an understanding of the subject matter, we can call such understanding romantic.

Of course, we can never be sure about the development of romantic understanding unless an evaluation takes place, which produces evidence that supports the presence of the specific characteristic elements of romantic understanding. Journal writing, for example, can be content analysed in order for those romantic elements in students' comments to be indentified. So while romantic teaching requires planning and delivery, in accordance with a romantic framework, romantic understanding requires evaluation. Paper-and-pencil tests can evaluate conceptual understanding, but they can also evaluate narrative understanding—provided the test items are such that students have the opportunity to use narrative—which, in turn, can help one identify romantic elements. However, journal writing may be considered one of the best ways not only to assess romantic engagement but also to identify romantic elements and hence to evaluate romantic understanding. That said, one has to admit that evaluating romantic understanding is not an easy and straightforward process, and unless, if ever, specific instruments are developed, collecting multiple data may be the best way to have reliable evidence of romantic understanding. In other words, students' understanding, even when the content analysis of their journals does not provide adequate data that support romantic characteristics, cannot be dismissed as 'unromantic', if teaching is based on a romantic framework.

Moreover, if our data do support understanding of content knowledge and, at the same time, romantic associations with contextual factors but not with the content itself, one should not be readily inclined to reject such understanding as 'romantic'. This argument has some weight in the case in which teaching takes place through storytelling. The reason is that the story provides a context that integrates content knowledge, and understanding, naturally and inevitably, derives from that integrated whole. This means that we can talk about romantic understanding, even when our data do not support explicit or even implicit romantic characteristics, provided, of course, that students have developed an understanding by listening to a certain story. For example, a student who understands a certain concept, according to data provided by an evaluation test, may say or write that he or she is just amazed by the courage, the passion, and the persistence of a scientist who is, in some way, associated with that concept. This student's understanding is most likely romantic, since the concept was imbedded in the plot of the story and, therefore, was connected to the life and work of that scientist.

One may very well also argue that some romantic characteristics can be 'tacit'. For example, a student may have made an association with a heroic quality but which is neither explicit nor implicit in what the student says or writes. The same could be said about wonder, although this particular characteristic can be observed in students' expressions. The point that is being made here is not to be taken as an argument against the possibility of evaluating romantic understanding. It should rather be taken as an argument that for some students, it may be really hard to identify some romantic features, even when their understanding is romantic. And that is why access to multiple data, as has already been noted, is crucial when it comes to identifying romantic features.

3.6 What the Research Shows

Even though there is some anecdotal evidence about the educational benefits and the effectiveness of romantic understanding, no empirical study can be found in the literature except the study conducted by Hadzigeorgiou, Klassen, et al. (2012). The purpose of this study was to investigate the effect of the Nikola Tesla story (see Appendices A and B) on grade 9 students' understanding of the concept of alternating current. This story, based exclusively on historical events, included all the romantic elements (see Appendix C). Three specific research questions were asked: (a) Does teaching through the Tesla story, that is, a story with all the characteristic elements of romantic understanding, encourage more engagement with science, in comparison with an explicit teaching method? (b) Does teaching through the Tesla story is more effective in helping students learn specific science content, in comparison with an explicit teaching method? (c) Is there any evidence to support the development of romantic understanding for the students who were taught through the Tesla story?

The sample consisted of students who attended private preparatory schools in the wider metropolitan area of a city in Greece. This means that the age range of the students was between 14 and 16 years. A total of 197 students were the participants in this study with an average age of 15.25 years. These students attended 19 private preparatory schools. It should be pointed out that those schools offered the same courses as the regular public schools. The main objective of the courses they offered was to better prepare students for the public school, while, at the same time, some courses were just remedial.

A quasi-experimental design with two groups was used, and the research study was conducted over a 10-week period, followed by a repeat of the final test 8 weeks thereafter. The first 4 weeks were spent teaching the fundamental prerequisite knowledge of electricity to all participating students, and the fifth and sixth week consisted of the first assessment of students' understanding through two tests. The following 3 weeks consisted of instruction in two different modes, the control group receiving direct instruction and the experimental group receiving indirect instruction through the use of the Tesla story.

The results show a significant difference between the control group and the experimental group, in terms of engagement, retention, and understanding. The empirical evidence found through an analysis of student's journals and traditional tests can be seen in Appendix D.

3.7 Concluding Comments: The Need for a Romantic Understanding of School Science

The difficulties in evaluating romantic understanding notwithstanding a careful design of a teaching framework can be effective in fostering romantic understanding. The results of the study by Hadzigeorgiou, Klassen, & Froese-Klassen (2012), as was discussed above, do document students' romantic understanding. Such evidence demonstrates that the development of romantic understanding in the context of school science education is indeed a possibility.

This evidence, however, raises the question of how one can explain the significant difference in performance of similar test groups. A possible explanation may be the storytelling approach itself—that the novel and innovative approach for the experimental group (i.e. Hawthorne effect) may have been a motivating factor to learn in and of itself. Another possible explanation is that the Tesla story, by providing context, caused students to focus their attention entirely on the introduced ideas and the actual phenomenon of study. This explanation correlates with research findings that confirm that 'when information is emotionally charged, we are more likely to pay attention to it, continue to think about it over a period of time, and repeatedly elaborate on it' (Ormrod, 1999, p. 420), and stories provide such opportunities for emotional stimulation. Another explanation may be that narrative understanding plays an important role in learning (see Chap. 4). Given the evidence that narrative understanding is an important factor in learning science (see Avraamidou & Osborne 2009; Hadzigeorgiou & Stefanich, 2001; Milne, 1998), one may opt for the third explanation. Most likely all three factors, that is, the innovation of the storytelling approach, the opportunity students had to use narrative understanding, and the Tesla story itself, as a romantic story that encourages romantic understanding, must have played a role.[13]

Regardless, however, of any explanation regarding the evidence from the study, it is important to note that a romantic teaching framework, based upon a 'romantic science story' (i.e. a story with all the romantic elements, according to Egan's conception of romantic understanding), such as the Nikola Tesla story, can indeed facilitate the acquisition and retention of science content knowledge, which is at the heart of conceptual understanding. The chapter that follows discusses the notion of storytelling, which can be used to foster romantic understanding.

---

[13] While the study yielded indisputably positive results, it is important to note that studies with a two-group quasi-experimental design have some inherent limitations. The first is the absence of a random assignment, so despite the ability of the students in the two groups being comparable and similar, with most of them attending schools where students are performing at the average or below average level, one cannot say, categorically, that the groups were the 'same'. Second, although background knowledge of the fundamentals of current electricity was tested, no pretest on alternating current was administered in the study. The researchers of this study are confident that the students did not have such knowledge; however, this aspect was not formally tested. Despite these limitations and also the interpretation one might give to the test results, these results, however, support the idea that the Tesla story played a role in student learning.

3.7 Concluding Comments: The Need for a Romantic Understanding of School Science

Klassen and Froese-Klassen (2014b), in their study on the role of interest in science education, point out that it is possible 'to re-interpret the cognitive tools of romantic understanding and designate humanization as the overall context for romantic understanding and the sense of wonder as the interest reaction to the heroic, […] the extremes of reality, […] and the contesting of conventions' (p. 140). In other words, what Egan (1997, 2005) has called cognitive tools appear to be very similar with some of the characteristic features of interest (even though the experience of wonder is much richer than that resulting simply from an interest, especially a situational one). It is for this reason that Klassen and Froese-Klassen stress that:

The insights provided by romantic understanding and its success in achieving improved student learning could add an enriching new dimension to the research on interest. (Klassen & Froese-Klassen, 2014b, p. 140)

Chapter 4
Narrative Thinking and Storytelling in Science Education

> *It is very likely the case that the most natural and the earliest way in which we organize our experience and our knowledge is in terms of the narrative form.*
>
> Jerome Bruner, in The Culture of Education, p. 121.
>
> *Teachers tell their students stories from the first day they start school and children's storybooks are better made and more engaging than they have ever been. Yet stories are an underused medium for learning. Pushed to the margins of the curriculum to stimulate art and drama activities, but forgotten or neglected when the study of more 'serious' subjects begins.*
>
> Kieran Egan, in Teaching as Storytelling, p. 5.
>
> *A first insight into modern narrative theory regards the fact that every good story has a consistent plot. Effective storytelling includes a carefully developed structure in the organization of the narrated facts […] The simplicity of a good story is deceiving. A simple plot is a piece of art. If we want to go beyond the pure repetition of stories already created, we have to deal with the theory of poetics.*
>
> Fritz Kubli, in Can the Theory of Narratives Help Science Teachers Be Better Storytellers, Science & Education, 10, p. 595.

It would no doubt be a truism to state that over the last two decades, there has been an emphasis on narrative in our culture (Avraamidou & Osborne, 2009; Schank & Berman, 2002). Nash (1991) had stressed the fact that 'Preoccupation with discourse…is no longer the private province of specialists in literature and language' (p. xi). In fact, whole movements have sprung up within the physical and the social sciences which seek to apply the techniques which are involved in literary and linguistic studies (see Green, Strange, & Brock, 2002).

Narrative is indeed central to our cognitive activities, to historical thinking, to psychological analysis and practice, and to political critique and praxis. It is also crucially important in the formation of an individual's sense of identity. And in educational research, narrative has been used as a medium of data representation and also as a guide for the development of methodologies (see Conle, 2000; Connelly & Clandinin, 2000; Fenstermacher, 1994; Green et al., 2002; Klassen & Froese-Klassen, 2014a; Nash, 1991). Thus it is quite reasonable that Coles (1989) has recommended 'a respect for narrative as everyone's rock-bottom capacity' and considers narrative a 'universal gift to be shared with others' (p. 30).

But why is there such an emphasis on narrative? First and foremost, because it is a fact that human beings tell and listen to stories. And this is because of an impulse that seems to be primordial. According to Bell (1991),

> The shaping of experiences by narrative, indeed the very impulse to tell stories, may suggest primordial but subliminal processes underlying even the apparently independent planes of reason or evidence. (p. 172)

If this hypothesis is correct, then it makes sense to use narrative and especially storytelling across the curriculum.

Teachers, of course, have always made use of stories at one time or another. However, that has been a mere accompaniment to their traditional mode of instruction and certainly an activity in line with the traditional academic or even learner-centred curriculum. But if narrative is to be rediscovered, as indeed it has been over the last two decades, then much more attention should be paid to the notions of orality and storytelling and their consequences for teaching and learning as well as for curriculum organization. This chapter discusses the ideas of narrative thinking and storytelling as tools for understanding the world and then focuses on the role of storytelling in science education.

4.1 Narrative Thinking

Human beings have a natural impulse to tell and listen to stories. Jerome Bruner (1985) has made the following bold claim:

> There are two irreducible modes of cognitive functioning-or more simply, two modes of thought-each meriting the status of a "natural kind". Each provides a way of ordering experience, of constructing reality and the two (though amenable to complementary use) are irreducible to one another. (p. 97)

By 'natural kind' Bruner (1985) means that each mode of thought, under minimal contextual constraint, comes spontaneously into being. The first kind he calls paradigmatic, and the second one narrative. The paradigmatic (or logico-mathematical) mode is concerned with the formation of hypotheses, the development of arguments, solutions to problems, finding proofs, and rational thinking in general. The narrative mode, on the other hand, is concerned with 'verisimilitude', that is, life-likeness and the creation of meaning. It seeks explications that are context sensitive and particular (not context-free and universal). The narrative mode is entirely divergent and employs literary devices, such as stories and metaphors, in order to evoke meaning. It is, therefore, a mode of thinking that is about people (i.e. human

4.1 Narrative Thinking

intentions, human actions, and human experiences) and not about thinking per se, which is what the paradigmatic mode is all about.

Although these two modes of human thinking are complementary, they are irreducible to one another. This means that they are both indispensable in understanding the world, for they both aim at making sense of the world. Their difference—and hence their irreducibility—is that while the paradigmatic or logical mode does this (i.e. making sense of the world) by distancing itself from reality and the human emotions, the narrative mode places primacy upon the human element, particularly human emotions (e.g. human motivations, endeavours, struggles, ambitions, hopes, anxieties, successes, and failures).

It is important to note that the narrative mode of thinking is used by people in everyday life; they use it to communicate, to make sense of their experiences, and to plan future actions. Even though all normal people become competent at thinking in the narrative mode, in comparison, fewer people become competent at thinking with the logical mode. It is no wonder that the narrative mode, according to Bruner (1991), is the default mode of thinking.

The narrative mode of thinking, however, is also important to science. Bruner (1986) pointed out that many scientific and mathematical hypotheses start their lives as stories and metaphors. This view is in line with the one held by Popper (1972), who argues that today's science is built upon the science of yesterday and that the older scientific theories were built upon 'prescientific myths' (p. 346), and shows the importance of the narrative mode, even for the case of science, which has, arguably, been built upon rationality, objectivity, and truth. Of course, the notions of objectivity and truth—bequeathed to us by modernist science that was rooted in a Cartesian/Newtonian conception of reality—have been challenged by postmodernist conceptions of science, as discussed later in this section. Popper's and Bruner's views, however, deserve particular attention, as they point to the crucial importance, if not the centrality, of stories in the context of school science education.

A story, of course, generated by the narrative mode will not necessarily convert into a scientific theory. The scientific maturity of a story always presupposes a process of conversion into verifiability; it cannot rest on its dramatic origin. In other words, scientific maturity presupposes the use of the paradigmatic (or logico-mathematical) mode. Although the narrative mode of thinking (and therefore a story or a myth) could be viewed as the cause of the construction of unreal or even impossible worlds, Bruner (1985) points out that 'the narrative mode is not as unconstrainedly imaginative as it might seem to the romantic' (p. 100). So, in science, we cannot construct any kind of world (reality) or all kinds of impossible worlds, since the paradigmatic (logico-mathematical) mode of thinking, which is a complementary kind of thinking, tests concepts through the use of evidence, arguments, and so on. Although scientific theories may have started their life as stories or myths, not all of those stories resulted in a scientific theory.

Bruner's (1985) hypothesis is nonetheless very bold, given that in his *Actual Minds, Possible Worlds* (1986), he breaks away from realism by arguing that nature itself is a hypothesis and that a scientific theory and a well-made story are 'two forms of an illusion of reality' (p. 52). It is quite evident that Bruner's notion of 'illusion of reality' raises an argument on metaphysical grounds: is reality a human construction?

Of course, with the rise of, and emphasis on, the constructivist epistemology, the idea of 'construction' has become commonplace. The idea that epistemologically there can be nothing known to which our ideas can correspond and it is precisely the construction of concepts and theories that aims at representing the world is shared among philosophers and cognitive scientists (Manicas & Secord, 1983; Matthews, 2015). Even if one were to cling to a realist ontology, there would be no reason why epistemologically one could not be a fallibilist. However, to claim that all reality is a human construction or invention appears to be too far-fetched.

Even according to the Piagetian epistemology, children construct cognitive structures that have to accommodate a reality 'out there'. Although the notion of transformation is a central one to Piaget's (1977) epistemology, which, of course, entails the constant modification of reality through the action of the epistemic subject, the interaction between epistemic subject and reality presupposes their existence prior to this interaction. This is in line with both Kitchener's (1983, 1986) and Bruner's (1986) idea that Piaget's constructivism is committed to some kind of metaphysical realism. Simply put, constructions for Piaget are representations of a real world to which children have to accommodate. And this is not surprising given that for Piaget the paradigmatic (or logico-mathematical) mode was the cornerstone of his epistemology. 'Operative thinking' (as opposed to 'figurative thinking', which is based on sensory experience), according to him, results in transformations, that is, changes in the position and in the form of the objects young children play with. But the objects were certainly there. They did not appear as a result of this transformation. It might be argued that there is 'a construction of reality', in the sense that a previous reality has been modified or changed.

However, the notion of 'construction of reality' through the narrative mode is much more challenging, as it raises a number of arguments about the notions of truth, rationality, and objectivity. And in the context of education, where the development of rationality has been considered an important aim—if not an ideal—the use of the narrative mode through stories, myths, and fantasy poses certain problems. Indeed, given that 'rationality and reality are closely intertwined in our mental lexicon' (Egan, 1988, p. 11), a story or myth could be viewed as the source or cause of the construction of unreal or even impossible worlds. While reason asserts the possible, the verifiable, and the real, a story could assert the impossible, the unverifiable, and the unreal. That is why Bruner (1985) emphasized that the narrative and the logico-mathematical modes are complementary, which means that the constructions made by the narrative mode are checked, in a sense, by the logico-mathematical mode.

Of course, the narrative mode of thinking introduces arguments that apparently cannot be settled here—in fact they are not settled within the area of philosophy (see Bereiter, 1994; Rorty, 1980; Toulmin, 1976). However, over the last three decades or so, there has been mounting evidence, especially from the area of history of science, that even truth is something that is negotiated by scientists (Kuhn, 1970). Paul Feyerabent's (1993) celebrated motto 'anything goes' does support the view that the distinction between reality and fiction is an unclear one.

4.1 Narrative Thinking

Narrative theory has indeed challenged us to view the structuring of scientific discourse similar to that in the human sciences (Altman, 2008). As Gough (1993) argued, the world 'operates more like novels tell us than what modern science describes'[1] (p. 621). Thus 'objective truth' and fiction are not as distinct as we have previously thought. Bruner (1985), who espoused such a view, writes:

> *Truth is stranger than fiction. Each is a version of the world and to ask which depicts the real world is to ask a question that even modern metaphysicians believe to be undecidable.* (Bruner, 1985, p. 100)

These ideas are in line with a postmodern perspective on science, which rejects the notion of objective truth and, instead, proposes the idea of 'science as progressive discourse' (Bereiter, 1994; Bereiter, Scardamalia, Cassells, & Hewitt, 1997). The key idea in this postmodern perspective is neither the abandonment of the notion of objectivity nor the view that scientific theories and ideas in general are human constructions. It is rather the shift from 'objectivity' to 'progressiveness', which will lead to better knowledge. Such a shift

> *[...] makes possible a conception of scientific progress that does not depend on an objective standpoint of judgement and a conception of scientific method that does not presuppose a realm of objective truth [...] The key assumption supporting the scientific enterprise, therefore, is not objectivity but progressiveness. It is not necessary to believe that science is approaching some objective truth, but it is necessary to believe that today's knowledge is better than yesterday's.* (Bereiter, 1994, p. 4)

However, even if one were to put aside the idea of scientific progress, one could still do away with the notion of objective truth. In the case of scientific models, for example, which are representations of physical reality, some are more powerful than others, in the sense that they (models) can explain and predict more phenomena. It would be meaningless to ask which model is 'more real' or 'more true'. Even in the case in which one adopted a realist ontology (i.e. one believed in the existence of a real world), one's constructions would still be viable models that help explain physical reality.

In light of this short discussion, Bruner's (1985, 1986, 1991) claim about the centrality of the narrative mode of thought appears to be a sound one. After all, both truth and fiction represent two versions of the world. Even in the case of science, as was pointed out above, we can talk about different versions of the world (i.e. about different models). Gough (1993), echoing Bruner, stressed the view that the distinct differences between the narrative forms found in scientific reports or in school textbooks, on the one hand, and in mass media, journalism, and fiction, on the other, are neither necessary nor desirable.

It is worth pointing out that there is evidence that a fictional story might be generated in order to assist the successful dissemination of new scientific discovery (Myers, 1991). Also according to Harre (1991), the authority of 'apparently neutral

---

[1] Gough (1993), in making reference to a poster that he had once seen and which read 'the universe is not made of atoms – it is made of stories', discusses the idea that the Bohr-Rutherford model is 'one fiction which can be fashioned from certain scientific facts [....] but it is not the only plausible story that these facts can be used to generate' (p. 615).

scientific evidence and argument' strongly depends upon narrative of heroism and virtue that are covertly propounded by scientists themselves on the grounds that they are members of an elite 'moral community'.

The construction of 'truth', of course, can also be approached from a perspective, which acknowledges the role that beliefs have played in this construction process. Indeed, if beliefs are 'the source of all knowledge' (Polanyi, 1958, p. 266) — is it not true that even 'truth is something that can be thought of only by believing it' (p. 305)? — then it is important to acknowledge that even in science, it was beliefs that guided the scientists' constructions (see Holton, 1996, for a discussion of the 'unconfessed, unconscious basic presuppositions and preferences' that scientists may choose to adopt, even in cases in which they are not led to do so by the available data or current theory). The role of beliefs in the development of major scientific theories is well known. For example, metaphysical beliefs about the universe played a major role in Einstein's thought, and Galileo held on to his conviction about the motion of the Earth despite his fear of imminent death. Thus, drawing a distinction between subjectivity and objectivity and between reality and fiction is not an easy task.

So although Bruner's hypothesis is very bold, it does shed light on the development of scientific knowledge, which cannot be explained solely in terms of paradigmatic (logico-mathematical) thinking. Both the 'irrational character' of scientific thinking (Di Trocchio, 1997; Feyerabend, 1993) and the idea that scientific theories start their life as myths (Popper, 1972) do support Bruner's theory about the two modes of thought and make us pay more attention to the narrative mode of thinking. Moreover, the creation of scientific ideas necessitated mental leaps, which were not possible through logical thinking (see Chap. 1).

There is evidence that the scientists' personal stories, that is, stories based on events from their everyday life, can help them think about their own work (Hadzigeorgiou & Stefanich, 2001; Martin & Brouwer, 1991, 1993). However, a good example is Niels Bohr himself, who, as he reports, conceived of the idea of complementarity from an everyday event involving his son who had stolen sweets from a tobacco shop. Bohr, in thinking about this particular event, became aware that his love towards his son and the punishment that he (the son) deserved were two sides of the same coin; although irreducible to each other, they were complementary (cited in Bruner, 1986). Bohr's account does provide support for the crucial, if not central, role of narrative thinking in the case of concept creation/construction.

Perhaps the centrality of narrative thinking is captured in the notion of 'mind as a narrative concern' (Sutton-Smith, 1988). This notion, coupled with the limitations of reason (i.e. logical thinking) and the primacy of the constructive property of the mind — a property that is not simply involved in phenomena of optical illusion, but a property that is fundamental to our understanding of the world — points to the importance of storytelling as a means to understand the world, through imaginative engagement with ideas, and hence to its importance as an effective teaching/learning tool. It has been pointed out that the quite recent rediscovery of the narrative mind necessitates not only our closer attention to the imagination (Egan, 2005) but also a recognition that learning to follow narratives in general involves 'the development of more significant intellectual capacities than has traditionally been recognized' (Egan, 1992, p. 63).

4.2 Storytelling: A Tool for Understanding the World

Before the invention of the printed word, humans relied exclusively on orality. Stories played an important role as a means of conveying knowledge and experience from one generation to the next. However, with the advent of literacy, storytelling almost disappeared (Ong, 1971). Orality was pushed to the background, even though it is, and has been, a human characteristic, most prominent among young children.

It is a fact that young children love to listen to a story—any kind of story—so much so that they literally hung on the storyteller's every single word. From fairy tales and Aesop's fables to myths about the ancient world and the origin of the cosmos, young children become emotionally involved with the story, namely, with its plot, the characters, and the ideas conveyed through that plot. Even though the claim that all knowledge comes in some story form (see Schank & Abelson, 1995) is an extreme one, the universal power of a story to convey information with an emotional meaning is indisputable (Bruner, 1990, 1991, 1996; Coles, 1989; Egan, 1986, 1999; see also Campbell, 1990 and Kolakowski, 1989).

Bruner's (1996) view that we swim in a sea of stories is quite interesting. According to him, 'we live in a sea of stories and, like the fish, who will be the last to discover water, we have our own difficulties grasping what it is like to swim in stories' (Bruner, 1996, p. 147). It is a fact that even though people are not capable of understanding 'the specific thought patterns of another culture', they have 'less difficulty understanding a story coming from another culture, however exotic that culture may appear' (White, 1981, p. 1).

But if stories—those we tell and those we listen to—are central to our life, then school life and education cannot be an exception (Schank & Berman, 2002). Indeed, the power of stories has long been recognized in education and considered an effective teaching/learning tool, since they can engage, perhaps more than any other medium, people's affective imagination and responses, but, of course, in different ways. (And this is the reason why very young kids, at least those from the Western culture, engage imaginatively with Cinderella and talking bears, while children of 9 or 10 years of age prefer stories involving heroes and the extremes of reality and human experience.)

However, there has been a renewed interest in stories in the last two decades, with the rediscovery of the 'narrative mind', and also with the re-emergence of orality, as a central characteristic of people's—especially young children's—lives (Egan, 2005). Orality has been seen not as a lack of literacy, but rather as a feature upon which the development of literacy can take place (Egan, 1999, 2010). Storytelling, apparently, is an excellent way to develop orality.

It is quite clear that this (storytelling) approach to teaching and learning, especially with young children, is very different from direct contact with real objects and events. Indeed, storytelling has nothing to do with concrete experiences (e.g. hands-on activities) and physical action (as the primary source of young children's knowledge). However, storytelling is a way, as was previously said, to create meaning. Fictional stories, in particular, are more concerned with affective meaning and affective imagination (see Bruner, 1990).

Egan (1986, 1988, 1999) has convincingly argued that the story metaphor is more appropriate in describing what we learn about the world, according to research based on the constructive nature of human sense making. Indeed the story receives the attention it deserves in Egan's recapitulation theory (see Chap. 3), since it is considered a 'cognitive tool' (Egan, 1997, 2005). The development of both 'mythic' and 'romantic' understanding necessitates the use of storytelling, even though the stories necessary for the development of these two kinds of understanding are different. While a story for the development of mythic understanding is structured around a binary opposite (e.g. good/bad, useful/useless, motion/rest, fast/slow) and includes mental imagery, humour, mystery, and rhyme, a story appropriate for developing romantic understanding is structured around a heroic quality (e.g. heroic characteristic or achievement) and includes such elements, as wonder, humanization of meaning, the extremes of reality and human experience, and revolt against mainstream/accepted ideas (see Chap. 3).

The story metaphor is also important from a postmodernist perspective on teaching and learning. From such a perspective, understanding the world involves a rejection of traditional dichotomies, like those between fact and fiction, reality, and epistemic subject. And this is why, from such a perspective, storytelling is considered an all-important pedagogical tool. As Gough (1993) has argued, 'An appropriate science pedagogy […] tacitly embraces […] the relatedness of the observer and the observed and the personal participation of the knower in all acts of understanding'. He goes on to say that a postmodern pedagogy 'implies a reversal of many taken-for-granted assumptions about the relationship between fiction and reality' (p. 621).

In discussing, of course, the notion of storytelling, it should be noted that, more often than not, the notion of 'narrative' is conflated with that of 'story'. While both narratives and stories are used to communicate ideas, stories constitute a class, a subset, within the narrative genre. And narratives should also be distinguished from other forms of communication, like exposition/description and argumentation. These distinctions are imperative. Defining a narrative, as Klassen and Froese-Klassen (2014a) point out, is crucial not only for clarification but also for ascertaining whether writing that purports to be narrative can indeed be characterized as narrative. The notion of narrative, according to a recent analysis by Klassen and Froese-Klassen, is quite broad, and it does not have a categorical definition in the literature (see also Avraamidou & Osborne, 2009), in contrast to the literary notion of story.

For pedagogical purposes it is crucial, but at the same time easy, to distinguish between the two notions: a narrative can be considered a flow of events associated with a theme/idea, while a story can be considered a sequence of these events to create meaning.² Despite the fact that, more often than not, the terms 'narrative' and 'story' are used interchangeably, the sequencing of events (very often with a causal relationship between the two) is the most distinguishing characteristic of stories. A collage, for example, can be used to generate a narrative but not a story, unless the events that the various photos illustrate or give rise to (through an interpretation) are deliberately placed in a sequence, so there is a beginning, a middle part, and an end.

Ricoeur (1981) writes that stories refer to the actions and the experiences, as a result of these actions, of real or imaginary (fictional) characters. However, it has been pointed out that there has to be 'a structured, coherent retelling of an experience or a fictional account of an experience' (Schank & Berman, 2002, p. 288). Thus, we can say that a story refers to a narrative account of events and experiences, as a result of the action of characters (who, more often than not, resolve a conflict or crisis and who may also perform a heroic act). This, very often brief, narrative account has a specific 'beginning/middle part/end' structure.

A story has been used as a technique for organizing events, facts, characters, ideas, and so forth, into a meaningful unit. This unit has a beginning that sets up a conflict or expectation, a middle part that makes up the plot, and an end that resolves the initial conflict. Unlike other forms of narratives, stories end. According to Egan,

> *One reason why stories provide affective meaning is that unlike the complexity of everyday events, they and. They don't stop at some arbitrary point. What makes them stories is that their ending completes and satisfies whatever was raised in their beginning and elaborated in their middles.* (Egan, 1988, p. 102)

A story, therefore, is a conceptual tool for providing coherence, continuity, and meaning to its contents. There is considerable evidence from empirical research that we understand more easily and remember better various pieces of information if they are integrated within the plot of a story, than when the some pieces are presented in a logical or conceptual sequence (Bruner, 1990; Egan, 1986, 2005; Schank, 1990; Schank & Abelson, 1995; Schank & Berman, 2002). A story, on the other hand, could be considered the means of translating knowing into telling (Avraamidou & Osborne, 2009), an idea that is important from an instructional point of view, especially in the context of science education, where abstract knowledge needs to be presented in a way that makes sense to the students.

What must be stressed is that, regardless of the type of story (i.e. myth, fable, folk tale, personal story, science fiction story, 'historical' story), the pedagogical importance of storytelling is quite undisputed, as regards its potential to capture the students' attention, to engage their imagination, and to convey effectively ideas and values (Haven, 2000; Simmons, 2006; Truby, 2007). Also, storytelling appears to satisfy all three elements of brain-based learning. According to Caine, Caine, McClintic, and Klimic (2005, p. 233), effective learning should be based upon the

---

²One may argue that a story requires at least two events. Thus, while the single event 'It was Newton who formulated his three famous laws of motion' cannot be a story, by adding another event such as 'Scientists have applied these three laws in order to describe, explain, and predict all kinds of motion on or near the surface of the Earth and in outer space', we can create a story like 'the story of motion'.

following: (a) relaxed alertness (i.e. a state of mind created in a low-threat atmosphere, which also creates a sense of community), (b) planned immersion (i.e. the creation of an environment in which students become involved with the objectives of the lesson), and (c) active processing (i.e. utilization of learning methods, which encourage reflection and integration of the information in a meaningful way).

Indeed, storytelling has, first of all, the potential to enhance the classroom atmosphere. We all know that a good story, especially if it is entertaining, can relax learners and reduce stress, while, at the same time, it can create a sense of community. Solomon (2002) has argued that a story is like a dialogue, which involves both the teller and the listener, even though the latter does not appear to be actively involved in the telling of the story. This happens because the listener tries to create meaning out of the events of the story.

Storytelling can also be very engaging (compared with other teaching methods), as students become emotionally involved with content knowledge on a deeper level (Richter & Koppet, 2000; Hadzigeorgiou, Klassen, & Froese-Klassen, 2012), and has many of the benefits of experiential learning due to high levels of the listeners' active engagement (Richter & Koppet, 2000). Moreover, storytelling can appeal to a wide range of intelligences (e.g. linguistic, logico-mathematical, interpersonal, spatial) and a variety of learning styles, such as auditory, visual, and kinaesthetic (Rose & Nichol, 1997).

But how can one use storytelling effectively in the context of science education? This is the topic of the next section, but here it is important to say a few things about the implications of storytelling for the curriculum, in general. These implications refer to the reconsideration of some principles that have guided learning and instruction for a very long time. These are as follows: (a) we proceed from the concrete to the abstract, and (b) we proceed from the known to the unknown. First of all, we all know that children's imagination is captured by such unfamiliar and unknown things and entities as witches, dragons, and talking rabbits.[3] Indeed these are intellectually more engaging and meaningful than their everyday experiences—their meaningfulness derived from mediation process between two binary opposite concepts (e.g. big-little, good-bad, oppressor-oppressed) (see Egan, 1999). Second, research reported by both Bruner (1990) and Egan (1999) provides evidence that children do understand, although intuitively, abstract concepts, such as oppression, freedom, cruelty, or kindness. Scientific ideas are no exception, as empirical evidence (although little) suggests (see next section). Storytelling can indeed be considered a good means for helping pupils understand science ideas, if these are embedded in the plot of a story, and also for helping them to convey their ideas (Egan, 1986, 1997; Hadzigeorgiou, 2006a).

---

[3] These are entities that capture the imagination of children from European and North American countries. I believe Egan (1999) refers to these children. Perhaps other entities capture the imagination of children from other non-Western cultures.

4.3 Storytelling in Science Education

It is a fact that the historical method in science education was proposed as an alternative to the 'logical method' that was dominant in science education in the 1960s. Central to the historical method was the use of the history of science as both an instructional medium and a pedagogical method (Brush, 1969; Klopfer & Cooley, 1963). Matthews' (1992) proposal for, and central contribution to, making the history of science an indispensable instructional tool has laid the foundations for the use of storytelling in science education per se, even though storytelling had been occasionally used by teachers in general.

However, if the narrative mode is 'more natural' than, and develops before, the logico-mathematical mode (see previous section), then storytelling can indeed be the starting point to teach science (in early childhood) and then, later on, act as a bridge between the two modes of thought. However, in science education (with the exception of early childhood science education), the narrative mode has not been very popular, even though it has been argued that 'forms of discourse in the natural and human sciences are themselves ordered as narratives' (Gough, 1993, p. 607).[4]

But if stories, as is well known, were used in the past to convey information of vital importance from generation to generation (Bruner, 1990), it is quite reasonable to use stories nowadays to convey information not only about historical events and our cultural roots but also information about science and more specifically information about both the content and the process of science. According to Martin and Brouwer (1991), storytelling can be pedagogically effective because 'the students themselves are involved in giving meaning to the stories' (p. 719). As they pointed out,

> *A story with just a few well chosen words, can draw in the reader or listener and create a world of shared experience. The story can at times communicate in a few words that which a dense, technical analysis might require many lines to accomplish.* (p. 708)

Lemke (1990) deplores, even in an indirect way, the presentation of science as a description of the way the world works and recommends its presentation as a human activity taking place in a social context. But this can be done easily through storytelling. In fact, Lemke points to the crucial role of storytelling in science education, even though he does not make an explicit reference to it:

> Science is projected as a simple description of the way the world is, rather than as human social activity, an effort to make sense of the world. Statements about the way atoms are or the earth is tend to be less interesting to many students than statements about who did what to come up with these unfamiliar ideas. (pp. 130–131)

---

[4] According to Ogborn, Kress, Martins, and McGillicuddy (1996), scientific explanations have a similar structure as stories. 'Firstly there is a cast of protagonists, each of which has its own capabilities which [...] might include entities such as electric currents, germs, magnetic fields, and also mathematical constructions such as harmonic motion [...]; secondly the members of this cast enact one of the many series of events of which they are capable; lastly these events have a consequence which follows from the nature of the protagonists and the event they happen to enact' (p. 9).

It is quite clear that a narrative or story about atoms (i.e. from Democritus and Dalton to the present-day scientists who study atoms in accelerators) or about the Earth (i.e. the cosmic creation story that recounts the formation of planets with the Big Bang) can be told in the classroom in order to make learning more interesting[5] and memorable.

Students, of course, at all levels of education, use narratives on a daily basis (e.g. to describe what happened at a soccer game, at a friends' gathering). And yet school science education is still dominated by 'expository text' (e.g. textbook writing, oral lectures, teacher guidance). Textbooks, for example, contain expository text that can be found in the various descriptions of physical entities and processes (e.g. of molecules, of the water cycle), in various sequences (e.g. of laboratory experiments), in comparisons and contrasts (e.g. of vectors and scalars, of simple and complex machines), in lists (e.g. lab safety rules), and in causes and effects (e.g. why the sky is blue) (Begoray & Stinner, 2005; see also Hadzigeorgiou & Stefanich, 2001). And it is a fact that the organizational structure of textbooks has proven difficult for many students.

Narrative texts or stories, on the other hand, have been occasionally used by science teachers (e.g. anecdotes, short stories) in order to illustrate an idea (e.g. Newton's apple to illustrate gravity, Archimedes' bathtub discovery to illustrate buoyancy). However, over the last two decades, storytelling has been receiving more and more attention. The literature on the subject testifies to this fact, even though this literature is not extensive (e.g. Avraamidou & Osborne, 2009; Clough, 2011; Hadzigeorgiou, 2006b; Hadzigeorgiou & Stefanich, 2001; Klassen & Froese-Klassen, 2014a; Millar & Osborne, 1998; Milne, 1998; Solomon, 2002). Science teachers, of course, can change an expository text into a narrative one, in order to take advantage of the instructional value of the latter.[6]

In reviewing the literature on storytelling in science education[7], one can find the use of exclusively fictional stories or historically based stories (see Avraamidou & Osborne, 2009; Hadzigeorgiou, 2006a, b; Klassen, 2006a, 2009). And according to

---

[5] The power of storytelling to make an idea interesting should be recognized. Even such an obvious idea, as 'A moving object does not come "naturally" to rest (i.e. the Aristotelian idea that falling objects will reach the ground and come to rest because that is where their natural position is), as there must be a force to do this (i.e. to bring the object to the state of rest)', can be made interesting through a story that provides the historical context in which this idea was embedded (see Appendix A, third story).

[6] **Expository text**: And with this simple powerful telescope, we can see many details when we use it to look up into the night sky. The Moon may look like a smooth ball of light covered with dark spots, but on closer look through this telescope, we can see deep valleys and great mountain ranges. Through the telescope we can now see all the different marks on the Moon's surface.

Narrative text: When Galileo looked through his new telescope, he could see the surface of the Moon, and so he began his first close look into the space. He slept during the day in order to work and see the Moon at night. Many people thought that the Moon was a small ball with a light of its own. Now that Galileo had a closer look through his new telescope, we realized that the Moon's surface had mountains and valleys.

[7] Most of the studies reported in the literature are descriptive (with or without exemplars), while there are also a few empirical studies (see Klassen & Froese-Klassen, 2014a).

this literature, the storytelling approach to science teaching and learning has ranged from the smallest stand-alone story element, such as the vignette or anecdote, to the largest story-like structure, such as a curriculum unit unified by a theme or storyline (see Klassen & Froese-Klassen, 2014a).

The plot of the story, depending upon instructional goals, curricular time, and other instructional elements (e.g. using storytelling to teach a unit on force and motion or a single lesson on the concept of wave and/or the nature of science), can be taken entirely from the history of science[8] (see Appendix A) and can be based on a single real historical event (e.g. the debate between Lord Kelvin and T. H. Huxley about the age of the universe) or even on an undocumented historical event (e.g. Newton sitting under the apple tree and thinking about the falling apple, Galileo dropping bronze balls of various weights from Pisa's Leaning Tower). While in the first two cases, the history of science should be portrayed as accurately as possible through the use of both primary (original) and secondary sources, in the third case, a plot may be further developed, for the purpose of making the story engaging, without reference to the history of science. However, in all three cases, the science content to be learned and the nature of science should be embedded in the plot of the story (Hadzigeorgiou & Stefanich, 2001).

What is important to bear in mind, with regard to creating stories for school science education, is that in the case in which teachers or science educators do not find an exciting story based on historical events (e.g. as is the case with the story about Otto von Guericke's Magdeburg experiment, the story about Galileo observing the swinging motion of the lamp while attending mass at the cathedral, or the Nikola Tesla story, which can be used for the introduction and teaching the concept of atmospheric pressure, pendulum motion, and AC electricity, respectively), they do a simple 'trick': they create themselves a story (or narrative), based upon a personal or common experience, which illustrates the human element and its relation to the science concept(s) to be taught (in a way that students become aware that these concepts, in fact their significance, are related to their own life). For example, an experience in an amusement park or playground can be the basis for a story about the concept of force, while an experience in a forest can be the basis for a story about the significance of photosynthesis. Or they create an entirely fictional story

---

[8] In utilizing history as the raw material for science stories, one can find two approaches: (a) the history of science is viewed in light of current knowledge, and (b) the history of science is viewed and interpreted in the light of the context and knowledge of the past time. While the first approach has been criticized on the grounds that application of present-day standards is inappropriate because historical figures and ideas can be placed and viewed only in the context of the past time, the second one has been criticized on the grounds that history cannot be interpreted when comparisons to the larger context cannot be made. Moreover, there is an argument that a focus on chronologies of events restricted to the local historical context is uninteresting to the non-specialist. There is also an argument that what is referred to as 'internal history', that is, the events describing the scientists' work and which (events) led to the development of an idea (written primarily by the scientists themselves), could provide a distorted view of the nature of science itself. Even though an 'internal history' can provide a first hand and thus an official account of the origin of scientific ideas, it nevertheless tends to romanticize the events and portray science as an inevitable consequence of the force of progress (see Klassen & Froese-Klassen, 2014a).

(see footnote 9 in this chapter for such a story, which introduces the concept of dissolution), for the main purpose of engaging students and especially young children with science. The students tend to identify with the character(s) of the story, who is (are) about of the same age as the students. The plot presents a challenging problem situation that the character(s) must solve. The students can help the character(s) by trying to solve the problem situation. Such fictional stories are just like the short 'mystery stories' used to introduce students to scientific inquiry as is discussed later in this chapter (see also Appendix G).

On the other hand, the teacher can use a story about a technological application (past or present)—based upon historical records—in order to introduce students to a specific concept. For example, a story about the radio or the radar can introduce students to such concepts as electromagnetic wave and electromagnetic spectrum, while a story about the airplane can be used to introduce students to the physics of aerodynamics (i.e. the phenomenon of flight, the concepts of air resistance and lift). Such stories, in addition to introducing students to certain science concepts and ideas, help students become aware of the fact that human life has always been affected, in one way or another, by some kind of technological application of scientific knowledge. This is the issue I take up later in this section when I discuss the purposes/functions of storytelling and more specifically the idea that storytelling can help humanize scientific knowledge.

4.4 The Characteristics of a Science Story

Making a good story, even one that is based exclusively upon historical events (taken from both original and secondary sources), is not an easy task. Jerome Bruner did remark that 'we know precious little in any formal sense about how to make good stories' (Bruner, 1986, p. 14). It also deserves to be pointed out that the interest in using storytelling in science teaching does not guarantee that the guidelines for the construction of planning/teaching frameworks are appropriate or clear. Klassen (2009) maintains that creating good stories is a challenging task, especially for science educators without a background in the humanities.

If a story is to accomplish its two fundamental tasks, namely, to communicate effectively science ideas and, at the same time, to involve the listener's feelings in relation to those ideas, then attention should be paid to how the story is written (or more precisely how it is 'crafted'). Even though 'storytelling' (i.e. the actual telling of the story) can play an important part in the process of engaging the listeners in the plot of the story, it is 'story making' that determines the listeners' engagement and how effectively the plot and the ideas in it become understood and retained.

Apparently, this is not easy to happen if the story simply provides a historical context. Some elements, in order to make the story engaging, need to be considered. These elements, according to the literature on the selection or creation of an engaging story, are coherence, problematization, mystery, characterization, mental imagery, conflict and problem resolution, challenging previous knowledge, irony, and raising questions (Altman, 2008; De Young & Monroe, 1996; Hadzigeorgiou &

Fig. 4.1 A framework for the development of romantic understanding (Hadzigeorgiou, Klassen, & Froese-Klassen, 2012)

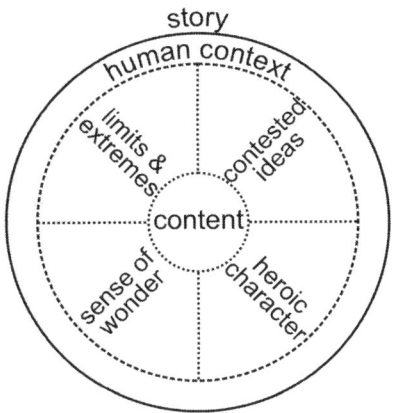

Stefanich, 2001; Klassen, 2009; Kubli, 2001; Norris, Guilbert, Smith, Hakimelahi, & Phillips, 2005). Egan's (1997, 2005) planning frameworks certainly consider these elements and provide useful guidelines for teachers without adequate knowledge of narrative theory and/or experience (see Fig. 4.1 in this chapter; see also Appendix B).

According to Avraamidou and Osborne (2009), the elements that play a significant role in a narrative and must always be included in it are *purpose* (to understand the natural and human world), *events* (a sequence of events that are connected to each other), *structure* (a temporal relationship of events with a beginning, middle, and end), *time* (the past), *agency* (human actors or entities that cause and experience events), *narrator* (the one who is either a real character or, alternatively, a sense of a narrator), and *reader* (the one who must interpret or recognize the text as a narrative). These elements are certainly important to bear in mind, but for the purpose of writing a story for instructional purposes, Klassen and Froese-Klassen's (2014a) list appears to be more helpful (see Table 4.1). For example, because understanding science ideas is the purpose, or one of the purposes, of telling stories, *consequential coherence* is a crucial element that needs attention when writing the plot of a story. Even though strict chronological sequence is not necessary (as the plot might contain flashbacks and flashforwards), the plot needs to have *consequential coherence*. Klassen and Froese-Klassen use the following two passages to demonstrate the notion of *consequential coherence (CC)*:

(A) A passage without CC: *Ohm took up a teaching position in Cologne; he performed an experiment to determine resistance; and he published his experiment.*
(B) A passage with CC: *Ohm understood the mathematical theory of heat; he applied the theory to electricity by analogy; and as a result, Ohm understood the mathematical theory of electrical resistance.*

In the case in which the story is based on a sequence of historical events, most of these elements are easy to identify. For this reason Hadzigeorgiou (2006b) paid more attention to the following elements: the main protagonists, wonder, mental

Table 4.1 An evaluative list of features for the construction of a science story

Character(s) taken from the history of science
Actions that are consistent with the historical record
Situations or states
Consequential coherence of the characters' actions
Past time
Plot structure with rising action and climax
Critical choice made by the main character
Appropriate science and NOS content

Source: Klassen and Froese-Klassen (2014a)

imagery of the major events, and science ideas (content knowledge and knowledge of the nature of science). However, in the case in which the story is written for the purpose of teaching students a specific science topic (e.g. force and motion, dissolution, gravity, sound) which is not based upon specific historical events[9], then attention should be given to all the elements provided in Table 4.1. It goes without saying that, in all cases, science content ideas must be embedded in the story. Whether the story is made up by the teacher or is based upon historical events, science content ideas can either be implicit in (i.e. emerge from) the plot or explicitly stated in the plot.[10]

In looking at Table 4.1, one may raise a question with regard to the absence of the notion of problematization (i.e. awareness of a problem), given that in the

---

[9] Such an example is the following fable with the two donkeys, which invites and challenges young children to think about the events in the plot and which introduces them to the physical process of dissolution. *Once upon a time, there were two donkeys. They had been asked to carry something to the local mill. The donkey who would reach there first would get more hay to eat. They did know, of course, what that something was; they just had to do the job. They both agreed to this job and went to get their cargo. They then realized that one of them had to carry a sack of salt, while the other one a sack full of sponges. They decided to flip a coin, because they could not decide who was going to get which sack. And they did, and the donkey who got the sack with the salt was not so happy. Anyway, they started to walk towards their destination, but the donkey who was carrying the salt was feeling tired; his walking demanded really great effort. The other donkey looked really relaxed and moved without any effort, as if he was not carrying any load at all. The time came though when they both stopped at the lake. They both felt that they had to cross it in order to reach the mill. As they started to get into the water, the donkey that was carrying the sacks of salt realized that he began to feel much lighter, while the other donkey carrying the sack filled with sponges began to feel heavier. They did cross the lake, but they arrived at the mill in different times.*

[10] In an attempt to demonstrate the dangers of AC power, Edison sponsored an electrical engineer to travel the country electrocuting animals with both DC and AC. Because the frequency of AC confuses the heart, animals that are electrocuted by AC die, whereas animals that are electrocuted by DC are stunned but survive. Edison used these so-called experiments to contrast the danger of AC with the relative safety of DC. We know that the effect of any type of electric current on a human being is very difficult to predict, as it depends on a number of factors (e.g. the condition of the skin, the amount of fluid in the body, and the point of contact). Tesla, however, had been experimenting with very high frequency currents, which, as he showed, did no harm. With his theatrical flair, Tesla could draw sparks to his own fingers and even walk through sparks without being hurt. He had realized that the high frequency of the current kept it on his skin. It was this strange effect, known as the skin effect, which made Tesla famous. He even sent sparks to the audience, making people realize that AC current, at least as used by him, was safe (Hadzigeorgiou, Klassen, & Froese-Klassen, 2012).

4.4 The Characteristics of a Science Story

context of science teaching and learning, a problem is crucially important for making students think. However, while it does explicitly appear on this list, problematization is included in the *situations or states*, since it is these situations or states that pose the problem.

Another important characteristic of an engaging story that is not often discussed in the literature is what is referred to as 'personal background information'. This can be biographic knowledge about the scientists themselves and their struggles (see next section in this chapter). This information can help students develop a kind of contextual knowledge and more specifically a kind that can be called 'knowing with', as opposed to the two most common kinds, namely, 'knowing that' and 'knowing how' (Hong & Lin-Siegler, 2012). The results of a quite recent empirical study, which reported higher achievement for a group of students, who listened to a story about Tesla's life and work in order to learn about AC electricity (Hadzigeorgiou, Klassen, & Froese-Klassen, 2012), are evidence of the role of 'personal background information'.

There is another question, though, that also needs to be answered: do stories based upon actual, historical events have greater value than made-up stories? Given that any historical event is open to the critique of interpretation (i.e. it may be a historian's interpretation and not an event that actually took place), this question, at least from pedagogical view, should not concern science teachers. Milne (1998) does make a good point regarding the authenticity (or realness) of some well-known historical events (presented as historical facts) that are frequently used by science teachers. The power of these examples[11] does not lie in their factual significance, but in the fact that they do tell a good story; they do capture the students' imagination. In so far as they serve their pedagogical purpose(s), they are valuable in science education. The next section deals with this specific issue.

In closing this section, the important message that needs to be understood is that making a story, especially an attractive and an instructive one, requires more than a sequence of (historical or fictional) events. The beginning of the story, its main part, and its end are all crucial. They each serve a purpose. The beginning is crucial to capture the pupils' imagination, to evoke a sense of wonder, and to provide a general background for the main plot of the story. The end is also important to provide a closure and a moral (e.g. a significant idea). For the main part of the story, the ideas to be learned should be made clear in the plot.

---

[11] Galileo dropping cannon balls from the Leaning Tower of Pisa; Galileo responding at the end of his inquisition trial, 'And yet it moves'; Kekule's dream and James Watt and the boiling kettle. It might be difficult in these science stories to separate history from fiction.

4.5 The Functions/Purposes of a Science Story

The rationale for using storytelling in science education is quite well known: it can make science more interesting and engaging (the lesson can be really entertaining), it can illustrate a concept, it can provide opportunities for contextualized and meaningful learning, it can make content memorable, and it can also help integrate science and literature (Hadzigeorgiou & Stefanich, 2001; Klassen & Froese-Klassen, 2014a; Kreps-Frisch, 2010). Moreover, storytelling can help students develop their imagination—a neglected but nonetheless crucially important intellectual ability in science education—and also their sense of anticipation. In regard to the latter, it has been observed that students are anxious to listen to the next event in the plot of the story, to the resolution of a conflict or controversy, and to the solution of a mystery or problem. A story, in particular, which presents, or simply includes in its plot, a mysterious event, situation, or phenomenon, makes the students so anxious that they cannot wait to find out how this mystery is explained (Hadzigeorgiou & Fotinos, 2007; Hadzigeorgiou & Stefanich, 2001). However, what is not stressed, or, at least, not explicitly mentioned—even though it is quite crucial—is the fact that a story can help students understand the problem situation that led to the development of scientific knowledge. Popper (1972) did point out that to understand a theory, one has to understand the problem situation from which it arises.

A story that presents, or simply includes in its plot, a problem that a scientist faced when she or he was trying to explain something and the proposed solution (e.g. the stability of the atom despite the energy radiated by orbiting electrons and Bohr's proposal of discreet electron orbits, black body radiation and Planck's idea of quantization of energy), a controversy (e.g. between Galvani and Volta regarding the nature of electricity) or an experiment (e.g. Galileo's experiment with free fall that proved the fallacy of Aristotelian theory of motion, Count Rumford's experiments that provided evidence against the idea of caloric), not only provides the context but also the problem that gave birth to a scientific concept. Even a scientist's intellectual struggle to accept an innovative idea (e.g. Einstein accepting quantum theory) may also make students, who have difficulty to understand a theory or an idea, focus on the conceptual problems that led to the formulation of innovative ideas. Perhaps it is assumed that the historical context is in and of itself enough to capture the students' imagination and help them become interested in the science lesson and that this interest or engagement is enough for understanding to take place. Or that the historical context necessarily helps them understand the problem situation that led to the development of a concept. Neither is always true and this is why storytelling, if it is to be effective in fostering understanding, should include both the historical context and the specific problem situation, which led to the development of a science concept (Hadzigeorgiou & Stefanich, 2001; Hadzigeorgiou, Klassen, et al., 2012).

A story about heat (see Wilson et al., 1965), for example, could help students understand the difference between heat and temperature, the idea of specific heat capacity, and the link between heat and motion, if its plot included:

- Joseph Black experimenting with a burner, an iron disc, and a pan of water and concluding that 'heat is obviously an amount of something, while temperature is the degree of hotness' (Wilson et al., 1965, p. 31).
- Count Rumford supervising the boring of a brass cannon in Bavaria and experimenting with the brass shavings and solid brass (in order to test whether equal amounts of hot brass shavings and hot solid brass cause equal or unequal increases in the temperature of the water in a bucket) and also demonstrating to the astonished bystanders that he could boil water in the absence of a fire (i.e. he had fitted a water-filled container around the tip of the cannon barrel, and the friction between the drill and the cannon increased the water temperature to such a degree that in about two and a half hours, the water literally boiled).
- Bernoulli's thought experiment involving a single molecule/particle moving around inside a sealed container and hitting the walls of the container and his mathematical proof (after he extrapolated to the case of billions of molecules/particles inside the container) that, in the case of a gas, increases in temperature cause increases in pressure in the container (well before Charles and Gay-Lussac were able to observe such a change).

Even though students could learn about these ideas, through a variety of methods, it is through storytelling that they are invited to take part in the drama and follow the logic, behind the development of scientific knowledge. Thus they come to share the same conceptual problems that puzzled the scientists of the eighteenth and nineteenth centuries. Such a story helps students see the conception of heat as molecular motion, not as something arbitrary that simply must be learned, in one way or another, but as a consequence of the scientists' endeavour to explain the natural world. I believe that such a story 'conceptually prepares' students for Mayer's and Joule's ideas regarding the equivalence between mechanical work and heat.

What follows below is a discussion of six specific functions that a story can perform, which deserve particular attention by the science education community. These are (a) the humanization of scientific knowledge, (b) the introduction of ideas from the nature of science, (c) the encouragement and development of romantic understanding, (d) the introduction to scientific inquiry, (e) the presentation of thought experiments, and (f) the development of environmental awareness. These functions certainly subsume the most frequently cited ones, like illustration, contextualization, engagement, and meaningfulness, but they nevertheless represent the pedagogical importance of storytelling from a more general educational perspective.

4.5.1 Storytelling as a Means to Humanize Scientific Knowledge

Humanizing, as the word itself suggests, must be equivalent to portraying scientists as human beings who experienced not only successes but also failures and who really struggled to achieve what we now know and take for granted. And the humanization of scientific knowledge must refer to the association of such knowledge with the human element. It has long been recognized that a story can help humanize science, by presenting the human context relating to the content to be learned (Hadzigeorgiou & Stefanich, 2001). If knowledge, as Egan (2005) argued, cannot be considered without reference to the hopes, ambitions, struggles, successes, and failures of those who created this knowledge—because the development of this knowledge is precisely tied to all these elements—then storytelling is the best, and perhaps the only, medium that can be used to present the human context in which knowledge was created and developed.

It is indeed through storytelling that students can learn, for example, about Galileo's heroic courage, who, although blind during the end of his life, continued to work on his revolutionary ideas, assisted by his son, or about Benjamin Franklin's passion for knowledge, who, during a severe thunder and lightning storm, climbed to the top of a church and conducted his famous 'key and kite' experiment, in order to show the connection between lightning and electricity. It is through storytelling that students can learn about such controversies, as those between Volta and Galvani, in regard to the nature of current electricity, or between Edison and Tesla, in regard to the superiority of alternating current (see Appendices A and B). And it is also through storytelling that they can learn about how an original idea, like the one Galileo had, namely, to incorporate pendulums into clocks in order to make them more accurate, was not recognized by Christian Huygens, who conceived of the same idea but after Galileo died.

These examples do show that storytelling, by presenting the human element inherent in the scientific process, actually presents the human aspect of the nature of science (i.e. the role that personal ambitions and interests play in scientific investigations and work in general), something that cannot be seen in the scientists' 'final products' (i.e. as these appear in journals and conference papers and generally in their published work).

Another element, however, that needs particular attention, with regard to the humanization of knowledge, is biographic knowledge. Such knowledge, if imbedded in the plot of a story, can facilitate science content knowledge acquisition. In other words, from an instructional point of view, it is more effective to combine science ideas with biographic knowledge (Mandler, 1984). It appears that biographic information helps increase the social presence of scientists in students' minds. Knowledge of the scientists' experiences, including the difficulties they faced and their struggles, helps develop in students a feeling of closeness to the scientists themselves, something that can facilitate learning (Hong & Lin-Siegler, 2012; McKinney & Michalovic, 2004). There is empirical evidence that if such biographic knowledge includes the scientists' struggles, then learning science is more effective. More specifically, there is evidence that students who were presented with struggle-oriented background information performed better compared with students who were presented with achievement-oriented personal background information (Hong & Lin-Siegler, 2012).

It is important to note that because storytelling has the capacity to develop a feeling of empathy in students (see Klassen & Froese-Klassen, 2014a; Solomon, 2002)—which is, perhaps, impossible for students to develop through other teaching approaches—storytelling can be seen as one of the best ways to humanize science education, especially physical science (Hadzigeorgiou, 2006b; Stinner, 1995).

However, it deserves to be pointed out that storytelling can humanize scientific knowledge in another way, namely, by helping students 'share' in the creation and development of scientific knowledge. This can be done if students listen to science stories and then they replicate experiments that have been conducted by famous scientists (e.g. Galileo, Newton, Volta, Faraday). The experiments, of course, must be carefully chosen so the students are successful in conducting them—even though failure in doing an experiment can happen. Thus, the students can see for themselves that science is indeed an activity that can be done even by students themselves.

Or they can listen to a narrative, which, at some point in its plot, poses a problem that needs to be tackled by the character(s) in the story. The students can either listen to the narrative and the solution the characters in the story find, or they can attempt to help the character(s) approach the problem, in which case, they may come up with the concept or idea that is supposed to be learned through the narrative. In both cases though they become aware that scientific knowledge is tied, in one way or another, to human life. For example, the teacher can use a narrative about a group of friends, who went up a mountain, but at some point they got lost, and they must now find a way to let the people living in the wider area at the foot of the mountain know that they got lost. The darkness, the fact that they lost their compass, and also the fact that there is no signal, and therefore their mobile phones cannot be used, make things really difficult. Nonetheless they must find a way to send a message to the people down below. This is an excellent opportunity for students to think of various ways to send a message to the people living at the foot of the mountain. In their attempt to categorize their ideas, they become aware that their message can be sent either through an object (i.e. an object with mass carries the message, as is the case of a written message in a bottle placed in a stream) or through something that has no mass, like a fire or the sunlight (reflected by a mirror). Thus, the concept of wave 'emerges out' of the problem posed by the narrative itself (Hadzigeorgiou & Stefanich, 2001).

By the same token, a narrative that presents the interrelationships of various technological applications, both past and present, and human life can be used to introduce science concepts and ideas. A good example is a story about lighthouses, which can be used to introduce students to the physics of optics and more specifically to the concepts of reflection and refraction and the prismatic lens.

Even though such concepts can be introduced through a variety of stories, the story about lighthouses helps students become aware that lighthouses were symbols of safety and hope, as they (lighthouses) determined the difference between a safe passage and a wreck. Such a story helps students understand how human life was once dependent both upon the lighthouse keepers and those scientists, who worked on the problem of long-distance visibility (i.e. how to make the light from the lighthouse visible from long distances in the sea), and has been found to be an engaging and effective way to teach a unit on light (i.e. the concepts of reflection, refraction, and the various kinds of lenses and their uses). Knowledge of optics, as high school students thought, was not something that was 'imposed' on them but rather something that emerged either as 'a prerequisite for understanding the problem of safe voyages, and especially the challenges that both housekeepers and scientists of the time faced', or as an 'illustration of the relationship of scientific knowledge and human life'. Awareness of that relationship was indeed a source of a sense of wonder for many students, as they themselves reported (Hadzigeorgiou, 2007). A teaching framework for a unit on light utilizing the 'lighthouse story' can be seen in Appendix H.

4.5.2 Storytelling for Introducing Ideas from the Nature of Science

It is a truism that any science story should portray the nature of science. Students should be helped, by listening to a science story, to become aware of this nature. This can happen if the plot (the events) of the story and the ideas embedded in it portray scientists[12] and their endeavour as they really are. In such case, students are not just helped to view science as a human endeavour; they are helped to see and understand the tentative, provisional nature of scientific knowledge and, most importantly, both the scientists' successes and failures (see Heering, 2010; Solomon, 2002). There is evidence that 'historical narrative is a 'good opener' to teach the nature of science' (Schiffer & Gueria, 2015, p. 409).

In regard to presenting science as it is, that is, with its successes, its failures, and its problems (i.e. practical, intellectual), a teacher does not just encourage engagement with science; he or she helps students to see the scientists portrayed in the story as real human beings (and not as superhumans with exceptional intellectual powers) and thus help them to identify with those scientists.

In regard to the tentative, provisional nature of scientific knowledge, the plot of the story should incorporate real events from the history of science that document

---

[12] The portrayal of a scientist as a lone-star genius or hero does not contribute to students' understanding of the nature of science. This image of a scientist can most likely discourage engagement in science. As Heering (2010) writes, the story of the lone genius is one of the various classical myths in the history of science. Indeed most scientists had assistants and collaborators who played an important, if not crucial, role for the success of the project.

(a) the change of scientific ideas over the course of time, (b) the existence of various (competing or complementary) theories for explaining a natural phenomenon, and (c) the irrationality of scientific progress.

Apparently, stories that can document the change of scientific ideas are easy to find in the history of science, which, in fact, testifies to the growth of scientific knowledge, as a result of the reconsideration, modification, or rejection of previous ideas (e.g. the story of motion, from Aristotle to Galileo and Newton; the story of heat, from Joseph Black and Count Rumford to Robert Mayer and James Joule; the story of gravity, from Newton to Einstein). It is more difficult, however, to document the irrationality of science and scientific progress. The reason may very well be that the notion of irrationality does not appear to describe science, which seems to be a rational endeavour. And it really is, unless one becomes aware of the fact that there are events in the history of science that testify to the downright rejection of 'crazy' or irrational ideas, which, however, ended up being very rational, since they now count as standard scientific knowledge.[13]

Even though an 'anarchist theory of science' (Feyerabend, 1993) can help one understand the irrational character of scientific progress, with younger students, this can become easier through stories that include historical events documenting the initial rejection of 'crazy' or irrational ideas, but which are now accepted as scientific ideas. These events can show that judging an idea as irrational or impossible (at the time of its conception) does not help scientific progress. The important idea that needs to be infused into the plot of the story is that the certainty with which certain scientists of the past defended the rejection of some ideas was eventually shattered.

Heering (2010) argues that in order for the story to help students understand the nature of science, four ideas need particular attention and, of course, need to be included in, or emerge from, the plot of the story: (a) scientific knowledge, while

---

[13] The history of science once again can provide many examples and therefore material to be incorporated into the stories (see Di Trocchio, 1997):

- In 1896 Lord Kelvin claimed that aviation was impossible (no thing heavier than air can fly).
- In 1907 Lord Kelvin claimed that the atom was impenetrable.
- In 1917 Robert Millikan claimed that humankind will never be able to utilize the energy released by a nucleus.
- In 1933 Lord Ernest Rutherford claimed that the idea of utilizing atomic energy is absurd.
- In 1937 Niels Bohr did not believe that atomic energy can prove practically useful.

It is interesting to note that even Einstein himself, who did not believe at first in the implications of his famous equation concerning the equivalence between mass and energy, had stated that it was unthinkable for humankind to be able to use those huge amounts of energy, as predicted by that equation (Di Trocchio, 1997). What is really ironic though is the fact that Max Planck (1933), who had said that the 'pure rationalist' has no place in the field of quantum physics—that is, a field based upon the very notions of uncertainty and unpredictability—advised Einstein to drop the idea to include gravity in his theory of relativity, because such an idea was indeed absurd (Miller, 2001). History, of course, has the last word, and in that particular case, we all know that that absurd idea ended up as a major idea of the general theory of relativity, that is, one of the greatest—perhaps the greatest—achievements of Western thought, and only storytelling can make these ideas and the events behind their rejection or acceptance alive before the students.

durable, has a tentative character, (b) people from all cultures contribute to science, (c) scientists are creative, and (d) science is part of social and cultural traditions. The stories about the development, for example, of the concepts of force and motion, gravity, heat, and light, do help students understand all of the above ideas. The story about the heat, in particular, can show that the development of scientific knowledge is the result of the contribution of many scientists (e.g. Joseph Black from Scotland, Benjamin Thomson or Count Rumford from America, Humphrey Davy from England, Jacque Charles from France, Robert Mayer from Germany, James Joule from America) and thus dispel the myth of the 'lone-star scientist'. However, such a story shows students another important aspect—perhaps an indirect consequence or rather by-product of scientific work—that is not often associated with the nature of science. This aspect refers to the recognition that a scientist deserves. For example, Joule's contribution to the discovery of the nature of heat had been recognized more in America and Britain than in Continental Europe, while exactly the opposite happened with Mayer's contribution. Of course, 20 years later that issue was finally settled, as it was recognized that both scientists deserved credit (Wilson et al., 1965). But with Daniel Bernoulli, things are different, since his pioneering work that made a major contribution to our understanding of heat as molecular motion remained unrecognized. His work was well ahead of his time, not, of course, because he was not recognized as a scientist, but because of the way science develops:

> Bernoulli's work was at least a century ahead of his time. His theories came in the period when caloric was the most popular explanation of heat, and he went unheeded, when, in his long-neglected work, Hydrodynamica, he turned his back on caloric by saying that "heat may be considered as an increasing internal motion of the particles. Indeed, Bernoulli's postulation was made more than 100 years prior to Joule's final statement on the equivalence of heat and motion. (Wilson et al., 1965, p. 53)

It should be noted that storytelling, perhaps more than any other approach, can help present to the students a more realistic image of scientific inquiry, since it portrays the scientists' life events and their experiences and how they (life events, experiences) are associated with scientific knowledge.[14] Although inquiry science (see next section) can help students become aware of the tentative nature of knowledge (i.e. by becoming aware of the fact that there are so many parameters involved in the investigation of a problem that they can never be absolutely certain of the result of their study or that there are so many factors to be considered in a given investigation that knowledge of all those factors is impossible and hence our knowledge is incomplete), storytelling represents a direct approach to learning about the nature of

---

[14] No doubt the Tesla story (see Appendix A) can promote knowledge of science as a process, as a human endeavour. More specifically, the story promotes the view that (a) science has a personal dimension, and (b) mainstream, traditional ideas have to be confronted and contested in order for scientific progress to take place. Thus the contribution of romantic understanding to one's understanding of science as a human endeavour is catalytic. Such understanding, in turn, is crucial for an understanding of the nature of science. Given the considerable attention over the past two decades (due to scholarship in the philosophy, history, and sociology of science) to students' understanding of the nature of science (NRC, 1996; Schwartz, Lederman, & Crawford, 2004), the idea to use storytelling in order to foster such understanding appears promising.

scientific knowledge—in line, in my view, with the one recommended by Lederman (1998). It is a direct approach because it explicitly presents ideas—in addition to the tentative nature of scientific knowledge—that only through storytelling can become understood by students, especially young children.

4.5.3 Storytelling as a Means to Introduce Scientific Inquiry

Inquiry, a teaching/learning method that goes back to ancient Greece and especially Socrates, has one distinctive feature: the posing of questions in an attempt to understand an idea, a situation. The most crucial aspect of the inquiry method is that 'it allows both student and teacher to become persistent askers, seekers, interrogators, questioners, and ponderers' (Olrich, Harder, Callahan, & Gibson, 2001, p. 320). And this is the reason why there has been an intense interest in what is now called 'inquiry science' or 'science as inquiry' (e.g. Bass, Contant, & Carin, 2009; Martin, Sexton, Franklin, Gerlovich, & McElroy, 2008; Stefanich, 2001). Indeed, The National Research Council has recommended inquiry science be 'a critical component of a science program at all grade levels and in every kind of science' (NRC, 1996, p. 214).

Central to inquiry science is the idea that students, like scientists, use a set of interrelated processes, by which they ask questions and investigate natural phenomena. These processes refer to the well-known science processes, such as observation, classification, prediction, formulation of hypotheses, and experimentation (NRC, 1996, 2007). The use of these processes requires students to move beyond rote memorization and become independent thinkers and problem solvers (Hadzigeorgiou & Garganourakis, 2010; Stefanich & Hadzigeorgiou, 2001) and even helps them with better retention of subject matter (Renzulli, Gentry, & Reis, 2004). The question, of course, is how to encourage students to pose questions in the first place. For it is these questions that will make them pursue investigations in order to answer them (questions) and thus understand how the natural world works. Fostering students' curiosity and their sense of wonder (see Chap. 6) is the most plausible answer. However, this is a very challenging task, and storytelling can help towards this goal, especially if the story is such that captures the students' attention and engages their imagination through mysterious situations and unexplained phenomena featured in its plot. Depending on grade level, students and young children alike can face the challenge of helping the central character of the story solve a mystery or problem.

For example, instead of asking students questions that are supposedly challenging, like 'what are the properties of white powders (such as salt, sugar, cornstarch, flour)', the teacher can use a story or narrative, through which he or she presents these 'puzzling powders', and the students can then play the role of a detective, who will try to solve the problem or mystery presented by the plot of the story. Thus students learn about the properties of those substances as a result of their investigations (see Appendix G for problems/mysteries in a narrative/story form). Also a question such as 'What do you think you might find on a forest floor or on a tree?' can be more engaging if it is included, in one way or another, in the plot of a story. In this particular case, the question can be asked indirectly or even directly by a character featuring in the story (Hadzigeorgiou, 2010).

The 'trick' here is to find a plot or create one, in which the central idea to be learned is not the answer to the teacher's question but the answer to a problem or a mysterious situation. This answer, however, will be the outcome of students' inquiry into the problem or mysterious situation. For example, such questions as 'What affects the temperature change of water heated by sunlight?' and 'How can we see what's inside the next door room without leaving our room?' can be posed by the plot of a story or narrative, and students, through their own investigations, can answer them and thus learn about how different colours absorb heat from sunlight and about mirrors and the law of reflection, respectively. The empirical evidence regarding the role of storytelling in fostering student inquiry is encouraging, both in terms of its impact on actual learning outcomes[15] and in terms of motivation.[16]

In the case, of course, in which a teacher finds a strange or mysterious situation or incident, which relates to the science content of the lesson and which helps evoke in students a sense of wonder, she or he can tell this incident as an introduction to the lesson and the students' inquiry. For example, a short narrative that presents such a strange and perhaps mysterious incident is what happened on February 2, 1901, that is, the day Britain mourned Queen Victoria's death: the booming of a cannon was not heard by people in the surrounding London area but instead by people about 150 km away (Stevens et al., 1966). This is a good way to introduce students either to the concept of sound, but through the strange phenomenon of refraction, or to the phenomenon of refraction per se. Other strange and mysterious incidents can also be used to introduce students to the concept of sound and other strange phenomena such as sound diffraction, interference, and absorption. Scientific inquiry will start once questions are posed and investigations are carried out, in order for students to answer their questions. For example, in the case in which the mysterious incident refers to sound absorption (e.g. in a snow-clad valley, no sound could be heard despite the fact the two friends, who had lost each other in last night's blizzard, were

---

[15] Corni et al. (2010) have reported that grade 4 children, who listened to the adventures of a lively and creative character and who helped this character solve problems, regarding the flow of water in an aqueduct and the filling of swimming pool, did develop problem-solving and other inquiry skills (e.g. interpretation, experimentation). They observed a shift from description to interpretation. Kokkotas et al. (2010) have also reported that grade 6 students who listened to a story about electricity developed inquiry skills, such as hypothesis exploration and formulation and interpretation. They also reported that students developed metacognitive skills, such as comprehension of new knowledge. Hill and Baumgartner (2009) have reported that high school students, who listened to the story of 'FloJo: The World's Fastest Woman', ran their own race experiment and thus learned about kinematics.

[16] Morais' (2015) study showed that elementary school children (ages 8–10), who listened to a story (which was followed by hands-on activities), enjoyed the story and the combination of story with hands-on activities. Similar findings have also been reported by teachers who used the story about the Galvani-Volta controversy, in order to teach about current electricity (Hadzigeorgiou, 2006b).

4.5 The Functions/Purposes of a Science Story

very near to each other and shouted at the top of their voice), students can inquire about how the landscape might affect the absorption of sound (i.e. the snow or other materials around there) and then test various materials, both porous (e.g. sponges, rugs) and non-porous (e.g. paper, plastic), in order to see which of them best absorbs the ringing sound of a noisy alarm clock (i.e. by covering the alarm clock with these materials) and also whether they are successful in completely containing the sound of the clock.

In the case in which the history of science provides the material for the stories,[17] the life and work of famous scientists (e.g. Newton, Galileo) can be used to introduce students to scientific inquiry. Such stories, depending upon curricular requirements (i.e. teaching a single lesson or a unit), can present either critical incidents from the scientists' lives (e.g. Galileo's experiments with free fall help students reject their Aristotelian/intuitive ideas about motion) or the evolution of ideas, as a result of the ingenuity of a number of scientists (e.g. the evolution of ideas regarding force and motion from Aristotle to Galileo and Newton or the evolution of ideas regarding heat from Black to Rumford to Mayer and Joule). In both cases, the stories can invite students to investigate the ideas that led to the development of science.

Two examples can help illustrate the case in which the story is based upon a single historical incident/event. The first one refers to Galileo attending mass at the cathedral and the second one to the Magdeburg experiment:

Example 1: A teacher, whose objective is to introduce students to the notion of pendulum and its laws of motion through inquiry, can tell a story involving Galileo—at the age of 19 as a student at the University of Pisa—in a moment of reverence in the cathedral at Pisa. Even though this story can be made interesting if various and most likely undocumented facts are included (e.g. his mind was on his experiments on motion, so he became interested in anything that moved during the mass, he was looking up in order to notice details of the cathedral's interior, he simply found mass attendance boring and was thus distracted), the critical incident is at the moment when Galileo observed the rhythmic swinging motion of a lamp suspended on a long chain and noticed that, while the arc of its swing grew

---

[17] The history of science is replete with such stories. See the second and third story in Appendix A. While the second story focuses on Tesla's main life events and ideas, the third one presents the ideas developed by Galileo and Newton. Such a story can be used as it is, perhaps complemented with a few incidents from the personal life of both scientists, for the teaching of a unit on motion, or can provide the material for shorter stories referring to a specific idea and/or experiment. For example, Galileo's experiments with free fall at Pisa can be used to introduce students to the law of free fall, while the famous, and most likely anecdotal, incident concerning an apple falling on Newton's head can be used to introduce students to circular motion (if the plot of the story includes Newton enquiring about why the Moon does not fall and using an analogy that compares the motion of the Moon with the circular motion of a stone tied to a piece of string). On the other hand, Galileo's *Starry Messenger*, which is a book on his explorations of the heavens and his discoveries about the Moon, can be read to the students, before they begin to inquire about celestial bodies and the use of the telescope and even introduce them to such concepts as data reliability and scientific evidence.

shorter and shorter, the time the lamp took to get from one side to the other remained constant. Not having a pocket watch—pocket watches were invented much later after Galileo died—Galileo measured the time of each full swing with his own pulse beat (i.e. by counting the number of his pulse beats). Thus Galileo made an inference that no matter how short the swings grew, their period (i.e. the time of a full swing) remained constant. In modern language we would say that Galileo discovered an important law, namely, that the period of the oscillation of the lamp is independent of the amplitude of the oscillation. This story invites students to ask questions and conduct experiments in order to determine the accuracy of Galileo's observations and inferences (e.g. they can use their own pulse and regular watches and compare them) and also the factors that influence the period of a swinging pendulum (Hadzigeorgiou & Stefanich, 2001).

Example 2: A teacher, whose objective is to teach about atmospheric pressure through inquiry, can tell a story involving Otto von Guericke and his experiment. Guericke was one of the two candidates, who aspired to become the next mayor of the small town of Magdeburg in Germany. And because this town was not famous like other German towns or cities—and its people were fed up with not being considered inhabitants of a great town—Guericke promised to make it famous. After the first candidate was booed for his ideas (e.g. to make Magdeburg the cleanest town in Germany), Guericke wanted to really surprise the citizens of Magdeburg. So he brought two identical hollow bronze spheres, about 60 cm wide, put them next to each other, in such a way that they fit really tightly together, and thus no air or water could flow in or out. When doubt was voiced, in regard to how those two bronze spheres would be making Magdeburg famous, Guericke surprised them by saying that no one, not even two groups of eight horses, each pulling on either side of the sphere, could be able to pull the two hemispheres apart. And, to their big surprise, the demonstration proved him right, and he became the new mayor of Magdeburg. This story invites students to ask questions and investigate the concept of pressure, using Magdeburg hemispheres found in science supply stores (Isabelle, 2007).

It should be noted that storytelling can be integrated into any constructivist teaching/learning model, and it best serves its purpose if it is used in the engagement phase or during the exploration phase or even in both phases.[18] The point that needs to be made here is that, while some exploratory hands-on activities can initiate the teaching/learning sequence, it is the story itself that introduces students to the concept or idea to be learned. And this learning will be the result of students' investigations after (or sometimes during) the telling of the story. The story about

---

[18] Given the fact that storytelling can create interest in science, as it can include all the features of interest, Klassen and Froese-Klassen (2014b) propose the story-driven interest approach (SDIA), which can be the basis for learning sequences that include, apart from the narrative context, practical and social contexts as well. They recommend the episode concerning Galileo attending mass at the cathedral, which, after it is narrated, is complemented with hands-on activities, based on students' questions, and with group discussions (see also Chap. 1, section on 'Imagination and Science Learning').

Galileo and the swinging lamp can be used as a starter in the engagement phase in order to introduce students to the idea of pendulum motion. In such a case, the students explore pendulum motion after the telling of the story (Hadzigeorgiou & Stefanich, 2001; Klassen & Froese-Klassen, 2014a). On the other hand, the story about the Magdeburg experiment can be told after the students have explored the behaviour of the Magdeburg hemispheres, through hands-on, so the story is used to complement the students' investigations by introducing the idea of pressure (see Isabelle, 2007).

Nowadays storytelling can facilitate scientific inquiry if it is integrated with technology. Whether we are talking about interactive stories, which are relevant to the specific course content, or about the use of technology per se as an aid to students' investigations, the potential of technology to enhance the process of storytelling needs to be taken into account. Given that storytelling can introduce students to interdisciplinary connections,[19] the use of technology becomes indispensible. Galileo's story, for example, starting with his birth in Florence in 1567, moving on to his observations of the sky and his experiments and ideas on motion, his sentencing in 1633 to death (because he was a heretic) and his death in 1642, and ending with the launching of the Galileo spaceship in the late 1980s and his subsequent posthumous recognition a few years later by the Catholic Church (because he was indeed right about the Earth moving around the Sun), can be used to introduce students to inquiry science and interdisciplinary work, with technology playing a vital role (see Appendix A, third story, for more information about Galileo's life and work).

4.5.4 Storytelling as a Means for the Development of Romantic Understanding

As was pointed out in Chap. 3, romantic understanding can be stimulated and developed if the elements of romantic understanding are placed in a human context, which is readily achieved by integrating them into the plot of a story. In this respect, a story provides the structure, the human context, and also the narrative context, which is indispensable for the creation of emotional meaning and the generation of narrative thinking.

In order to facilitate the development of romantic understanding, the challenge for science education is to identify heroic qualities that relate to the specific content

---

[19] For example, a story about Leonardo da Vinci can be used to introduce students to interdisciplinary connections between art and science. Also a story about Antoine Lavoisier; his discoveries about the existence of minerals in water; the composition of water that we cannot make chemical substances disappear, only to change the form; his work as a tax collector; and his subsequent sentencing to death during the French Revolution (Arnold, 1997) can be used in interdisciplinary approach. However, interdisciplinary connections can be introduced through stories which focus on the mystery and development of certain concepts, like 'time' and 'energy'. In the case of the former, for example, starting from Father Time Cronus, as described in ancient Greek mythology, the concept of time can help introduce students to interdisciplinary connections regarding history, biology, chemistry, physics, earth science, and language arts.

to be taught and organize this content within a narrative structure. The heroic qualities, in particular, can capture the main narrative thread and thus help maintain students' attention from the beginning to the end of the lesson. The narrative should be crafted in such a way that the elements (i.e. the characteristic features) of romantic understanding are included in the story. Thus the plot of the story should ensure that (a) a sense of wonder is evoked, (b) some limits and extremes of reality of human experience are exposed, (c) heroic character traits are applauded, and (d) conventions and conventional ideas are contested (see Fig. 4.1). The human context can be derived from the history of science, which will ensure that real events and ideas are woven into the plot of the story. This human context will facilitate the humanization of meaning, which is a central characteristic of romantic understanding, as was discussed in Chap. 3.

It must be pointed out that the planning frameworks shown in Appendix B do not include all the romantic elements. Their focus is on the human context and the experience of wonder, and, therefore, they can be used to humanize the teaching of physics (Hadzigeorgiou, 2006b, 2007; Hadzigeorgiou & Garganourakis, 2010). Of course, the rest of the romantic elements can be included in the planning framework, if an association between these elements and the science content to be embedded in the story becomes possible, as is the case with the Tesla story. Indeed, this story, according to the literature concerning Tesla's life and work on AC electricity[20] (see Appendix A), includes—as it stands—all the romantic elements (see Appendix C), and, as such, it represents a good example of a romantic story.

4.5.5 Storytelling as a Means to Introduce Thought Experiments

From an instructional point of view, finding an effective way to present thought experiments during a science lesson is crucial. It is true that encouraging students to design their own experiments has its own benefits, apart from the development of students' imagination (see Chap. 1). However, there are experiments that have played a decisive role in the development of science, and these experiments can be recounted in the class in the form of a story or narrative.

Notwithstanding the logical argument component of thought experiments,[21] the narrative component cannot, and should not, be dismissed. Given the complementarity of the logico-mathematical (or paradigmatic) and the narrative modes of thinking, it would be hard to argue that the conception (generation) of a thought experiment is based exclusively on logic. And there have been a number of scholars, who point out the narrative element inherent in thought experiments (see Klassen, 2006b). But even if one were to side with the view that thought experiments are

---

[20] Based upon the following sources: Cheney (1981), Cheney and Uth (1999), Johnston (1983), Jonnes (2003), Lomas (2000), O'Neil (1992), Seifer (1998), and Tesla (1982).

[21] Norton (1996) has shown that any thought experiment can be presented as a logical argument.

4.5 The Functions/Purposes of a Science Story

nothing more but logical arguments,[22] the narrative component should still be considered for pedagogical purposes. Klassen, in his study of published thought experiments, pointed out that such experiments 'are capable of becoming narratives' (p. 89), even though one cannot justify the claim that narrative structure is a major feature of them.

Certainly, this is not the place to discuss—let alone settle—the philosophical and scientific debates concerning the logical argument vs narrative component of thought experiments. However, it is easy enough for us to understand that in the case in which a thought experiment is told in class, it inevitably becomes a narrative (i.e. it has a narrative structure). In other words, a thought experiment is always conveyed in a narrative form. And this is important from a pedagogical and instructional point of view.

In defining thought experiments as mental re-enactments of natural phenomena, for the purpose of clarifying science concepts or answering students' questions about science, and in considering that learning sequences may also be structured as re-enactments, Klassen (2006b) points out that it is reasonable to view thought experiments as learning sequences. Thus, what must become clear here is that even though the instructional goal is the understanding of the logical argument (i.e. the science component) of the thought experiment, it is the narrative (i.e. the human component) that will help students create meaning.

As examples of thought experiments as mental re-enactments in the class, one can refer to Galileo's thought experiment, involving the frictionless motion of a ball down an incline and then up another incline, forming an angle with the first incline (see footnote 3, Chap. 1), and to Einstein's famous visualization of himself pursuing a beam of light in a vacuum. In the case of Galileo's experiment, the students imagine, as the thought experiment is being told in class, the gradual increase in the angle formed between the two inclines, until the second incline touches the horizontal ground, and imagine the ball moving forever in a straight line with constant speed. And in Einstein's case, the students imagine him observing a beam of light as an electromagnetic field at rest and then imagine him thinking and coming to the conclusion that such a thing, as an electromagnetic field at rest, cannot exist, as it would violate Maxwell's equations. It is evident that, despite their simplicity and their brevity, these two mental experiments can become narratives for the purpose of helping students understand the law of inertia (Newton's first law) and the impossibility of a speed greater than that of light, respectively.

---

[22] The best example to illustrate this logical argument is Einstein's simple thought experiment, involving a visualization of him riding alongside a beam of light and moving with the speed of light. If he did that, he would observe an electromagnetic field at rest. But since there is no such thing as an electromagnetic field at rest, then he cannot move with the speed of light.

4.5.6 Storytelling as a Means for Raising Environmental Awareness

Despite the contested relationship between science education and environmental education (e.g. Ashley, 2000; Gough, 2002, 2008)—a relationship that does not provide a fertile ground for the development of environmental awareness through school science—postmodern perspectives on science and science education do make the relationship between science education and environmental education more compatible (see Jenkins, 2009). Therefore it makes sense to pursue the development of environmental awareness even through science[23] and consider the potential role of storytelling (Hadzigeorgiou & Skoumios, 2013).

As far as environmental education is concerned, storytelling has been considered an effective strategy, in cases in which direct experience is impossible and also in cases in which the consequences of our experience are negative or undesirable (De Young & Monroe, 1996; see also Hadzigeorgiou, Prevezanou, Kabouropoulou, & Konsolas, 2010). In science education though, there is a question about the kind of stories we tell students. Indeed, what kind of stories can help raise environmental awareness in the context of science education? This is a crucial question given that in science education, storytelling is used mainly as an (effective) means of teaching scientific ideas. Whether or not the plot of the story helps humanize science, by providing a human context, the primary aim of storytelling is the teaching of science content knowledge, along with ideas from the nature of science (Hadzigeorgiou, 2006b; Klassen, 2006a). Even if an argument, which most science teachers could raise, namely, that while stories in which such concepts such as energy and its transformations, plants and animals, the water cycle, and the atmosphere are embedded can help raise environmental awareness, what about the rest of the science concepts? Two strategies appear appropriate: (a) infusing wonder about science ideas and their interrelationships into the plot of story and (b) recounting the story of the universe.

In regard to the infusion of wonder, regardless of whether students' object of study is an environmental issue (i.e. water pollution, global warming, acid rain), a socio-scientific issue (i.e. genetically modified food, transportation, overpopulation),

---

[23] Certainly, it is not only through socio-scientific issues that environmental awareness can be raised. Many teachers, for example, can use imagery and more specifically Google Earth, to help their students see a larger picture and thus develop a larger perspective of the state of the environment. This activity, in fact, can contribute to the development of a global awareness, which is seen as a prerequisite for environmental awareness (Selby, 1998) The GAIA (Global Awareness, Investigation, and Action) project, for example, which aims to inspire middle and high school students worldwide to become involved with environmental research and collaborate on a local, regional, and global level, appears an excellent way to raise environmental awareness. The aesthetic appreciation of nature may also be another avenue. The 2012 NASA 'The Earth as Art' collection, consisting of images of the planet taken from observation satellites over the last 40 years, as well as images from several environmental satellites, can help raise awareness, by presenting the diversity and beauty of the planet and also by revealing features and patterns, which are not visible to the naked eye.

or a natural phenomenon (i.e. a flash of lightning, the water cycle, photosynthesis, aurora borealis) or an object or entity from the natural world (i.e. a piece of rock, a flower, a glass of water), it (object of study) should help evoke a sense of wonder. This is easier said than done, but what should be borne in mind is that a sense of wonder can be more easily evoked if the narrative or story incorporates the science ideas and their interrelationships with human life. For example, in teaching about the water cycle, the idea that 'the water we drink is the same water that dinosaurs drank thousands of years ago' should be embedded in the narrative or story. Similarly, in teaching about photosynthesis, an idea that can help evoke wonder is that of all living organisms, it is only plants that can support human life.[24] The evidence that even young children can raise their awareness of the need to protect trees, if ideas about the importance of trees are presented through storytelling, is very encouraging (Hadzigeorgiou et al., 2010).

On the other hand, in considering the fascination that human beings have shown towards, and their attempt to understand, the universe as a whole (Toulmin, 1982), a global perspective, developed through the 'the cosmic creation story', that is, a story about the evolution of human life since the Big Bang (Swimme, 1988), can make an important contribution to environmental awareness by helping students gain a larger perspective that includes the natural environment.[25]

4.6 Storytelling in Science Teacher Education

At the beginning of this chapter, it was mentioned that narrative has been used for the development of methodologies. Given the power of stories to engage the emotions and help the listener to create meaning, storytelling can be a very useful tool in teacher education. Even though this tool can be used in science teacher education—just like in school science education—for such purposes as student-teacher involvement, to help student teachers bridge the gap between their prior ideas and new information, to relate to the instructor as a person (e.g. Kreps-Frisch, 2010; Kreps-Frisch & Saunders, 2008), and to better understand science ideas by writing their own stories about science ideas (Kreps-Frisch 2010), its potential to convey values about teaching (Zuckerman, 1998) and its role in helping student teachers

---

[24] It is this idea and not just the description of the chemical process of photosynthesis (i.e. only how oxygen is released from the plants' leaves) that must be quite explicit in the plot of the story that students will listen to (Hadzigeorgiou et al., 2010). By the same token, the fact/issue that the corn used to produce ethanol used by an SUV in one day can feed as many as a hundred people for more than a week can help evoke a sense of wonder about human behaviour, which can have an effect on human life. It is evident that it is the sense wonder about science ideas and socio-scientific issues that has the potential to foster and raise the students' awareness of their significance.

[25] It is the bigger picture of the planet and a global perspective that can help promote a larger concept for the natural environment. And this perspective is indeed needed, since the natural environment is a much larger concept than what some people may think (i.e. for people whose environmental concern stops at their yard fence).

construct practical/professional knowledge should also be recognized (Carter, 1993; Connelly & Clandinin, 2000; Clandinin & Connelly 1996).

By practical knowledge we mean the knowledge that teachers have and apply to situations they encounter during their teaching profession. According to Zuckerman (1999), practical knowledge refers to 'the insights derived from one situation that have an immediate application to other situations' and therefore 'is concerned with discriminating the relevant details in complex or ambiguous situations' (Zuckerman, 1999, p. 235).

Zuckerman (1999) found that stories can help student teachers construct (a) rules of practice (e.g. be receptive to criticisms, keep a journal to reflect on the areas that you need to improve, ask other teachers for suggestions), (b) practical principles (e.g. teaching involves the execution of many complex tasks), and (c) images (e.g. teaching as a juggling act, humour as a useful tool in the classroom, sharing passion for understanding). She also found that listening to and discussing stories foster a sense of community.

4.7 What the Research Shows

It is a fact that there is a lot of anecdotal evidence regarding the effectiveness of storytelling. There also exists published work on the theoretical underpinnings of narrative (e.g. Avraamidou & Osborne, 2009; Norris et al., 2005) and on the value of teaching through storytelling (e.g. Hill & Baumgartner, 2009; Linfield, 2007; Seeley & Gallagher, 2014). However, some empirical/experimental studies reporting on the effectiveness of storytelling can also be found, even though the number of such studies, in comparison with theoretical, descriptive, and prescriptive studies, is quite limited (see Klassen & Froese-Klassen, 2014b).

Starting with early childhood education, a very interesting finding is that ideas far removed from children's everyday environment, like volcanoes and stars, can be understood if appropriately presented. For example, Kalogiannakis and Violintzi (2012), in an intervention study with preschool children in Greece, reported that children's understandings about volcanoes can be changed. Their strategy was to tell children the ancient Greek myth of *Chimera* and then to interview them and compare their views about volcanoes with those held before they listened to the myth. Gordon (2013) also provided evidence of the effectiveness of storytelling with very young children. According to Gordon's intervention study with 4- and 5-year-old children in Australia, storytelling can foster understanding of concepts from the areas of astronomy and astrophysics. During a 9-week intervention, Gordon read to children such stories as *Once Upon a Star* (focusing on the death of a star in a supernova explosion), *Once Upon a Planet* (focusing on the role of dust in the formation of rocks, mountains), *Nuclear Fission, Oh No* (focusing on the accretion of a new star—our Sun), and *The Making and Breaking of Rock* (focusing on the crust of the early Earth). This study provided evidence that even complex concepts, such as those involved in astronomical phenomena, can be understood if presented through an appropriate form. Even though the influence of storytelling per se cannot be assessed, since discussions and dramatization[26] were also part of the invention strategy, the role of storytelling in helping children understand natural phenomena needs to be acknowledged.

In the context of elementary science education, Banister and Ryan (2001) used storytelling to teach elementary school children in England about the water cycle. They created the *Great Journey of William Water* story, in which they gave a water molecule human attributes. They reported that the story enhanced retention and that anthropomorphism can aid understanding and it is not necessarily an obstacle to learning. Corni, Gilberti, and Mariani, (2010) and Kokkotas, Rizaki, and Malamitsa (2010) have also reported very positive learning outcomes. More specifically, according to Corni et al. (2010), who used Pico's stories with grade 4 children in Italy, storytelling fostered emotional, cognitive, and heuristic involvement. Pico's stories (i.e. animated stories in slide show format about the adventures of some friends in various situations) helped elementary school children to understand such concepts as pressure difference and flow. For example, Corni et al. have reported that *the swimming pool story*, which poses various problematic situations regarding a swimming pool that has to be filled with water to the right level, encouraged children to help the character of the story solve this problem through inquiry and hands-on science. The children used toys and simple models (e.g. of the swimming pool, pipes, and aqueduct) in order to investigate the problem. Corni et al. summarized their results as follows: (a) children became involved with problem-solving, (b) there was a transition from description to interpretation, (c) they identified the relevant variables, and (d) they generalized the meanings of the relevant ideas. And according to Kokkotas et al. (2010), storytelling helped grade 6 elementary school children in Greece to understand concepts from electricity and electromagnetism and also develop both inquiry skills (i.e. as hypothesis formulation, interpretation) and metacognitive skills (i.e. comprehension of new knowledge, as a result of the storytelling strategy).

In the context of secondary education, positive effects have also been found. Erten, Kiray, and Sen-Gumus (2013), in a study with 80 Turkish high school students (ages 11 and 12), reported that storytelling helped them with understanding ideas about the nature of science, such as 'there is more than one scientific method', 'scientists use their imagination in their research', and 'scientific research is not limited to one field'. Erten et al. (2013) also reported that student's ideas about science changed from a positivist philosophy towards a heuristic philosophy, while *the Draw-a-Scientist Test* showed that stories also helped students change their stereotypical images of scientists.

Hadzigeorgiou (2006b), in reporting on the results of a pilot study with grade 9 students in Greece, pointed out the positive comments made by the teachers who

---

[26] The death of a star during a spectacular stellar explosion was illustrated with confetti, representing dust and gas. Children playing with confetti (e.g. using a black satin sheet to hold the confetti together and then either throwing confetti in the air to represent a stellar explosion or collecting it in one place on the satin sheet to represent the making of a star).

used a story regarding the Galvani-Volta controversy (see Appendix A). More specifically, the teachers reported increased student involvement, particularly of female students, who developed an interest in the topic of the story and the science content knowledge (i.e. current electricity). This is in line with more recent findings concerning the nature of stories, namely, that stories presenting the scientists' struggles have a greater effect on students than stories presenting the scientists' successes (Hong & Lin-Siegler, 2012).

In a quasi-experimental study with grade 9 students in Greece, Hadzigeorgiou, Klassen, & Froese-Klassen (2012) used a sample of 197 students (i.e. 95 in the experimental group and 102 in the control group) in order to compare the effect of a scripted story, based on the life and work of Nikola Tesla, on student's understanding of fundamentals of current electricity (i.e. fundamental ideas about alternating current, its production, and transmission). The results of this quantitative study (based on an analysis of students' journals and on immediate and delayed tests), despite any of its limitations, provide evidence of the effect of narrative/storytelling on science learning and also evidence of the importance of encouraging romantic understanding in science, at least in the context of teaching the concept of alternating current. In comparison with a group of students who were taught the idea of alternating current (i.e. its production and transmission) explicitly, that is, in a lecture format complemented with dialogue and visual aids, but without a context, the students who, through storytelling, were encouraged to understand this concept 'romantically' (see Chap. 3) became more engaged with both the content and the context of the story and also performed better on the test which assessed their understanding of alternating current. The statistically better performance of the experimental group and the fact that students took out-of-school time to make journal entries, without receiving any credit, are encouraging for further research into the role of storytelling and also into the role of the romantic framework in the learning process (see Appendix D).

Similar findings were reported by Arya and Maul (2012), who designed an experimental study with a comparatively large sample of 209 seventh and eighth grade students in the USA. They compared the effects of exposure to a written science text focused on scientific discovery and the effects of traditional instructional forms that delivered the same science content as did the written science text. (The creation of the texts was controlled so as to isolate the presence of the discovery of narrative structure (independent variable).) Significant better performance of the experimental group on both immediate and delayed posttests on understanding science content provided evidence for the crucial role of the narrative text.

4.8 Concluding Comments: The Need for Storytelling in Science Education

Granted that narrative thinking is a natural form of thinking and the human mind imposes narrative structure to events and ideas in order to understand and create meaning, storytelling can play a more prominent role in school science education. Nowadays narrative thinking and storytelling acquire an additional significance,

4.8 Concluding Comments: The Need for Storytelling in Science Education

given the increasing importance of informal education/free-choice learning and the function of television as a storyteller in this free-choice learning context (Dhingra, 2006).

In considering also the role of attitudes towards science in the context of science education and the fact that gender differences in regard to attitudes towards science are not ability related (Wang, Eccles, & Kenny, 2013; see also Ceci, Ginther, Kahn, & Williams, 2014), the role of storytelling in developing positive attitudes towards science should be given more serious thought. These attitudes are indeed crucial, not only in the context of early childhood education, as is usually thought, but also in the more general context of attracting the 'outsiders' (Brickhouse, 1994, 2001), especially those female students with both high verbal and high mathematical abilities, who unfortunately do not select a career in a mathematics-based science field.

On the other hand, the role of storytelling as an effective means to learn science in the early years should also be recognized. Stories that very young children find interesting facilitate, according to Fleer (2013), a 'flickering between real and imaginary worlds', necessary though for the construction of scientific knowledge at this early age. Stories at this early age, in addition to their potential as teaching/learning tools—which help bring science to life and elicit students' questions—can be seen, according to Seeley and Gallagher (2014), as 'safe places', where magical things happen, new identities are taken on, and new meanings are negotiated. Thus, stories can be seen as the best places for creative learning. And this creative learning has the potential to lead to an aesthetic experience. The question, however, is whether 'creative learning' refers to an innovative method of learning and hence to the teacher's creativity or to learning that taps and develops the students' creativity. Storytelling, of course, can be a creative activity for both the students and the teacher, but there are certainly other, perhaps more effective, strategies whereby students can develop their creativity. This is the topic of the next chapter.

Chapter 5
Creative Science Education

> *It is the tension between creativity and scepticism that has produced the stunning and unexpected findings of science.*
>
> Carl Sagan, in Broca's Brain: Reflections on the Romance of Science, p. 73

> *The formulation of a problem is often more essential than its solution, which may be merely a matter of mathematical or experimental skill. To raise new questions, new possibilities, to regard old problems from a new angle requires creative imagination and marks real advances in science.*
>
> Albert Einstein and Leopold Infeld, in The Evolution of Physics: The Growth of Ideas from Early Concept to Relativity and Quanta, p. 92

> *Every great advance in science has issued from a new audacity of the imagination.*
>
> John Dewey, in The Quest for Certainty, p. 294

The fact that curriculum documents worldwide make explicit reference to creative thinking as a worthwhile aim of education reflects the great importance we attach to creativity.[1] If the world, as we know it today, is the result or the product of the creative thinking of few individuals, and if progress in any human endeavour and field of study is due exclusively to the development of new ideas and new ways of seeing reality, then it makes sense to make creative thinking a curricular goal. Science is one of the disciplines that can make a contribution to the achievement of this goal. The *Creative Little Scientists* project (2011–2015),[2] which aimed to foster creativity in early childhood in several European countries, is evidence of the priority given to

---

[1] An ERIC search revealed that well over 1 million articles have been written about creativity in the contexts of education and learning and a little over 150,000 about creativity in the context of, or relating to, science education. Yet the above question is quite timely, now that creativity is increasingly considered a crucial ability for the future. As we enter a new era, creativity is not just becoming increasingly important (Pink, 2005), but it seems that 'our future is now closely tied to human creativity' (Csikszentmihalyi, 1996, p. 6). Gardner (2010), in his *Five Minds for the Future*, has argued for the crucial role of creativity, as a one of the five cognitive abilities that leaders of the future should seek to cultivate.

[2] See about the project at www.creative-little-scientists.eu

creativity in general and creativity through science in particular, especially in early years education.

However, there is empirical evidence that students do not appreciate the creative thinking required in doing science and that they do not view science in general as a creative endeavour (see Schmidt, 2011). This is somehow paradoxical, given that creativity is inextricably tied to the nature of science itself (McComas, 1998a, 1998b; Schwartz, Lederman, & Crawford, 2004). One of the most important ideas about the nature of science, shared by scientists and science educators, is that scientific knowledge is indeed the product of creative thinking (Osborne, Collins, Ratcliffe, Millar, & Duschl, 2003). On the other hand, there is evidence that teachers' conception of scientific creativity and their views about how to foster it are limited or inadequate (e.g. Hong & Kang, 2010; Lin & Lin, 2014; Newton & Newton, 2010).

It is true that there is quite a lot of rhetoric around creativity in the context of science education in the form of pedagogical slogans, such as 'creative science', 'creative problem-solving', and 'creative inquiry' (see Kind & Kind, 2007; Schmidt, 2011). What, however, must be understood is that the above slogans regarding creativity in science education may remain just slogans if we keep on paying lip service to them (and the ideas they represent) and if we tend to identify creativity simply, or mainly, with the generation of novel ideas without appreciating, for example, the special role of imagination (Holton, 1996) and the role of content knowledge in creative thinking (see Rowlands, 2011). That is why one has to critically look at what scientific creativity is, before implications for science education are drawn and certainly before activities that supposedly make students more creative are designed and implemented.

It is the purpose of this chapter to discuss the notion of creativity in the contexts of science and science education and then propose a number of activities/strategies that encourage creativity, and more specifically imaginative/creative thinking, through the learning of school science. What Jean Piaget has said about creativity provides food for thought for all those involved in the field of education in general:

> *The principal goal of education is to create men who are capable of doing things, not simply of repeating what other generations have done—men who are creative, inventive, and discoverers. The second goal of education is to form minds which can be critical, can verify, and not accept everything they are offered. The great danger today is of slogans, collective opinions, ready-made trends of thoughts. We have to be able to resist individually, to criticize, to distinguish between what is proven and what is not.* (Jean Piaget cited in Ginsberg & Opper, 1969, p. 5)

5.1 The Notion of Creativity

Despite the complexity of the notion of creativity and the difficulty to conceptualize it, there is an agreement among the experts that creativity is the ability to produce work that is both novel (i.e. original, unexpected) and appropriate (i.e. useful) (Sternberg & Lubart, 1999). Boden (2001) defined creativity as one's 'ability to come up with new ideas that are surprising yet intelligible, and also valuable in some way' (p. 95). J. P. Guilford though was the first who tried to come up with a model/theory for creativity and indeed the first to measure creativity with a paper-and-pencil test. In Guilford's (1950) theory of creativity, divergent thinking skills are central to the creative process. These skills, according to Guilford (1967), are 'fluency' (i.e. the ability to produce many ideas or solutions to a problem), 'flexibility' (i.e. the ability to generate different types of ideas/solutions), and 'novelty' (i.e. the ability to generate innovative/uncommon ideas/solutions). However, this psychometric approach was found to be too narrow, as creativity is a complex and multifaceted notion, and for its development we should take into account cognitive, motivational, and generally affective factors, as well as personal and social factors (see Sternberg, 1999).

In Amabile's (1996) theory, creativity is conceptualized as a confluence of three skills/factors, namely, (a) creativity-relevant skills (across domains), (b) domain-relevant knowledge and skills (domain specific), and (c) motivation. While creativity-relevant factors include cognitive style, personality traits, work style, and knowledge of strategies for generating novel ideas, domain-relevant factors include factual knowledge and skills relevant to the particular domain. It is generally recognized that creative people who produce novel ideas and novel products in general, in any domain, possess deep knowledge about that domain (Nickerson, 1999). It is also believed that the myth of genius appears to be just a myth (see Weisberg, 1993), and this makes sense if one considers what Einstein has said about creativity, namely, 'The secret to creativity is knowing how to hide your sources' (www.quoteinvestigator.com).

In regard to the third factor in Amabile's (1996) theory, that is, task motivation, intrinsic, and not extrinsic, motivation plays a central role, as external constraints may very well inhibit the creative process. Indeed, it appears that affective factors play an important role in creativity. A 'playful attitude', humour, and a fun atmosphere (see Hennessey, 1995) appear to help one engage in a given task. It has also been suggested that creativity stems from 'the conscious desire to make a positive change in something real' (Feldman, 1988, p. 288). The motivation derived from seeing the creative products of other people is crucial, according to Feldman, for one's creativity.

Creativity experts, however, have identified various types of creativity. Boden (2004) makes a distinction between 'h creativity' and 'p creativity', the former standing for 'historical creativity' (i.e. when something, like a new idea, a new theory, a new discovery, is historically new) and the latter for 'personal creativity' (i.e. when something is new in a personal sense regardless of whether that something is not new to others). Craft (2001) also makes distinctions between extraordinary, 'big C creativity' (BCC) and ordinary, everyday, 'little c creativity' (LCC).

In distinguishing between ordinary (LCC) and extraordinary creativity (BCC), one may be tempted to think of the latter as a purely personal ability, since it involves imaginative leaps and, sometimes, sudden insights. But since it (BCC) exists within a sociocultural system, consisting of three interacting elements (i.e. a domain or culture that sets symbolic rules, a person who brings new ideas into the domain, and a community of experts who will validate the produced novelty) (Feldman, Czikszentmihalyi, & Gardner, 1994), the purely personal dimension of creativity, even in the case of some rare individuals, who make new discoveries and invent new scientific theories, seems to be complemented with a social dimension. It is for this reason that creativity is increasingly considered a socio-related issue (Miell & Littleton, 2004; Ricchiuto 1996). In such a case, however, a question can be raised about the originality of the ideas put forward by scientists. This is discussed in the next section.

5.2 Creativity in Science

The idea that science is a creative endeavour is indisputable. Scientific ideas are creations of the mind. As has already been mentioned, 'Physical concepts are free creations of the human mind, and are not, however it may seem, uniquely determined by the external world' (Einstein & Infeld, 1966, p. 33). The invention, of course, of concepts and theories, more often than not, requires extraordinary imaginative leaps, but it is also true that even everyday scientific work, for example, problem finding and problem-solving, hypothesis formation, and modelling, requires imaginative/creative thinking, although the latter is not usually associated with novelty. Moravcsik (1981, p. 222) has pointed out that scientific creativity

> [...] can explain itself in comprehending the new ideas and concepts added to scientific knowledge, in formulating new theories in science, finding new experiments presenting the natural laws, in recognizing new regulatory properties of scientific research and the scientific group, in giving the scientific activity plans and projects originality, and many other areas.

But if creativity is one's 'ability to come up with new ideas that are surprising yet intelligible, and also valuable in some way' (Boden, 2001, p. 95), then novelty and value should be the two conditions or characteristics of scientific creativity too. And according to these two characteristics, scientific creativity can be identified with 'historical creativity' and/or with 'personal creativity'. It can also be identified with both extraordinary, 'big C creativity' (BCC) and ordinary, everyday, 'little c creativity' (LCC). It seems reasonable to identify LCC with 'normal science' and BCC with 'revolutionary science', according to Kuhn's (1970) terminology, since LCC is akin to imaginative and what Craft (2001) calls 'possibility thinking', which is a kind of thinking that takes place in everyday life.

In the light of the social dimension of creativity, the originality of one's ideas is somehow questionable. Whitehead's (1957, p. 116) view that 'Everything of importance has been said before by somebody who did not discovered it' certainly reflects the sociological view that the source of ideas is not the individual per se and would be 'more correct to say that s/he participates in thinking further what others have thought before her/him' (Mannheim 1972, p. 3). However, such views raise an issue with regard to the originality of the ideas put forward by scientists.

To pursue this issue further is beyond the scope of this chapter, but it is nonetheless important to consider it when approaching creativity in the context of school science (see next section). Suffice it to say here, it makes sense to view creativity as a mental ability emerging from a social context, which is, anyway, compatible with the social dimension of science itself. In actual fact, the view that science is a social activity—'constitutively social'—as Woolgar (1993, p. 13) put it, says much more than science has a social dimension. It rather conveys the view that the very nature of science is social. There is indeed evidence that creativity emerges from interacting scientists (Latour & Woolgar, 1986). The image of the 'lone star' scientist, working in the lab and experiencing a sudden inspiration and insight, thus solving a problem on his/her own, although not completely a myth, is very rare, at least nowadays. Interactions among scientists and among groups of scientists play a catalytic role in the creation of knowledge (see Feldman, Czikszentmihalyi, & Gardner, 1994; Simonton, 2004).

However, in talking about scientific creativity, the personal dimension of creativity, which is associated with the aesthetic element of science, needs to be considered. In actual fact, the philosopher and historian Thomas Kuhn has stressed its importance in scientific revolutions: 'Aesthetic considerations can be decisive. Though they often attract only a few scientists to a new theory, it is upon those few that its ultimate triumph may depend' (Kuhn, 1970, p. 156). The history of science provides evidence that aesthetic factors did play a major role in theory construction and in influencing scientific practice in general (Hadzigeorgiou, 2005b).

It should be noted that the common ground shared by art and science was recognized after a shift from a positivist epistemology took place, particularly with the advent of the theories of quanta and relativity. Indeed, that idea that science might have a greater commonality with art than was originally thought in a more positivist era has been seriously considered by a number of scholars (see Tauber, 1996). Miller's (2001) work on the life and work of Einstein and Picasso also revealed parallels between the two men, thus providing an insight into how the shift from positivism influenced both artistic and scientific creativity.

The similarities, however, between artistic and scientific creativity should not be linked only to advanced research and to such areas as relativity theory and quantum physics. The similarities between art and science (see Chap. 7) also point to similarities between the two types of creativity. What must be stressed through here is that there are three main differences between artistic creativity and scientific creativity. First, while in artistic creativity 'something travels', in the sense that people, who did not participate in the creative act (which produced a piece of art), feel delight and inspiration, and are carried away, in science the delight and inspiration are closely tied to the act and context of the scientific discovery itself, that is, to the scientist who made the discovery or developed a scientific theory (see Medawar, 1967, p. 172). Second, in science there is always the process of verification, which does not exist in artistic creation. Third, in scientific creativity, domain-specific knowledge plays a decisive role (see Gardner, 1993a; Nickerson, 1999; see also Weisberg, 1993), as does polymathy (Root-Bernstein & Root-Bernstein, 2004).

Simonton (2004), in using as examples the *Principia*, the *Republic*, *Hamlet*, *The Last Supper, and* the *Fifth Symphony*, makes Newton's scientific creativity quite distinctive from all the rest, since, as his (Simonton's) argument goes, any literate person can get an understanding of Plato's logical argument in the *Republic* or an idea of the dramatic development in *Hamlet* and also what is graphically conveyed in *The Last Supper* or musically expressed in Beethoven's *Fifth Symphony*, but makes no sense of what Newton in the *Principia* (Mathematical Principles of Natural Philosophy) actually put forward. It is perhaps this reason why scientific creativity is culturally valued more than artistic creativity.

From neuroscience's perspective, Dietrich's (2004) work on creativity types is very informative, since it can help us differentiate between scientific and artistic creativity. He classifies creativity into four major types: deliberate and cognitive, deliberate and emotional, spontaneous and cognitive, and spontaneous and emotional. Scientific creativity requires sustained attention and focus on an idea or problem and is therefore the result of a deliberate cognitive function (taking place in the prefrontal cortex), although there are occasions when spontaneous cognitive creativity makes one solve a problem. Artistic creativity, on the other hand, is associated with spontaneous and emotional functioning (taking place in the amygdala). Scientific creativity, more often than not, is based on logic, in the sense that 'Once a scientist masters the logic of science and the substance of a particular discipline, creativity is assured' (Simonton, 2004, p. 6). In other words, what is usually considered a mystical inspiration, a product of some extraordinary ingenuity, may very well be the product of logic. Although logic is one of the sources of scientific creativity—the other being genius, chance, and 'zeitgeist', according to Simonton—it is nonetheless an important one. There is consensus that scientific creativity is always matched by rationality, with experiments always playing a crucial role (Schwartz et al., 2004). Scientific ideas, as was pointed out previously, are always subject to verification through experimental testing.

The comparison that has just been made between artistic creativity and scientific creativity is quite crucial, given the possibility of an interdisciplinary connection between the two in the context of education. This issue I will take up in Chap. 7, but what must be stressed at this point with regard to scientific creativity is that, while art may be considered more imaginary than science, in the sense that in science logic is always a complement to imagination, the question about whether science is less or more creative than art is hard to answer. Further, both can be creative, and in fact transformative, in the sense that both can contribute to our change of outlook on the world (see Miller, 2001).

This comparison may lead one to identify scientific creativity mainly with two general abilities, that is, imaginative and logical thinking. Both intellectual abilities, in actual fact, are considered necessary, although not sufficient, for the generation of novel ideas. Yet creativity, as an emergent ability, is the result of a complex interplay of several factors, such as intellectual abilities (i.e. problem finding, seeing problems in novel ways), prior, domain-specific knowledge, personality traits (i.e. self-efficacy, risk taking, a tolerance for ambiguity), motivation, and environment (Sternberg, 2006). This complexity, coupled with the way and the circumstances under which scientific ideas burst forth (Gardner, 1993b, 1997), makes scientific creativity an unpredictable ability or event, for it is also true that the subconscious and unconscious have also played a decisive role in the generation of novel ideas and problem solutions. According to the literature, dreaming and daydreaming have resulted in sudden and unexpected insights—even illuminations (see Vernon, 1970; Kind & Kind, 2007; see also Zenasni et al., 2011, for the role of tolerance of ambiguity).

The notion of 'extracognition', which refers to those things that are not purely cognitive, can also be useful in the case of studying scientific creativity. Extracognition includes (a) beliefs; (b) feelings of harmony, beauty, and style; (c) intuition; and (d) values in regard to one's choice of a field (Shavinina & Ferrari, 2004; see also Runco, 2004). According to a study, which looked into biographical and autobiographical data regarding scientific geniuses and compared those with data derived from interviews with gifted adolescent students of physics and mathematics, common extracognitive features included (a) beliefs in truth and in the power of ideas, (b) feelings of direction, and (c) preference for harmony and beauty (Shavinina & Ferrari 2004).

The age of those involved in the creation of novel and revolutionary scientific ideas is a factor that needs some consideration with regard to scientific creativity, as is the fact that scientists, more often than not, are 'deliberately creative' (in the sense that they deliberately look for novelty that can be useful to society).[3] The fact that the scientists who transformed both their disciplines and the way people see the world were very young—i.e. in their early or mid-20s (Simonton, 2004)—is something that may very well have implications for science education, namely, giving creativity a more prominent role, especially in early years of education.

---

[3] In regard to the scientists' age, some examples can be illustrative. Marie Curie was about 30 when she began working on radioactivity, and by the time she was 45, she had won two Nobel Prizes. William Lawrence Bragg, the youngest-ever Nobel Laureate, received, at the age of 25, the Nobel Prize for his work on X-rays and crystal structure. Werner Heisenberg also did his pioneering work on quantum mechanics in his mid-20s, as did Albert Einstein, who published, some of his most important papers at that age. As for the secret of life, this was unravelled by James Watson, who codiscovered the structure of DNA when he was only 25 (Simonton, 2004). It is quite evident that these scientists transformed both their disciplines and the way people see the world. And this transformation of outlook may be more important than the improvements, marginal or not, in people's daily life, due to the application of the novel scientific ideas themselves (Peters, 1988). But what is important and may very well have implications for science education is that of all sciences only physics is the field with the youngest pioneers, preceded only by poetry (Simonton, 2004).

As for the scientists being deliberately creative, again two examples, from two different eras, can help illustrate the point. Nikola Tesla was looking for new ways to harness the energy of the ionosphere. Tesla was deliberately trying to harness the naturally occurring electricity in the ionosphere and then broadcast it back to relay stations that could then transmit free energy all over the planet. The fact that his creative vision was not realized is another story. However, out of that work emerged ideas in regard to the wireless transmission of electrical power. More recently, physicists Andre Geim and Konstantin Novoselov, in visualizing graphite as billions of layers of carbon atoms laid on top of each other, and in being interested in isolating a few of those layers, or maybe just one layer, took a block of graphite (i.e. the same material used for the centre part of a pencil), stuck tape to it, and then pulled the tape off. Not only did the tape method work, but in October of 2004, the two scientists announced that they had managed to create single sheets of carbon just one atom thick and in 2010 received the Nobel Prize!

It is therefore apparent that scientific creativity is the outcome of a complex interplay of factors that cannot be predicted. Yet despite the complexities inherent in science, as a field of inquiry, and despite the complexities in approaching it as a creative endeavour, a reliable picture of it, as was discussed in Chap. 1, should include the idea of science as a creative endeavor, in which imagination plays a central role. This reliable picture is important to consider when approaching scientific creativity in the context of school science and science education. The reason is it can provide the background and rationale for designing activities, which have the potential to foster creativity in school science education. In considering, of course, the fact that school science education can often be more procedural than creative—McComas' (1998b) analysis of science textbooks did reveal a 'procedural' emphasis, as far as the portrayal of the nature of science was concerned—the implementation of such activities can at least help change the school atmosphere into one that is more creative.

5.3 Creativity in Science Education

The answer to the question 'what does creativity in science education mean?' may seem quite straightforward, for one can readily say that creativity in the context of science education refers, or should refer, to what the science teacher does (i.e. she/he stimulates and encourages creative thinking) and/or to the opportunities the students have, independently and/or as a result of what their teacher does, for creative thinking. What is not straightforward, however, is the extent to which creativity in the context of the science classroom can or should reflect scientific creativity, for it sounds reasonable, one might say, that creativity in the context of science education should reflect, as much as possible, the notion of scientific creativity. There is a view that any approach to scientific creativity in the context of school science should be both 'authentic' in scientific research terms and meaningful and appropriate to the students' needs and abilities (Kind & Kind, 2007). However, the idea of 'authenticity' may be misleading.

Although 'scientific creativity' should reflect what real scientists do, the differences between scientists and students (especially young students), as well as the nature of the tasks encountered by them, need to be taken into account. Students have neither the scientists' conceptual framework nor the time to pursue a topic for a long time, unless of course this has been arranged (i.e. through participation in a project that poses no immediate restrictions on time). Moreover, the deliberate pursuit of novelty by scientists may be totally absent from students, who may very well do things, including scientific inquiry, because they *have to*. The nature of the problems scientists encounter is also another issue (i.e. ill-structured problems admitting multiple solutions) to be seriously considered. This point I take up further down this section of the paper.

A point also that needs to be made here is that although we know what real scientists do, we cannot say the same about what and how they think. Given that there is no 'universal scientific methodology', scientists can approach and solve the problems they face in their research in many possible ways (see Simonton, 2004). Moreover, as Medawar (1979) observes, 'Scientists are people of very dissimilar temperaments doing different things in very different ways' (p. 3). In other words, scientific creativity emerges from experiences extremely unique to the individual scientists. Even if some students had conceptual frameworks similar to those of scientists, the nature of creativity, as an emergent intellectual ability, would make the comparison between students and scientists unrealistic. All these arguments make 'authenticity' an unreliable, if not invalid, criterion whereby one judges scientific creativity in the context of school science education.

The issue of 'authenticity', however, becomes more complex if one considers it in the context of scientific inquiry. Furthermore there is also a crucial question: how authentic can inquiry science be? Are students really free to explore or are somehow guided by their teachers to follow a step-by-step procedure (i.e. collect and analyse) that quite often can take the form of 'a recipe' for inquiry (see Asay & Orgill, 2010)? The main flaw with inquiry science, as Kind and Kind (2007) have observed, is that the freedom and openness existing in real science are rarely achieved in the everyday reality of the science classroom, and, more often than not, teachers inevitably 'frame' student inquiry, by facilitating and providing most of what is required in the investigation. There is some evidence to argue, as they report, that scientific inquiry does not offer any guarantee for fostering students' scientific creativity. In fact, this evidence suggests that 'any claims that 'scientific creativity' is developed through inquiry science are certainly spurious' (Kind & Kind, 2007, p. 27). Moreover, it is questionable whether the fact that different groups of students come up with different ways to answer questions and approach problems (Barrow, 2010) can be taken as an indication or a criterion whereby one can judge scientific creativity.

Yet these words of caution do not imply that inquiry science cannot be creative. Three examples can help illustrate the point here. In a science class, a teacher provides the groups of students with certain materials, like batteries, wires, and light bulbs, and asks them to make the light bulb light. In another classroom, students are asked to make a model of a house and investigate how illumination (with light coming from the sun) within the house can increase. And in another classroom, students are asked to investigate and then come up with an explanation how substances like sugar and salt affect the evaporation rate of the water in a container. Here one can see three different activities with the first requiring little imagination and divergent thinking—trial and error suffices to get the simple system working—the second requiring imagination and divergent thinking in order for students to try possible factors that might affect the illumination in the house (i.e. the position of windows, the colour of curtains, the arrangement of furniture), and the third requiring the creative proposition of a theory to explain why adding substances, such as salt or sugar, to water causes the water to evaporate more slowly at a given temperature. (Students will have to visualize water molecules forming bonds with sugar or salt molecules, so it will take extra energy to break those bonds, thus slowing down the evaporation process.) Of course, different variables here, like distilled water and tap water, may be included in the investigation, making the whole process even more creative (see Barrow, 2010). And despite a lack of conclusive evidence resulting from a critical evaluation of some inquiry science approaches to scientific creativity, programmes that offer extended opportunities for project work over a longer period of time, demanding student commitment and ownership, appear to be more promising as far as the development of creativity is concerned. Such programmes appear to meet 'creativity' criteria to a greater extent than traditional inquiry teaching (Kind & Kind, 2007).

A word of caution, however, should also be said about art. True, art is an excellent tool to help students learn science (Ashley, 2011; Merten, 2011). For example, students who make a collage, illustrating the water cycle or the states of water, are helped to learn science content. Moreover, for some students, such an activity can be a great stimulus for learning. But this is different from saying that this activity necessarily helps students develop their scientific creativity, which can be more successfully fostered through activities that 'explicitly' encourage divergent/imaginative thinking (e.g. creating an analogy for the water cycle or the states of water, writing about the daily life of a photon). However, if students participate in an art-based science activity that requires imaginative/divergent thinking (e.g. representing a science idea through an artistic form of their own choice or through two different forms of visual art) or if the activity that students select (e.g. collage making) is carried out with an artistic spirit, that is, with imaginative engagement and the desire to produce a novel, uncommon, or useful in some way work (which, in fact, shows how important affective and other factors, as was previously discussed, are), then art can indeed help with the development of creativity.

The point that is being made here is that, while participation in an art-based science activity does not necessarily tap the students' imagination and creativity, the possibility for students to have a truly creative experience through their involvement with an art-based activity needs to be recognized too. This point, along with the various factors influencing, according to the literature, creativity (e.g. motivation, content knowledge, social milieu), is crucially important in the context of planning for activities that are considered creative or that supposedly help foster the students' creativity.

However, even in the case in which an art-based activity does not appear to be very imaginative and creative, one has to recognize that the initial involvement with such an activity can increase the possibilities for students to become engaged at a deeper level and hence the possibilities for a creative moment and an aesthetic experience too (Hadzigeorgiou & Fotinos, 2007).

If science is indeed 'a holistic enterprise that may be influenced by art, music, dance, yoga, meditation, imagination, wonder and may other things' (Lunn & Nobel, 2008, p. 803), then art should be considered, through the possibilities it can provide for aesthetic experiences, an excellent avenue to an aesthetic kind of understanding, which is documented in the literature on both science and science education (Girod, 2007a). Jackson's (1998, p. 33) argument provides not only a justification of art and science connections in the curriculum but also an answer to the question regarding how one can induce an aesthetic/transformative understanding of school science:

> The arts do more than provide us with fleeting moments of elation and delight. They expand our horizons. They contribute meaning and value to future experience. They modify our ways of perceiving the world, thus leaving us and the world itself irrevocably changed.

The possibility for students to see things and ideas in novel and unusual ways (Gardner, 1993a, 1993b, 2010; Sternberg, 2006) is something that needs particular attention. Further, the development of this ability, or at least the effort to foster it, can be considered an important goal regarding scientific creativity and should complement two more (traditional) goals, such as (a) the generation of multiple ideas (i.e. solutions to problems, answers to questions) and evaluation of those, which are worthwhile to be pursued further, and (b) making associations between semantically remote or seemingly unrelated ideas, events, and phenomena (Craft, 2001; Sternberg, 2006). It is apparent that the above three goals concerning creativity in school science are both compatible with the general notion of creativity (Sternberg, 2006; Gardner, 2010) and scientific creativity in particular (Kind & Kind, 2007; Gardner, 1993b, 1997; Schmidt, 2011; Simonton, 2004) and also realistic in the context of science education, in the sense that activities aiming to achieve them can be designed and implemented (Hadzigeorgiou & Fotinos, 2007; see next section).

Apparently, the above goals necessitate a distinction between an innovative teaching approach and an approach that provides opportunities for creative thinking. Helping students to think creatively in the context of school science is certainly very different from both teaching them creatively (i.e. by implementing an innovative approach) and teaching them about the nature of science, in order to help them become aware of, and appreciate, science as a creative endeavour. It is also helpful to distinguish between learning to be novel in the context of everyday life is very different from learning to be creative in science (Rowlands, 2011). It is for these reasons that the distinction between three frames, namely, 'creative teaching', 'teaching about creativity', and 'fostering students' scientific creativity' (see Kind & Kind, 2007), is important to consider. Such distinction focuses our attention on what we really do or on what we really want to do.

The above goals also necessitate a special attention to imagination. There is empirical evidence to support the view that people who have opportunities to operate in imagined worlds become more creative. And although the evidence that imaginative skills in science education are transferable to other areas is not convincing, there is good reason to believe that 'imagination offers the promise of making scientific creativity more concrete and helping to identify a potential starting point for further research' (Kind & Kind, 2007, p. 25).

It is interesting to note that, according to De Cruz and De Smedt's (2010) analysis, based on theoretical models and experimental results of the cognitive sciences, scientific creativity, like other forms of creativity, is structured and constrained by prior ontological expectations, and distant analogies[4] can be a powerful epistemic tool that helps one to overcome these constraints.

In discussing, however, the role of creativity in science education, the role of teacher creativity should also be considered. Indeed, if the having of an aesthetic/transformative experience is an instructional goal, then the teachers themselves must be creative too. Given that such an experience presupposes at least some initial involvement with the ideas of the lesson, these ideas must be imaginatively engaging. 'Crafting science content' and 'crafting dispositions', as Girod, Rau, and Schepige (2003) recommend, require teachers not only with a good grasp of science content knowledge but also with a creative mind and disposition, as they try to present science ideas in a way that students feel the mystery, the wonder, and the drama involved in their (ideas) conception and development and also as they try to develop their students' creative imagination through questions that make them wonder about the potential of these ideas.

With regard to fostering in students a sense of wonder, it is usually assumed that a question or a challenge in general (e.g. a problem) will automatically make students wonder. However, as discussed in Chap. 6, the experience of wonder requires the science teacher to (a) 'craft' science ideas out of the traditional school curriculum (e.g. the idea of atom as 'the most sociable entity in the universe', the idea that heat is always present in electrical circuits as 'the incorrigible and ubiquitous thief of electrical power') and (b) formulate questions that have the potential to make students wonder (e.g. 'is the sand we see on a beach different from a rock or a big mountain?', 'how can sound—not light—help us to see what we cannot see?', 'what would a water molecule say of its travel during the water cycle?'). It is evident that both of the above require, among others, the teacher's creativity.

5.4 What the Research Shows

Although the literature on creativity in science education is limited, there are some studies which fall in one of the following two categories: (a) studies on teachers' or scientists' views about creativity and its relationship to the nature of science, and science in general, and (b) studies on students' creativity. In regard to the first category, Osborne et al.'s (2003) Delphi study involving 25 scientists from a variety of fields found that there is a consensus among the scientists that science is an imaginative/creative endeavour: 'science is an activity that involves creativity and imagination as much as many other human activities […] Scientists, as much as any other profession, are passionate and involved humans whose work relies on inspiration and imagination' (p. 702). The teachers—a subgroup of scientists—also thought that creativity is an important aspect of science.

---

[4] Current research on analogies in scientific understanding focuses on near analogies, that is, analogies in which the source and the target concept or domain are close.

However, it is interesting to note that some studies found that teachers had inadequate conceptions of scientific creativity. Newton and Newton (2010), using a phenomenographic analysis of primary school teachers' views of what constitutes creativity, found that teachers had a narrow conception of scientific creativity. The teachers focused on practical investigations and judged creativity in terms of what stimulates interest and on-task discussions. Also their views regarding the development of creativity were inadequate. Their main criterion for fostering creativity was fact-finding and practical activities.

Lin and Lin (2014), in using open-ended and Likert-type questions, found that while the teachers in Taiwan were aware of the key features of creativity and many of them had a number of ideas about how to foster it, they did overlook such aspects as problem finding and the link between art and science. On the other hand, Hong and Kang's (2010) comparative study—they used two groups of 44 South Korean and 21 US science teachers—showed that even though individual teachers' conceptions of creativity were quite limited, as a whole, their views were consistent with the international literature on the subject. For example, problem-based and project-based inquiry was considered a very creative activity. In regard to the process of fostering creativity, three reasons were prevalent among the teachers: (a) pressure for covering content, (b) difficulty in the assessment, and (c) class size. In regard to the differences between the two groups, as far as fostering creativity is concerned, Hong and Kang found that, while South Korean teachers believed that the most important thing the teachers can do is to provide opportunities for students to think creatively, US teachers thought that environmental and emotional support was what matters the most.

In regard to studies on student creativity, Shea, Lubinski, and Benbow (2001), in their longitudinal study with talented 12–14-year olds, found evidence for the importance of visualization when it comes to choosing a career later in life. They tested spatial visualization (i.e. students' ability to visualize and manipulate their visualizations) and verbal abilities, with the purpose of predicting students' career paths later in life. Their conclusion was that students with strong spatial abilities (relative to verbal abilities) were more likely to follow a career in mathematics, science, and engineering (see also Furnham et al., 2011, for the role of individual differences as predictors of creativity in art and science).

The role of creativity in the context of student-generated analogies has also provided evidence for the positive role of the latter in the learning process. In a study with 14-year-old American students, Pittman (1999) found positive connections between the students' ability to generate their own analogies in drawing tests and their scores on a written test. On the other hand, Spier-Dance, Mayer-Smith, Dance, and Khan (2005) also found positive learning effects in the context of chemistry lessons, where undergraduate students were asked to create their own analogies for halogen oxidation. An interesting finding was that the learning benefits of generating analogies were greater for lower-achieving students.

In Diakidoy and Constantinou's (2000–2001) study, Greek Cypriot undergraduate physics students were asked to generate as many responses as possible to three open-ended physics tasks (i.e. with more than one solution). The assessment of the creative solution of those tasks was based upon Guilford's three divergent thinking abilities, that is, 'fluency' (i.e. the number of solutions), 'flexibility' (i.e. the number of different types of solutions), and 'originality' (i.e. how innovative the solution was). They found that creativity was context dependent but did not correlate strongly with subject matter knowledge.

In China, Lin, Hu, Adey, and Shen (2003) tested the influence of the *Cognitive Acceleration Through Science Education* (CASE)[5] programme on students' creativity with a sample of 11–15-year-old students. They used Hu and Adey's (2002) model for assessing scientific creativity for secondary school students,[6] and their results showed that students who had been taught CASE material increased their creativity test score (on five out of seven items). Such a finding suggests, according to Lin et al. (2003), that analytical thinking skills and reflection enhanced the students' metacognitive abilities, which, in turn, influenced their ability to think creatively.

Studies that report on the use of dramatization in science education (see Chap. 7) have also revealed that children's creativity, especially if they are asked to write the script and/or create their own production, can foster science learning. For example, there is evidence that seven Australian fifth grade students, who were seeking extended learning, learned advanced scientific concepts through the creative process of script writing and production of a science play called 'hectic electric'. Aided by two parents and a mentor, the students wrote the script, submitted it for a science drama competition, and finally won the first prize in the primary school category. This project provided evidence that creativity can help students develop self-efficacy and higher-order thinking skills, both of which are necessary for learning science (Nicholas & Ng, 2008).

With young children, whose creativity has its source in play, there is also evidence that creativity, play, and science are closely related. The studies by Siry and Kremer (2011) and Fleer (2013), with preschool children, have provided evidence that the back and forth movement between young children's imaginary constructions and the real world can indeed help them learn science. Children's creativity played a central role both in the context of learning about the rainbow (Siry & Kremer, 2011) and in the everyday context of cooling a bowl of porridge (Fleer, 2013).

---

[5] The CASE programme was designed to increase students' scientific reasoning, especially analytical skills rather than creative thinking.

[6] Hu and Adey's (2002), based on Guilford's theory of creativity, developed a model for testing scientific creativity in the context of secondary school science. Their assessment focused on three factors, namely, *process* (referring to imagination and thinking), *trait* (referring to fluency, flexibility, and originality), and *product* (referring to technical product, scientific knowledge, science phenomena, and science problem). Even though their model offers the possibility for assessing 24 'factors', Hu and Adey focused on seven items, namely, unusual use, problem finding, product improvement, creative imagination, problem-solving, science experiment, and product design. Administered to 160 UK students, they found both a high internal consistency and inter-scorer reliability.

Weiping, Adey, Jiliang, and Chondge (2004), who applied Hu and Adey's model to compare Chinese and British adolescents' creativity, found that creativity develops in stages, with a levelling off occurring at age of 14, something that agrees with the results of the Guilford-Torrance test.

The recent study by Cremin, Glauert, Craft, Compton, and Stylianidou (2015) also provides evidence for the role of creativity in learning science in the early years. Drawing on data from the EU project *Creative Little Scientists* (2011–2014), they explored the teaching and learning of science and creativity in early years education. Using a deductive-inductive approach, they analysed 71 cases (with three episodes per case) and found synergies between 'play and exploration, motivation and affect, dialogue and collaboration, problem-solving and agency, questioning and curiosity, reflection and reasoning, and teacher scaffolding and involvement' (p. 404). Apparently, such findings, along with the observed differences in practice between preschool and primary settings, are quite encouraging for further research into the role of creativity in learning science in the early years.

5.5 Fostering Creativity in the Science Classroom

The discussion thus far begs the question: how can we best foster students' scientific creativity? This discussion may have been illuminating, yet the evidence about the effectiveness of certain teaching strategies (i.e. imagery/visualization, inquiry science, integrating art with science) is inconclusive. An analysis, in fact, of research studies in science education points to the fact that 'creativity in school science is at a much lower level than is required even to begin approaching an answer' (Kind & Kind, 2007, p. 2). Moreover, research based on the life stories of scientists (e.g. Einstein, Maxwell, Faraday, Watts, Feynman) questions the development of students' imaginative skills by formal schooling (see Chap. 1). As Shepard (1988) has pointed out, 'their development occurs before, outside of or perhaps in spite of such schooling – apparently through active but largely solitary interaction with physical objects of one's world' (p. 181). All these make the development of students' creativity a real challenge. This challenge becomes even greater if we consider the fact that creativity is grounded in knowledge, and therefore science teachers should help students build content knowledge but 'without killing the creativity' (Boden, 2001, p. 102).

However, the above question can still be answered in the affirmative, if we consider what we know about creativity in general, namely, that individuals can become creative through the confluence of several factors, including the environment and the challenges it offers (Sternberg, 2006). Moving beyond the rhetoric, by carefully analysing and reflecting on current practices that supposedly help students develop their creativity, can be more fruitful than simply following the trend and adopting and implementing those practices. Even if one believes that 'creativity is to a large extent a matter of innate ability, talent, and capability' (Moravcsik, 1981, p. 227), one can nevertheless acknowledge that 'there are many ways in which science education can locate, foster, encourage, practice, and enhance traits, attitudes, and skills in the students so that whatever creativity the student has, it is more effectively converted into achievements and accomplishments' (p. 227).

Having already made reference to ideas from the nature of science that have a bearing on creativity (i.e. the social nature of science, its aesthetic dimension, the idea of scientific inquiry, and the role of imagery and imagination in science), and considering evidence from research on the effect of temporal and spatial distance on creativity (Shapira, & Liberman, 2009), it goes without saying that a notion of creativity in the context of school science should certainly take all these ideas into account and should also be compatible with the general notion of creativity. A few points that are crucial for designing activities can be reiterated here:

First, the fostering of creativity presupposes a strong conceptual framework. In other words, science content knowledge is a prerequisite for thinking and hence a prerequisite for creative thinking. Students should be as knowledgeable about science (i.e. content knowledge) as possible. Vygotsky (2004) has pointed out that creative thinking is based upon past experiences and knowledge. This, of course, sounds quite reasonable. The more ideas about any topic a student already has, the more new ideas are likely to occur to him or her. And more new ideas simply mean more opportunities for creativity.

Second, creativity in science education is about divergent/imaginative thinking. Encouraging creativity in the context of school science means encouraging idea generation in a nonthreatening and critique-free environment. This means that, in order for students to be creative, all ideas need to be heard and not ridiculed, no matter how crazy they may sound. Di Trocchio (1997) provides ample evidence that what was once thought to be a crazy/illogical idea was finally accepted by the scientific community (i.e. the transmission of electromagnetic waves over long distances, the splitting of the atom, the general theory of relativity).

Third, imagery and visualization should have a central place in the science curriculum and teaching. As Mathewson (1999) pointed out, visual-spatial thinking is an overlooked aspect of science education. The evidence concerning the role of imagery, although with some caution (see Kind & Kind, 2007), is encouraging.

Fourth, because creativity, more often than not, presupposes engagement with ideas, objects, and situations, special attention should be given to the notion of 'aesthetic experience'. Such an experience, especially when accompanied by a sense of wonder, increases the possibilities for deeper engagement in science and for inspiration (Hadzigeorgiou, 2005a).

Fifth, thinking about future events and possibilities (i.e. temporal distance) and also about far away events and people (i.e. spatial distance) is a strategy that can be incorporated into teaching activities (Hadzigeorgiou, Kabouropoulou, & Fokialis, 2012).

Sixth, the social nature of science, as has already been discussed, points to activities that provide students with opportunities to interact in a social setting, for the purpose of thinking imaginatively and creatively. That is, creativity, without completely excluding individualized activities, should be fostered within a sociocultural milieu.[7] This milieu includes both the culture of scientific inquiry and the culture of the school classroom, and both cultures can play a role in developing students' creative thinking (Hadzigeorgiou et al., 2012).

---

[7]An argument about whether individual creativity is superior or inferior to social creativity is hard to defend or maybe meaningless in the sense that there is an interplay between the two.

5.5 Fostering Creativity in the Science Classroom

However, a few words about evaluating the results of our teaching should also be said here. Given the nature of creativity, and scientific creativity in particular, as was discussed, no one can predict the emergence of creativity. And this can be a problem. This is not to say that creative thinking cannot be recognized. On the contrary, we can tell when we see it. The problem is that, even if one teaches deliberately for creative thinking, one cannot expect to assess it when one wants to assess it, as a result or as a consequence of his/her teaching. The problem becomes more complicated if one considers the fact that, regardless of the opportunities students have for creative thinking, the testing situation may not provide a reliable means to assess creativity, for the test itself may be felt as a constraint on a student's freedom and hence on his/her creativity. And the inauthentic situation in which it is taken may also be a constraint on his/her ability to think creatively. According to the literature, freedom and authentic situations can be considered, among other factors, preconditions for creativity, since they both relate to motivation, purpose, exploration, and confidence (Gardner, 1993b; Mumford, 2003; Sternberg, 2006; see also Simonton, 2004).

So the best we, as science educators and science teachers, can do is provide an environment that increases the possibilities for creativity to emerge. In the light of what has been discussed so far, some activities can be considered more appropriate for fostering scientific creativity in science education. Those are activities that are more likely (a) to provide opportunities for imaginative/divergent thinking and (b) to lead to aesthetic experiences. These activities are compatible with the three goals regarding creative thinking in science education (see previous section). Further the achievement of the first two goals require divergent/imaginative thinking, while the third one, in addition to such kind of thinking, presupposes a sense of wonder and aesthetic experiences in general.

Compatible with these goals and also with what has been said thus far about scientific creativity, the following activities, although not a recipe, have the potential to foster students' creativity in the context of school science:

- *Creative problem-solving* (e.g. measuring the height of a building using a barometer or tennis ball, measuring the surface of an irregular shape using a mechanical balance, predicting the fate of the earth after the total disappearance of the sun, calculating the density of a proton, of a black hole)
- *Creative problem-solving in the STS context* (e.g. how electrical energy can be linked to environmental problems, how technology might affect the environment in the future, how we can produce electrical power in the future, how we might approach the sudden invasion of bacteria from space, how interdisciplinary connections between scientific fields can help with the problem of water and food shortage in the future)
- *Creative application of science ideas* (e.g. the construction of a flashlight, electric motor, electromagnet from simple materials; the construction of a floating platform or raft by using empty plastic bottles; the production of a plan for spending a week on a desert; the construction of a weight loss plan, considering various factors such as taste, availability, and cost; the making of musical instruments using only veggies)

- *Creative classification* (e.g. to produce at least three classifications of the plants found in a certain region, to list as many electrical appliances as possible, and then to classify them using at least two criteria)
- *Creative writing* (e.g. an essay or narrative/story about a day in the life of a proton, a day without gravity, about the journey of a water drop in the water cycle, the possible journey of a water molecule from the time of dinosaurs to the present day)
- *Creative script writing for a theatre play about a physical process or phenomenon* (e.g. about the 'social life' and 'marriage' of atoms and the formation of molecules and chemical compounds, about the possible experiences of a gas molecule that was accidentally trapped inside a box containing molecules of another gas, about the motion of electrons in a wire while they collide with other atoms and in the process heat is produced)
- *Script writing for a theatre play about the life and work of famous scientists* (e.g. Galileo, Newton, Faraday, Curie, Bohr, Einstein[8])
- *Writing a 'personal science story'* (e.g. a narrative or story about a personal experience illustrating certain science concepts, like forces and motion, electricity, density, energy, entropy)
- *Creative science inquiry* (e.g. investigating possible factors that might have an effect on the illumination of a room, ways to produce electricity for the house in a case of emergency, ways to heat water in the absence of metal containers, measuring time without a watch)
- *Devising thought experiments to understand phenomena* (e.g. free fall inside an accelerating elevator, motion in the absence of gravity, motion in an electric field, molecular motion inside a sealed container as a result of various temperature and pressure conditions)
- *Hypothesis creation* (e.g. hypothesizing about the causes of global warming and greenhouse effect, about the connection between magnetism and electricity, about the forces exerted on moving subatomic particles)
- *Creating analogies to understand phenomena and ideas* (e.g. the phenomenon of resonance, the ideas of energy, nuclear fission and fusion, annihilation, chemical bonding, mixing of substances)
- *Challenging students to find connections among apparently unrelated facts and ideas*[9] (e.g. between sound, oil containers, and tax evasion; between Newton's laws, a nurse, and a soccer player; between gravity and executions in the Middle Ages; between light, electrons, and a surgeon; between water and electricity; between a glass of wine, the age of the universe, and the evolution of stars; between the sinking of Titanic and hydrogen bonding; between a thief, the police, and the speed of light; between a layer of rock, an atom, and a stopwatch; between a girl on a swing, a parachutist, and an orange)

---

[8] In this case the script is based on actual life events, but students can enrich it with imaginary situations, events, and dialogues that do not alter the historical reality.

[9] Such a challenge can take place at the beginning of the lesson, in order to motivate students to think creatively and at the same time become aware of the various interconnections of phenomena and ideas.

- *Mystery solving* (e.g. to explain the disappearance of something, like a certain volume of liquid; to find the identity of a substance out of a variety of similar-looking substances, like sugar, salt, and baking soda; to find something that is missing, like a beam of light)
- *Detective work* (e.g. to detect misconceptions about science ideas, like the kind of trajectory and floating behaviour of an object, or even to find whether the science behind certain phenomena, like a foggy window glass, a frozen lake, the kind of motion of a car, can reveal the innocence or guilt of a person)
- *Approaching the teaching and learning of science through the visual arts* (e.g. using photography to make a collage to present the various states of water and the results of a study of a topic such as the effect of modern technology on everyday life, using technologies to construct scientific models, using drawing to represent a phenomenon, such as photosynthesis)
- *Pursuing a project of personal interest* (e.g. the evolution of various technological applications—obsolete, current—and speculation about future ones, the creation of an alternative periodic table)

The role of visual arts and especially that of dramatization are discussed in Chap. 7, due to the attention they have received over the last decade. The role, however, of science fiction, which is missing from the above list of ideas, deserves particular attention, especially in the case of gifted students. Given that creativity, as Vygotsky (2004) pointed out, presupposes a rich knowledge base—the more one knows, the more creative one can be—and the use of the imagination, then science fiction can be seen as an excellent way to approach the development of students' creativity. Indeed, science fiction (a) helps students acquire a rich content knowledge, due to the variety of topics covered by the science fiction literature, and (b) encourages their imaginative engagement as science fiction topics are intellectually intriguing and challenging.[10] Stutler (2011) makes reference to Dabrowski's 'theory of overexcitabilities', which suggests that 'intellectual, emotional, and imaginational linkages are the basis for highly creative intelligence', such as in the case of gifted students.

5.6 Concluding Comments: The Need for Creative Thinking in Science Education

Whether or not one agrees with the idea that the development of creativity should be the first most important goal of education is irrelevant when it comes to the notion of scientific creativity. Given that imagination and creativity are considered central to the nature of science, a good science education cannot help but foster students' imaginative skills and creativity. And because creative thinking is necessary for inquiry and problem-solving—even problem-solving in the context of everyday

---

[10] The issues that can be raised about science fiction have already been discussed at the end of Chap. 1.

life—creativity can help make science education 'more functional'. This is an important message to those who lament over the state of science education.

If all learning is a possibility (Brent, Sumara, & Luce-Kapler, 2008; Hadzigeorgiou, 2005b), the development of students' creative powers through science learning is a possibility too. Maslow's (1968) distinction between 'special talent creativeness' and 'self-actualizing creativeness' can provide food for thought for those who are willing to make school science not just an adventure but a creative endeavour too, so that students' creative acts will be both acts of self-expression and acts of self-actualization. Although 'special talent' creativity has played a catalytic role in the development of our civilization, in the context of education, 'self-actualizing' creativity appears to be a much more realistic notion and also a more fruitful and promising one.

Carl Rogers' (1961) definition of creativity as 'the emergence in action of a novel relational product growing out of the uniqueness of the individual on the one hand, and the materials, events, people or circumstances of his life on the other' (p. 351), when applied to education, makes the distinction between the various degrees and/or kinds of creativity (as found in the literature) unimportant and orients us to the process of wholehearted engagement, as the first necessary condition of creativity (see Csikszentmihalyi, 1996). Perhaps, if approached from this perspective, creativity in school science can open new vistas for both teachers and students and can provide an answer to the perennial problem of students' engagement in science.

However, given the problematic nature of this engagement process, one point deserves to be made explicit here: A delicate but nonetheless important distinction needs to be made between those situations or questions that are supposedly challenging (i.e. we tend to think that they will challenge students to use their creative imagination) and situations or questions that are really challenging, in the sense that they encourage imaginative engagement. So if our goal is to foster creativity, we need to realize that not all situations and questions will result in student engagement and thus in creative thinking. Asking children, for example, what they might find on a forest floor, or even how to make a musical instrument by using various vegetables or a flashlight from simple everyday materials, can be challenging for some children or students, but not for all of them. We suppose that such questions will automatically make students use their creative imagination, in their attempt to think of all possible entities—living and nonliving—on the forest floor or of the possible veggies and materials in order to make a musical instrument and a flashlight, respectively. However, before students start to inquire and think in order to answer the teachers' questions, they have to be imaginatively engaged with the topic of the question. In other words, we can increase the possibilities for creative thinking, if we increase the possibilities for imaginative engagement (see Robinson, 2001). The work done by the *Imaginative Education Research Group* (IERG), directed by Kieran Egan at Simon Fraser University, is promising, at least as far as the role of the imagination in learning is concerned. More research is certainly needed, especially with regard to creativity, but, nonetheless, such research needs to be based upon a theoretical framework, which gives primacy to imagination.

5.6 Concluding Comments: The Need for Creative Thinking in Science Education

Imagination, of course, and creativity are linked to the arts. And this is quite reasonable, given that, oftentimes, these two notions are discussed, analysed, and researched in the context of the arts. And yet the crucial role of wonder in fostering creativity should be seriously considered too. Although this role may be implicit in artistic approaches and activities, it needs to be made more explicit in the context of school science, as the experience of wonder about phenomena and ideas can make students creative or, at least, more creative.

Given the drop in creative thinking past grade 5—even though up until fifth grade students were found to be increasingly open-minded and curious and also able to produce unique responses (Kim, 2011)[11]—as well as the fact that the enthusiasm to learn science drops after grade 4 (Kirikkaya, 2011), the possibility of evoking wonder as a means to creativity should be given serious thought. In fact, Cant (2014) believes that seeing the world through the lens of wonder can make schools more exciting and creative places. The next two chapters, which deal with the role of wonder and the role of art in school science education, can help one understand both wonder and art as prerequisites for student engagement and, from the perspective of this chapter, the role that wonder and art can play in creative thinking.

---

[11] The findings of a study, whose sample consisted of 300,000 students from kindergarten through grade 12, are overwhelmingly disappointing with regard to creative thinking. By and large, children's thinking beyond fifth grade, that is, in middle and high school, was found to be conformist (Kim, 2011).

Chapter 6
'Wonder-Full' Science Education

> *Nothing is too wonderful to be true, if it be consistent with the laws of nature.*
>
> Michael Faraday, in Bence Jones' *The Life and Letters of Faraday*, p. 253.
>
> *In the end, science, as we know it, has two basic types of practitioners. One is the educated man who still has a controlled sense of wonder before the universal mystery whether it hides in a snail's eye or within the light that impinges on that delicate organ. The second kind [...] is the extreme reductionist who is so busy stripping things apart that the tremendous mystery has been reduced to a trifle, to intangibles not worth troubling one's head about.*
>
> Loren Eiseley, in *The Star Thrower*, p. 151
>
> *We all have a thirst for wonder. It's a deeply human quality [...] There's wonder and awe enough in the real world. Nature's a lot better at inventing wonders than we are.*
>
> Carl Sagan, in www.azqotes.com/author/12663-Carl_Sagan

Wonder, in the context of school science education, is one of those notions that is either taken for granted—we all know it is important, so we need not talk about it—or simply associated with the emotional dimension of scientific knowledge, which is not the 'real stuff' of science. Science learning is about knowledge and understanding, even though such learning is influenced in one way or another by emotions. The value, of course, of wonder has been defended by philosophers and scientists alike. Toulmin (1976), for example, in line with Aristotle, writes that wonder is the engine of all intellectual inquiry, while MIT physicist Victor Weisskopf considers wonder the seed of knowledge (Weisskopf, 1979).

It deserves to be noted that Harvard psychologist and Hobbes professor of cognition and education, Howard Gardner, in agreement with Levi Strauss, believes that all psychological/social scientific understanding has to begin with a phenomenological approach—although it should not stop there—thus appraising indirectly the value of wonder (personal communication, 2009). This value, of course, has been seen by some science educators (e.g. Gilbert, 2013; Goodwin, 2001; Hadzigeorgiou, 2001, 2007, 2014; Millar & Osborne, 1998; Ritz, 2007; Silverman, 1989, 2003;

The original version of this chapter was revised. An erratum to this chapter can be found at DOI 10.1007/978-3-319-29526-8_8

Stolberg, 2008). And yet the role of wonder in science education has been downplayed or, worse, has been totally overlooked.

The reason why wonder has not received the attention it deserves by mainstream science education may be found in the following facts (Hadzigeorgiou, 2007, 2014): First is the emphasis on the social element of learning science (and hence the emphasis on discourse and cooperative activities) (Lemke, 2001). Second is the fact that it is not only difficult for both pupils and science teachers to completely abandon empiricist and logical positivist philosophies of science, especially when engaged in laboratory work (Monk & Dillon, 2000), but also because one can adopt a constructivist approach to the teaching and learning of science without even bothering about the issue of wonder (Hadzigeorgiou, 2005a, 2005b). Third, the recent emphasis on citizen science and the public understanding of science has led to the consideration of a pragmatist conception of school science education (see Jenkins, 1999, 2002; Roth & Lee, 2004). Fourth, wonder is inherently passive, and, as such, it might be an obstacle to curiosity (Hadzigeorgiou, 2007). And fifth, the notion of wonder itself is quite problematic since it can be associated with science fiction (Barron, 1987; Kelley, 1972) and also with magic, miracles, and even incomprehensibility (Silverman, 1989).

All the aforementioned reasons, however, stem from lack of awareness of the relationship between science and wonder, and especially from a misunderstanding of the nature of the latter. It is the purpose of this chapter to explore the nature of wonder and the relationship between science and wonder and then discuss the value of wonder in science education, by making reference to its role as a prerequisite for engagement with school science, as a source of students' questions, and also as a prerequisite for significant learning.

6.1 The Nature of Wonder

The nature of wonder is no doubt both complex and elusive. It does not take one long to realize that wonder, as a state of mind, can be associated with mystery, awe, perplexity, astonishment, surprise, amazement, admiration, and bewilderment. Yet the experience of a sense of wonder cannot be reduced to the experience of any one of the aforementioned elements. Mere surprise or even mere astonishment is not wonder anymore than is mere admiration or bewilderment. Also the difficulty one has to conceptualize wonder may be found in the fact that it is used interchangeably with the notion of curiosity (e.g. see Silverman, 1989, 2003), due to the circular nature of both notions.

Even though we all had—most likely at some point in our childhood—an experience of wonder, it has been difficult to articulate our feelings or state of mind.[1]

---

[1] I have always remembered the first time I visited, during a trip as an elementary school student, a cave and came face to face with stalactites and stalagmites. The excitement and the feelings of surprise, astonishment, and bewilderment have also been unforgettable. It was very recently, how-

And the fact that the notion of wonder is difficult to conceptualize can be seen in the various definitions or aspects of it, as found in the literature.

Silverman (1989), for example, differentiated between 'wonder in the sense of curiosity' and 'wonder in the sense of the magical, miraculous and incomprehensible' (p. 44). Wonder in the latter sense 'is like a narcotic and destroys curiosity and anesthetizes the intellect' (p. 44). Goodwin (2001), also, in distinguishing between two aspects of wonder, that is, 'wondering about' and 'wondering at', identified the former with curiosity (which 'reflects the activity of scientists') and the latter with our capability of wondering. It is this capability, according to Goodwin, which 'reflects the human response to discoveries and understandings' (p. 69). A 'wondering if' aspect has also been identified by Goodwin, although this appears to be similar to the 'wondering about' aspect.

Related to Goodwin's (2001) 'wondering at' the world are Stolberg's (2008) three categories of wonder, namely, 'physical wonder' (which is induced by interaction with natural objects and/or phenomena), 'personal wonder' (which is induced by interaction with human beings and/or their achievement), and 'metaphysical wonder' (which is induced by any kind of interaction, but the experience of wonder leads to a shift in perspective). It is evident that any one kind of wonder (i.e. physical, personal, metaphysical) may or may not induce a 'wondering about' attitude towards the world, depending on a number of factors (i.e. worldview, opportunities, and help to reflect on the original stimulus). It is also evident that all the aforementioned distinctions point to a passive aspect and an active aspect of wonder. The former can be identified with an emotional response to something (i.e. a phenomenon, an idea) or even with something magical, miraculous, and even incomprehensible, which may or may not lead to a shift in perspective, while the latter with curiosity.

Yet wonder is not the same as curiosity. The Oxford dictionary (11th edition) gives the following definition of wonder: a feeling of surprise and admiration caused by something beautiful, unexpected, or unfamiliar. It is clear that this feeling is not the same as curiosity. Hove (1996), in explicating the notion of wonder, identifies wonder with 'the emotion caused by the perception of something novel and unexpected or inexplicable' and with 'the state of mind in which this emotion exists' (p. 442). However, he also identifies it with an 'astonishment mingled with perplexity

---

ever, while studying research findings on motivation and attitudes, that I realized the full import of that experience in the cave. In reflecting upon my then reaction to seeing stalactites and stalagmites, I became aware that I had experienced a sense of wonder, which, in turn, became a source of interest and curiosity. From the moment I laid my eyes on those "ice- or brown-sugar-looking" rocks, I began to observe them closely, to touch them—despite the warnings against doing just that—even to taste them, and then to ask the tour guide and my teacher about their origin, age, and composition. My curiosity did not stop at the cave. Upon returning home I vividly remember my search in the encyclopaedia and my discussions with friends during the following days about what I had learned about stalactites and stalagmites. That experience in the cave was no doubt a transformative one with an impact upon my subsequent education, and recently, it served as a basis for an action research project on making physics more engaging for lower secondary school students (Hadzigeorgiou, 2012). The results are discussed in this chapter.

or bewildered curiosity' (p. 442). For him being simply curious about something without being astonished by that something cannot by itself evoke a sense of wonder. Curiosity, according to Burke's (1990) philosophical analysis, is the simplest human emotion and it should be differentiated from the state of astonishment during which 'the mind is so entirely filled with its object that it cannot entertain any other' (p. 53). Santayana (1955) also associates astonishment with the notion of the sublime, which in turn relates not to curiosity but to such notions as awe and wonder.

What deserves to be stressed here is a delicate distinction between curiosity and wonder: Curiosity is actually the drive to investigate or study something, while wonder is a state of mind or feeling. Moreover, wonder has an aesthetic dimension, which can be totally absent from curiosity. This aesthetic dimension implies that astonishment and admiration can be both present in the experience of wonder. For example, there is empirical evidence that students who watched a film, whose plot included Tesla demonstrations and experiments on the wireless transmission of electrical energy, experienced not only bewildered curiosity about such experiments or demonstrations, but also a feeling of astonishment, while, at the same time, they expressed their admiration for what Tesla did and Tesla himself (Hadzigeorgiou & Garganourakis, 2010). And the students who were astonished to learn that during a head-on collision the force a very small car exerts on a very big truck is equal in magnitude with the force the very big track exerts on the very small car—even though they knew which vehicle was going to be damaged!—expressed their admiration for Newton's ingenuity to conceive of the law of action and reaction (Hadzigeorgiou, 2012).

Taylor (1998), in associating admiration with wonder, talks of the 'poetic' nature of the latter. For Taylor the emotional response to what is being perceived stems primarily from the wholeness of the object of perception. While curiosity, according to him, is a scientific impulse that strives to dominate nature:

> *Wonder is poetic and is content to view things in their wholeness and full context [...] When a flower is taken apart and examined as pistil, stamen, stem and petals, each part is seen exactly and a certain curiosity is satisfied; however, curiosity is not wonder, the former being the itch to take apart, the latter to gaze on things as they are.* (Taylor, 1998, p. 169)

The association between admiration and wonder is present, according to Dawkins (1998), in the 'poetic' nature of science itself. Although science and poetry represent two different ways to experience the world (i.e. a sunset can be described by both a physicist and a poet, and here we have two different descriptions), one can nevertheless speak of the poetry of science, as Dawkins does, in the sense that one can feel admiration and astonishment at phenomena and ideas. One can admire and be astonished at the beauty of natural phenomena and at the unexpected connections among such phenomena. Moreover, one can feel admiration and astonishment at the fact that scientific understanding opens up new ways of looking at things and leads to new discoveries and understandings.

One, of course, could identify astonishment and admiration, that is, the poetic nature of wonder, with a 'wonder at' attitude, which, certainly, is not curiosity. It should be stressed though that even a 'wonder about' attitude should not be identified solely with curiosity. A person, for example, can wonder about how to proceed in approaching or solving a problem, without his/her curiosity being aroused. In such a case, there are first an awareness of a problematic situation and, second, feelings of perplexity, doubt, and uncertainty. This simple situation shows that wonder and curiosity are two different notions, and it is misleading to use them interchangeably.

That wonder is not the same as curiosity can be also seen from the fact that while curiosity is about things to which 'answers can be given by reference to procedures that are commonly acknowledged, wonder points to something beyond the accepted rules' (Opdal, 2001, p. 331). Going beyond the accepted, the usual, and the ordinary, however, does not mean that wonder, in order to be evoked, requires unusual and extraordinary objects, phenomena, or situations, as is usually the case with curiosity (Berlyne, 1960; Bruner, 1996, p. 114; Loewenstein, 1994). In fact, wonder, unlike curiosity, can be evoked even through simple and usual situations, as empirical evidence suggests.[2] Such evidence (i.e. that wonder can be evoked through simple, ordinary situations and can make one see something usual and ordinary as unusual and extraordinary) justifies Martin Heidegger's view:

> *Unlike curiosity which presupposes that there is a distinction between the usual and the unusual, ordinary and extraordinary, wonder is an attunement in which one finds the usual to be extraordinary.* (Stone, 2006, p. 208)

Heidegger, in his *Being and Time*, is quite clear about the difference between wonder and curiosity: 'curiosity has nothing to do with observing entities and marvelling at them' (Heidegger, 2008, p. 216). This marvelling cannot take place when one is curious, because, he pointed out, once curiosity 'obtains sight of anything, it already looks away to what is coming next' (p. 398). But wonder experienced as a feeling of astonishment can make the mind, as was previously said, to be 'entirely filled with its object that it cannot entertain any other' (Burke, 1990, p. 53). Thus, it is wonder, not curiosity, that makes one dwell on a phenomenon, an idea, and this is precisely its educational value. Engaging students in science presupposes this dwelling on phenomena and ideas. Certainly, curiosity, aroused from something unusual and extraordinary, has its place and value in science education, and education in general, but its different nature from wonder needs to be acknowledged.

---

[2] For example, a science teacher can help foster a sense of wonder at and about the force of gravity by helping students become aware (through a simple and very usual situation involving a magnet holding a paper clip) that gravity is indeed the weakest of all forces, since a tiny magnet can hold a paper clip despite the fact the whole earth is pulling down on it. According to empirical evidence, initially the demonstration meant nothing to the students, who saw only two ordinary objects, that is, a paper clip and a magnet. It was only after they became aware (through questioning) that the magnet attracted and held the paper clip, in spite of the fact that the whole planet was pulling down on it, and that they felt surprised and, in fact, astonished and started to wonder at and about the force of gravity. As one female student commented 'Although I knew that gravity was the weakest of all forces and I could see that in the numbers on that table about the relative strength of all forces in Nature, it was after that simple, and very easy-to-do experiment that I understood it better […] It is really remarkable and very strange now that I know that the force of gravity is very-very weak' (Hadzigeorgiou, 2006a, b, 2007).

That curiosity is the 'thief of wonder', as Stone (2006) put it, can be seen in that 'utilitarian curiosity' prevalent even in inquiry science, in cases in which 'the students pursue answer after answer as though making a tick on the tourist's place list' (Piersol, 2014, p. 15). That is why it is crucially important to distinguish between 'utilitarian curiosity' and 'bewildered curiosity' for it is the latter that relates to wonder (see Appendix G for narratives introducing a sense of mystery/bewildered curiosity).

In comparing curiosity and wonder, it should also be noted that the component of awareness, which is always present in wonder, may be totally absent from curiosity. For example, in the case of a child asking questions about the sky, there is a difference between the question 'why is the sky blue?' and the question 'If stars fall all the time, then why is the sky always full of stars?', in the sense that in the former case the child can be simply curious, while in the latter she/he becomes aware that his/her knowledge is either incomplete or mistaken. Of course, this is a very simple example, but whether we are talking about a demonstration (i.e. of the weakness of the force of gravity, of the invisibility of light) or a verbal expression (i.e. matter is 99.99 % empty space; there can be motion at extremely high speeds in a straight line in the absence of a net force), a kind of awareness must always be present. In actual fact, it is this awareness that makes something usual and ordinary to be seen as unusual and extraordinary. Harvard psychologist Howard Gardner believes that wonder makes sense only when awareness of some kind is present and that is why he considers wonder inappropriate for very young children (personal communication, 2009).

There may be, of course, an argument that a distinction between a 'wonder question' and a 'curiosity question' has only theoretical interest, as it would be impractical to try to distinguish between the two, for it is true that most of the time it is not an easy task to perceive that delicate distinction between those two kinds of questions (i.e. the question 'why is the sky blue?' may be a 'wonder question' or a 'curiosity question'). However, there are times that one can identify a 'wonder question'. When a student asks 'Why is it so cold up on the top of a mountain, since we know that hot air always goes up?', she/he is experiencing a sense of wonder, since she/he becomes aware that her/his knowledge is either incomplete or mistaken. So although both curiosity and wonder are important and a good science education should foster both of them, the role of wonder, as a potential source of students' questions, needs to be acknowledged and valued (see section on 'Wonder in Science Education: What the Research Shows' for examples of 'wonder questions' (Sect. 6.3)).

Dewey's (1998) distinction between three stages or levels of curiosity, that he calls organic, social, and intellectual, can be quite useful here since it is level three or intellectual curiosity that can be identified with wonder. As Dewey argues, at the organic stage, very young children are simply curious about anything—their curiosity being an expression of abundant organic energy—and at the social stage their curiosity is developed under social stimuli, but their motive is not an explanation but an 'eagerness for a larger acquaintance with the mysterious world' (p. 38) in which they are placed. At this social stage, in other words, young children's curiosity is about facts about the world, which though is 'not evidence of any genuine consciousness of rational thought' (p. 38). This is certainly different from the stage at which children become aware of something, and their curiosity is transformed into an interest in finding out answers for themselves. Apparently, it is this kind of curiosity that can be identified with wonder.

It is interesting to note that the association between wonder and awareness is present in Whitehead's notion of 'stage of romance'. The stage of romance, as the stage of 'first apprehension' of the subject matter of any school subject, precedes the stages of precision and generalization and has as a central element the experience of a sense of wonder through the awareness of unexplored or unexpected connections among facts, events, and ideas. As Whitehead (1957) pointed out, at the stage of romance, there is a feeling of 'excitement consequent on the transition from the bare facts to the first realization of the import of their unexplored relationship', and also a realization of 'unexplored connexions with possibilities half-disclosed by glimpses and half-concealed by the wealth of material' (pp. 17–18).

An illustration of Whitehead's notion of 'stage of romance' is astrophysicist Carl Sagan's experience, as a young child, in New York. Sagan, according to what he recounts, experienced a sense of wonder, which was associated first with the awareness that stars were that aspect of his environment that was different from all the rest—something that also made him wonder about their nature—and second with the awareness that they (stars) were suns, just like ours, and very far away. His comments on that kind of awareness are quite instructive and worth quoting:

> *It was in there. It was stunning. The answer was that the Sun was a star, except very far away. The stars were suns; if you were close to them, they would look just like our sun. I tried to imagine how far away from the Sun you'd have to be for it to be as dim as a star. Of course I didn't know the inverse square law of light propagation; I hadn't a ghost of a chance of figuring it out. But it was clear to me that you'd have to be very far away. Farther away, probably, than New Jersey. The dazzling idea of a universe vast beyond imagining swept over me. It has stayed with me ever since [...] I sensed awe. And later on (it took me several years to find this), I realized that we were on a planet – a little, non-self-luminous world going around our star. And so all those other stars might have planets going around them. If planets, then life, intelligence, other Brooklyns – who knew? The diversity of those possible worlds struck me. They didn't have to be exactly like ours, I was sure of it.* (Sagan, 1995, p. 25)

What needs, of course, particular attention in regard to Whitehead's ideas about the stage of romance and 'the realization of the import of unexplored relationships' is the association of wonder with the awareness of the significance of certain phenomena and ideas. Verhoven (1972) has in fact argued that wonder reveals the infinite significance of things and also urges us to respect that which it reveals. And, as such, wonder can help us view things in a new light (Hove, 1996). Indeed, wonder has the capacity to defamiliarize 'the familiar through a refreshed way of looking upon it' (Abrams, 1971, p. 379). Michael Faraday, for example, who is reputed to have said that 'Water is to me, I confess, a phenomenon which continually awakens new feelings of wonder as often as I view it', explicitly talked about a refreshed way of looking upon water, as a result of his experience of wonder.

In light of the foregoing discussion, wonder can be viewed as an intellectual attitude or state of mind that can have several and diverse sources: situations, phenomena, and ideas that give one the opportunity to admire, to feel a sense of mystery, and to be surprised, astonished, bewildered, and perplexed. However, awareness of some kind should also be present. This means that wonder has two components: an emotional and a cognitive component.[3] Therefore wonder, in order to be considered a 'pedagogical tool', should also result in the following kinds of awareness:

- Awareness that one's knowledge is incomplete or mistaken
- Awareness that there is more to be learned
- Awareness that some phenomena exist at all
- Awareness of unexpected connections among phenomena and among ideas
- Awareness of the beauty of natural phenomena

Such awareness, more often than not, comes as a 'shock' to the one who experiences it and can be quite powerful, for it makes one conscious of what one is learning, or even of what one has already learned, and, at the same time, makes one view the world in a new and different light.

This awareness, obviously, is crucially important in education, for it is after a 'shock of awareness' that students can perceive abruptly something and also see unexpected connections among phenomena and among ideas. The educational philosopher Maxine Greene writes:

> *A great part of our everyday life is not lived consciously, and since nothing makes an impression, the world seems bland, muffled, and vague. Now and then, however, there are exceptional moments, moments of response to 'shocks of awareness'.* (Greene, 1978, p. 185)

What Greene says applies to education. Indeed, for many students, the world of science and education in general appears bland, muffled, and even vague. And there may be students who have learned science but are not really conscious of what they have learned.

Aristotle's notion of *aporia*, whose meaning is best captured by the English word *wonder*, derives from a-poria, which literally means 'no path', that is, a state of mind, which makes one aware that one does not have a path to proceed further. When a teacher, for example, asks his/her students to ask a question, he or she asks them to express an *aporia*—not to be curious about what is being, or has been, taught—so he or she can make sure that they understood what he or she taught them. Students express their *aporia* because they do not know how to proceed (i.e. explain and/or predict something), as a result of lack of knowledge and understanding. Apparently it is the awareness of their ignorance (i.e. their lack of knowledge and understanding) that makes students be in a state of aporia, that is, in a state of wonder. Socrates, by evoking in his audience a sense of wonder, made them aware of their ignorance, that is, their lack of knowledge and understanding (see Matthews, 1997; Piersol, 2014).

The idea of 'shock of awareness' is central to the process of cognitive conflict since the aim of the instructional process is to challenge existing misconceptions. However, the development of awareness goes well beyond the cognitive conflict

---

[3] Perhaps the expression 'I feel a sense of wonder' makes one associate wonder with emotions rather than cognition.

approach and encompasses science learning in general. Awareness, for example, that the electrical resistance of our own skin determines in certain circumstances our chances to survive death; that gravity is an extremely weak force, much weaker than the attractive force a tiny magnet can exert; and that of all organisms on the planet only plants are responsible for maintaining life, is very different from knowing and applying Ohm's law in order to solve problems; from knowing that gravity pulls down on all objects, which accelerate at the same rate; and from knowing the chemical equation of photosynthesis and the substances that are involved in it, respectively. These examples do show that the notion of awareness goes beyond the cognitive conflict approach, as it (awareness) also enables students to see things differently. They also show that the awareness inherent in the experience of wonder presupposes imaginative engagement. Dewey (1934), in fact, had argued that 'When old and familiar things are made new in experience, there is imagination' (Dewey, 1934, p. 267).

It is quite interesting to note that a shift or change of outlook, as a result of the experience of wonder, is also in line with a view of knowledge inspired by complexity theory and more specifically by the notion of 'strong emergence'. Knowledge, from such a perspective, helps people perceive a new reality.

> *Knowledge does not bring us closer to what is already present. Rather it emerges into that which is unthinkable from the ground it precedes [...] Emergent knowledge, in other words, moves us into a new reality, which is incalculable from what came before.* (Osberg & Biesta, 2007)

There is no question that such a perspective on learning, as a change of student's perception of science ideas and phenomena, represents a great challenge for science education. Such a challenge had been posed by Richard Feynman (1969), who, in his address to the National Science Teachers Association, as is well known, had pointed out that the world looks so different after learning science. This is what science teaching should aim at, but the question is how can such change be encouraged and fostered in the context of school science. This is discussed in detail in the sections on the role of wonder in science education and on its pedagogical implications. However, the role of wonder in professional science, which is discussed in the next section, can help shed some light on the relationship between the two, for it is crucially important that one have an understanding of what wonder is and the role it plays in science, before one attempts to design activities that supposedly have the potential to foster it in the context of science education.

6.2 Wonder in Science

The relationship between science and wonder is well documented in the literature, and, as Holmes (2009) points out, the romantic science period (see Chap. 3) was indeed the age of wonder. Even though the image of the 'lone star scientist', spread during the romantic period,[4] is a myth (see Heering, 2010) and science is now considered a social activity—'constitutively social' as Woolgar (1993, p. 13) put it—the fact that science is also characterized by a strong personal element cannot be dismissed. Central to this personal element is not only the constructive nature of thinking and understanding but also the experience of wonder. This personal element of science has also been described as 'aesthetic' (Fisher, 2003; Root-Bernstein, 2002; Tauber, 1996). Central to this aesthetic element is the idea of beauty, which, in turn, can be directly linked to the experience of wonder (Girod, 2007a, 2007b; Hadzigeorgiou, 2005a; Hadzigeorgiou & Fotinos, 2007).

What needs to be pointed out is that the experience of wonder can be one of the greatest rewards of science: 'The procrustean oversimplification of fundamentalist reductionism […] cannot embrace the practice of science itself […] whose chief reward is the experience of wonder' (Polkinghorne, 1998, p. 2). In *The Star Thrower*, Loren Eiseley (1978) speaks of two kinds of practitioners in science. One is the 'extreme reductionist who is so busy stripping things apart that the tremendous mystery has been reduced to a trifle', and the other is she/he 'who still has a controlled sense of wonder before the universal mystery whether it hides in a snail's eye or within the light that impinges on that delicate organ' (p. 151).

It is true, of course, that the experience of a sense of wonder does not become evident like other elements central to scientific inquiry (e.g. intellectual, ethical). Yet it does emerge when scientists speak autobiographically about their work and the work of other scientists (see Root-Bernstein, 1996). What should be pointed out is that there is a confusion between what scientists do and what they actually report. 'At the heart of the unsolved problem concerning scientific thinking is the confusion of the form and content of the final translations with the hidden means by which scientific insights are actually achieved' (Root-Bernstein, 2002, p. 61). Unfortunately, the exclusion of the element of wonder from scientific reports 'discourages and delegitimates its expression or even admission by students and amateurs to having experiencing it' (Hein, 1996, p. 285).

Although curiosity is an important scientific attitude and a motive for research and inquiry, scientists, when speaking about their work and the work of other scientists, associate wonder not with curiosity but with such elements as mystery and awe (Root-Bernstein, 1996, 2002; see also Midgley, 2000a, 2000b; Wilson, 1986, p. 10). The association of wonder with mystery and awe provides evidence that those who experience a sense of wonder respond emotionally to their object of study and that this emotional response may be considered a prerequisite for engaging with that object of study. This association of mystery, awe, and wonder can be found in the literature and should be noted: 'Mystery generates wonder and wonder generates awe' (Goodenough, 1997, p. 13), and 'Our sense of wonder grows exponentially: the greater the knowledge, the deeper the mystery' (Wilson, 1986, p. 10). Perhaps Einstein's famous phrase epitomizes the role of mystery as a source of awe and wonder.

---

[4]The spread, however, of science to society and the establishment of scientific institutions and societies are a legacy of that period.

6.2 Wonder in Science

> *The fairest thing we can experience is the mysterious. It is the fundamental emotion which stands at the cradle of true art and science He who knows it not and can no longer wonder, no longer feel amazement, is a as good as dead, a snuffed-out candle.* (Einstein, 1949, p. 5)

The above phrase, linking mystery and wonder[5], is quite famous and is often quoted in books, articles, and speeches. However, it was Newton who implicitly expressed a sense of the mysterious and thus a romantic attitude towards knowledge:

> *I do not know what I may appear to the world, but to myself I seem to have been only like a boy playing on the sea-shore, and diverting myself in now and then finding a smoother pebble or a prettier shell than ordinary, whilst the great ocean of truth lay all undiscovered before me.* (Cited by Mandelbrote, 2001, p. 9)

It is interesting to note that Newton himself, whose physics became the focus of sharp criticism, and in fact came under attack by the Romantics, was motivated by a sense of mystery and spiritual wonder. Recent scholarship provides evidence that:

> *Newton was not the first of the age of reason. He was the last of the magicians, the last of the Babylonians and Sumerians, the last great mind which looked out on the visible and intellectual world with the same eyes as those who began to build our intellectual inheritance rather less than 10,000 years ago.* (White, 1997, p. 3)

There have been, of course, a number of scientists who have implicitly and even explicitly talked and wrote about the value of wonder in science. Michael Faraday, for example, is well known for his talent to amaze the audience with his demonstrations and scientific explanations about the wonder and beauty of candle flames, which tap all the known laws of the universe (see his *The Chemical History of Candle*). However, two famous scientists do stand out: Richard Feynman and more recently Richard Dawkins.

Feynman is famous for his love for science, particularly physics, for his ability to instruct and entertain his audiences, and also for his witty comments especially on science and its nature, as these have appeared in a number of books, most of them bestsellers.[6] His view about the aesthetic element of science, which (view) is quite explicit about the beautiful and wonderful things of science, and about their capacity to inspire people, can provide science educators and teachers with much food for thought:

---

[5] One has to bear in mind that wonder and mystery are associated, no matter their source or origin. For example, whether one observes a rainbow or a waterfall and becomes aware of their beauty and feels a sense of mystery about their origin, or a battery that is always running out of energy each time it is connected in a circuit, despite the fact that the battery was always fully charged, one's sense of wonder has its source in one's sense of mystery. In the first case, the sense of mystery has its source in the aesthetic perception of the rainbow or the waterfall. In the second case, there is mystery simply because there is an unexplained situation. In both cases there is a sense of wonder. However, while the first case is associated with a wonder-at and a wonder-about attitude, the second case is associated only with a wonder-about attitude or simply with curiosity.

[6] His most famous books are *Six Easy Pieces*, *What Do You Care What Other People Think*, and, of course, *The Feynman Lectures on Physics*.

> *The world looks so different after learning science. For example, trees are made of air, primarily. When they are burned, they go back to air, and in the flaming heat is released the flaming heat of the sun which was bound in to convert the air into tree. [A]nd in the ash is the small remnant of the part which did not come from air, that came from the solid earth, instead. These are beautiful things, and the content of science is wonderfully full of them. They are very inspiring, and they can be used to inspire others.* (Feynman, 1969, p. 319)

However, it has been Oxford evolutionary biologist Richard Dawkins who can indeed be considered the most fervent exponent of the importance of wonder in the practice of science. He believes that wonder is 'one of the highest experiences of which the human psyche is capable' (Dawkins, 1998, p. xii) and makes the following comment:

> *Yes, we must have Bunsen burners and dissecting needles for those drawn to advanced scientific practice. But perhaps the rest of us could have separate classes in science appreciation, the wonder of science, scientific ways of thinking, and the history of scientific ideas, rather than laboratory experience [...] Far from science not being useful, my worry is that it is so useful as to overshadow and distract from its inspirational and cultural value. Usually even its sternest critics concede the usefulness of science, while completely missing the wonder.* (Dawkins, 1998, p. 10)

In the context of contemporary science education, with an emphasis on standards, and with proposals for sociopolitical and instrumentalist/utilitarian conceptions of school science (see Hadzigeorgiou & Schulz, 2014; Schulz, 2009, 2014), what Feynman and Dawkins said about science is worth considering seriously. Indeed, it has been pointed out that a pragmatist/utilitarian conception of science does not contribute to our appreciation of both the cultural value and the beauty of science (Dawkins, 1998). Moreover, 'the practical mission to advance science alone seems grim, focusing as it does on the political and economic importance of science and its role in a technological world' (Hein, 1996, p. 285). It is for these reasons that attention should be paid to the notion of wonder and an attempt should be made in order to reclaim its value in school science education.

MIT physicist Max Tegmark is a case in point, as his experience with wonder can be truly instructive. Tegmark (2015) recounts how his intellectual struggle with a 'wonder question' played a role in his decision to choose a career in physics and also pursue an advanced study in quantum mechanics and theoretical physics[7]: 'If we know that subatomic particles can be in different places at the same time, then why can't people, who are made of such particles, as well?' This question has also played a central role in Tegmark's professional research, which has been described with such feelings as enthusiasm, fun, and excitement—all of them having their source in nature's weirdness, strangeness, absurdity, and beauty.

---

[7] This is my own interpretation of what he recounts in his *Our Mathematical Universe*.

6.3 Wonder in Science Education: What the Research Shows

Despite the dearth of studies on the experience of wonder, the very few studies found in the literature fall in two categories. First, there are studies with pre-service teachers and their views about the experience of wonder and, second, studies with primary and secondary school students. With regard to the first category, two studies by Stolberg (2008) in England and Gilbert (2013) in Australia deserve to be mentioned. With regard to the second category, three studies conducted by Hadzigeorgiou (2012) and Hadzigeorgiou and Garganourakis (2010) in Greece and by King, Ritchie, Sandhu, and Henderson (2015) in Australia deserve particular attention as they provide evidence of the specific roles that wonder can play in the teaching/learning process.

As regards the first category, Stolberg (2008) explored whether wonder-based reflections can be a source of inspiration. He studied 140 English pre-service primary teachers' views about what experiences of theirs were considered personal sources of wonder. The pre-service teachers described 240 separate situations/events, in which they personally experienced wonder. These were categorized as personal (induced by interaction with human beings and/or their achievement), physical (induced by interaction with natural objects and/or phenomena), and metaphysical (induced by any kind of interaction which leads to a shift in perspective). The analysis of extended interviews with 15 pre-service teachers showed the crucial role and value of wonder for emotional engagement in scientific inquiry.

Gilbert (2013), on the other hand, explored whether wonder can increase Australian pre-service teachers' interest in science and develop positive views with regard to science content knowledge. Four case studies demonstrated a shift in the pre-service teachers' desire to learn science content knowledge after their participation in activities that helped to evoke in them a sense of wonder. They did report that before the intervention, they used to detest that science content knowledge.

With regard to the second category, Hadzigeorgiou (2012) reports on an intervention study, in the form of action research, with grade 9 students. The study involved two grade 9 classrooms consisting of 27 and 30 students, respectively. The students of the first classroom served as the treatment group, while the students of the other classroom as the control group. Three units, from the 9th grade syllabus, namely, matter, force and motion, and light, were selected. For each of those three units, ideas as possible sources of wonder were identified (see Appendix E) and were presented to the students of the treatment class verbally, through demonstration experiments and/or hands-on activities.[8] For comparison purposes, the students of the control class were taught the same science content, but there was no attempt on the teacher's part to foster a sense of wonder.[9]

---

[8] On all occasions though, time was given to the class so that wonder could be evoked (i.e. through questioning, taking time to reflect on what the teacher said or on what they had observed).

[9] Both groups were taught by the same teacher, according to her regular teaching schedule. The main difference between the two teaching approaches was the initial attempt on the teacher's part to evoke a sense of wonder in the students of the treatment group. For example, in the case of

Observation and especially student optional journals were the main instruments of the research. A quantitative analysis of journal entries made by the students of both classrooms provided evidence for higher involvement for the students—both males and females—of the classroom where the teacher evoked a sense of wonder. Also an analysis of students' comments provided evidence that wonder was experienced as astonishment mingled with a 'shock' of awareness. Moreover, two paper-and-pencil tests administered at the end of the school year provided additional evidence that wonder had an effect on students' ability to remember 'wonder-full' ideas and also an effect on better understanding of, at least, three phenomena. This empirical evidence of better retention and understanding is evidence of the role of wonder as an 'attention catcher' and generally of the role of affective factors in the learning process (see Appendices E and F, which give the ideas that were selected as potential sources of wonder and 'important ideas' that students of both groups remembered at the end of the school year). Most importantly, however, this study provided evidence that the experience of wonder can help students change their perception of both natural phenomena and science ideas and also their perception of science as a school subject.

Hadzigeorgiou and Garganourakis (2010) have also documented the positive role of wonder in the learning process. Their study explored the extent to which situations that evoke a sense of wonder can promote scientific inquiry. Given the intense interest, curiosity, and wonder that some 11th grade students had begun to develop after seeing the movie *The Prestige*,[10] a science teacher used this movie with the whole class, as a source of wonder. The teacher asked students to keep an optional journal, in which they could write the questions they wanted to answer and also make entries of things they considered important (i.e. impressive facts and ideas, what they thought they learned during their investigations). It was found that of the 28 students (15 males and 13 females) in the class, 19 students, that is, more than 60 % of the class, became involved in self-directed inquiry at home. More specifically, through an analysis of students' journals, observation, informal discussions, and paper-and-pencil tests, it was found that students (a) became involved with Tesla's life and work, thus developing an interest in current electricity, (b) better

---

Newton's third law, the teacher started the lesson by presenting a photo illustrating a head-on collision between a huge truck and a very small car and asking students to predict the magnitude of the forces exerted on the vehicles. After students' (by and large wrong) predictions, the teacher challenged those predictions by telling students that Newton had proposed an axiom, according to which the force exerted by a huge truck on a very small car is equal in magnitude to the force exerted by the small car on the truck. In the other classroom the teacher simply presented the law, as a simple statement and used a variety of examples (contexts) to illustrate it. By the same token, in the case of presenting Bohr's atomic model, the teacher tried to focus the students' attention on the fact that an atom is mostly empty space and on some strange consequences and paradoxes that follow from that fact. In the other classroom, the teacher simply presented the atomic model and stressed its importance in explaining a range of phenomena.

[10]This film is about two stage magicians of Victorian London who competed with each other for the best stage act ever. One of them went as far as Colorado Springs, in the USA, to find Nikola Tesla (who at the time pursued his interest in and experiments on the wireless transmission of electrical power) and asked him for a stage trick that could help him beat his rival.

understood Ohm's law in their attempt to find out the degree of damage or injury that an electric current can cause, and (c) learned about ideas that were not part of their science curriculum, such as the skin effect and the biological effects of AC and DC electricity. Moreover, some students began to consider the possibility of a career in electrical engineering.

More recently, King et al. (2015) explored 8th grade students' emotional response during their participation in science activities on a unit of energy. Using multiple sources of data, they were able to identify the students' emotional responses. A case study with two students showed that 'choosing activities that evoked strong positive emotional experiences focused students' attention on the phenomenon they were learning' (p. 1886). Such findings agree with what Hadzigeorgiou (2001, 2012) and Hadzigeorgiou and Garganourakis (2010) have previously reported.

The above studies have provided evidence that wonder can play three specific roles in the context of science education: (a) a prerequisite for engaging students in school science, (b) a source of students' questions, and (c) a prerequisite for a change in students' perceptions of science ideas and phenomena. What follows is specific evidence for these three roles.

6.3.1 Wonder as a Prerequisite for Engaging Students in School Science

That feelings of surprise and astonishment make students focus their attention on the phenomenon or idea they are learning has been found by both Hadzigeorgiou (2012) and King et al. (2015). According to Andersson and Gullberg's (2014) study, capturing unexpected things was found to be one of the key factors that empower young children in their learning of science.[11] There is also evidence that students who experienced a sense of wonder were more engaged with science, as they spent more time making journal entries, in comparison with students who were taught the same ideas but their teacher made no attempt to evoke wonder. Not only the number of entries but also the number of comments and questions written in the journals was much higher in the case of the students who experienced a sense of wonder. It is quite interesting to note the differences that emerged between the 'outsiders', namely, females and low achievers, in the two classes (treatment and control), even though there was no difference between the academically able students in those two classes (Hadzigeorgiou, 2012).

And evidence from a study with kindergartners, who participated in hands-on activities, also suggests the same thing. Young children literally gravitated towards the objects that helped evoke in them a sense of wonder (e.g. emptying water from

---

[11] The other factors are paying attention to and using children's previous experiences, asking challenging questions that stimulate further investigation, and creating a 'situated presence', that is, remaining in the situation and listening to children and their explanations (Andersson & Gullberg, 2014).

one glass into another by using a piece of towel cloth without moving or tilting the glasses, changing the colour of a flower by putting it into a glass containing coloured water, lifting a pile of books using one's breath). Those children spent more time with the activities than was initially thought by both the teacher and the researcher. Moreover, the 'wonder activities' used in their classroom were real motivators for some children who had never concentrated on anything in the past. It was observed that those activities did help them become more focused and even begin to participate in other activities with interest and intention (Hadzigeorgiou, 2001).

6.3.2 Wonder as a Source of Students' Questions

Given the crucially important role of students' questions,[12] the challenge for science education is to identify the sources of the questions. More often than not, this process of asking questions is associated with curiosity, which is considered central to scientific inquiry and science education (AAAS, 1990, p. 173). Yet, no explicit reference to wonder and to its role in promoting scientific inquiry is made. Having already made a delicate distinction between curiosity and wonder, one could certainly ask the following question: which is more important for students, to wonder about natural phenomena or to be curious about them? If 'the beginning of science is wonder' (Silverman, 2003, p. 387), it would be preferable for students to experience initially a sense of wonder at and about their object of study.[13] For example, in Hadzigeorgiou and Garganourakis' (2010) study, students' most frequent questions (posed by more than 80 % of the classroom and written in students' journals) were 'wonder questions', which had their source in their surprise, astonishment, perplexity, admiration, and awareness that their knowledge of electricity was incomplete and that some phenomena can exist at all (Hadzigeorgiou & Garganourakis, 2010):

---

[12] Questions play the role of a link between thinking and learning (Chin & Chia, 2004; Chin et al., 2002; Gardner, 1991; Good & Brophy, 1995) and also lie at the very heart of scientific inquiry (NRC, 1996, 2009), which, as evidence suggests, results in better retention of ideas (Hadzigeorgiou, 2012; Renzulli, Gentry, & Reis, 2004).

[13] One, of course, could argue that what really matters in the end is not the source of students' questions but the questions themselves and whether these questions foster scientific inquiry. What should be stressed is that, if wonder, experienced as surprise or even astonishment, can make the mind to be entirely filled with the object of study (Burke, 1990), the experience of wonder increases the possibilities for students to ask questions about the object of study itself, in which they show a genuine interest. Moreover, the idea of wonder, as a potential source of students' questions, sounds more realistic, since it can be evoked even through familiar and ordinary situations, through phenomena and ideas that are taken for granted (Hadzigeorgiou, 2007), whereas curiosity presupposes novel, unusual, strange, and even extraordinary phenomena and situations if it is to be aroused (Berlyne, 1960; Bruner, 1996; Loewenstein, 1994). This is why the arousal of curiosity remains always a challenge. This is not to say that fostering or evoking wonder is not a challenge. It is indeed, but no unusual and extraordinary phenomena and situations are required.

- How could unwired light bulbs become luminous at the touch of Tesla's hand?
- How could light bulbs be planted in the ground light?
- How could Tesla himself walk through sparks totally unharmed?
- How can these experiments be explained? Can they be replicated?
- Who was really Nikola Tesla?
- Why has Tesla been marginalized by history and science textbooks?
- Why are Edison's and Marconi's names better known than Tesla's?
- Why hasn't Tesla been given credit for his inventions?

It is important to note that more than 80 % of the classroom asked 'wonder questions', which, though, became the precursors for 'curiosity questions' (i.e. 'How safe is it really to "play" with electric current?', 'Which is safer for humans, direct or alternating current?', 'How much current will kill a person?', 'Can we make, and how, our skin resistance so large that we will never run the risk of an electric shock?'). Whether or not one accepts such a distinction, the fact is that the source of all questions was students' initial sense of wonder. If, however, one were to consider the evidence that wonderment questions (in contrast to information questions) stimulate discussion at a higher level of cognitive functioning and play a major role in meaningful learning (Chin, Brown, & Bruce, 2002), then the distinction between the two kinds of questions seems to be a valid one. Indeed, according to Hadzigeorgiou and Garganourakis (2010), it was the initial wonder that stimulated discussion and facilitated meaningful learning.

Hadzigeorgiou's (2012) study, on the other hand, produced evidence that not only astonishing phenomena, as was the case with Tesla's experiments, but also ideas (i.e. matter, gravity, action and reaction, law of inertia) and everyday phenomena (i.e. free fall, seeing objects around us) can be sources of students' questions. For example, students asked 'Why is gravity such a weak force?' and 'How can we explain the fact that atoms are mostly empty space, and yet our hands do not go through the table in front of us?' after their teacher attempted to evoke a sense of wonder about the concepts of gravity and matter, respectively. It is quite interesting to note that not a single student from the classroom that was taught the law of action and reaction asked a question about that law, in sharp contrast to the 12 students (40 %) from the other classroom, who were taught that law through the teacher's attempt to foster a sense of wonder. These 12 students wondered about how Newton arrived at such a law, how he thought about it, how strange that law is, etc. These questions were not the result of learning about the law itself (as was the case with the students of the 'traditional classroom') but the result of their experience of wonder through a paradox or mystery (i.e. how is it possible for a tiny car to exert on a huge truck, during a head-on collision, a force that has the same magnitude as the force that the huge truck exerts on the tiny car?) that made them aware that their knowledge was either incomplete or erroneous. The admiration towards Newton himself, who put forward such a law, was very explicit in some students' comments, which, in turn, shows that wonder can indeed have an aesthetic dimension (Table 6.1).

Table 6.1 A sample of students' questions as written in their optional journals after an initial experience of wonder in their classroom (Hadzigeorgiou, 2012)

Why is gravity so small a force?
If matter is really empty space, then why my hand doesn't go through the table?
If all waves require a medium to travel through (we cannot have sea waves without water and sound without air), then why does light travel in vacuum? What is so different about it?
How is it possible that in using the mathematical formula measuring the time of free fall I get what actually happens and when I think about it, without using that formula, I understand that heavier objects fall faster than lighter ones?
How did Newton arrive at the law of action and reaction?
Can we apply in some ways Newton's law of inertia in order to reduce petrol consumption in our cars?
If matter is mostly empty space and if light is invisible, can scientists create an invisible man who can walk through walls, buildings, etc.?
Is it possible that the world we live in is just an illusion and the real world is the one that physics teaches us?

6.3.3 Wonder as a Prerequisite for Significant Learning

If 'we can typically store and retrieve information with highly emotional content more easily than we can recall relatively non-emotional information' (Ormrod, 1999, p. 431), then wonder can be seen as a prerequisite for learning science. Indeed wonder can charge emotionally information thus resulting in better retention and easier retrieval of that information. There is empirical evidence (see Appendix F) that a considerably larger number of science ideas were remembered by a larger number of ninth grade students (more than 50 % of the classroom remembered at least ten ideas) who experienced a sense of wonder at and about those ideas, in comparison with students of another ninth grade class who were taught the same ideas by the same teacher but who did not attempt to foster a sense of wonder in those students (Hadzigeorgiou, 2012).

Of course, one may very well argue that retrieval of information does not guarantee understanding. Yet there is a question concerning a piece of empirical evidence: Why did students, who experienced a sense of wonder, perform statistically better on written tests (which assessed conceptual understanding of certain science ideas) in comparison with students of very similar background and academic record, who were taught by the same teacher but did not experience a sense of wonder (Hadzigeorgiou, 2012)? A possible explanation of this fact is that wonder made students focus their attention entirely on the introduced idea and the actual phenomenon of study. According to the literature, 'when information is emotionally charged we are more likely to pay attention to it, continue to think about it over a period of time, and repeatedly elaborate on it' (Ormrod, 1999, p. 420).

However, wonder can make students aware of what they are learning, or even of what they have already learned, which suggests that wonder has a metacognitive dimension. The following are comments made by students in their journals:

Gravity

- *Although I knew that gravity was the weakest of all forces and I could see that in the numbers on that table about the relative strength of all forces in nature, it was after that simple and very easy-to-do experiment that I understood it better.*
- *That one pen falls simultaneously with a bunch of ten pens is something that I could never imagine. Now I understand gravity. What a strange force!*
- *It is really remarkable and very strange now that I know that the force of gravity is very weak [...] When I'm thinking about bodies falling freely, I am thinking about gravity as a very very weak force.*

Light

- *I had never thought that light is invisible. I know that light is a wave like sound although it does not require a medium to travel through. But I always thought that light is something we see.*
- *I was really astonished at seeing, with my own eyes, that light is indeed invisible, and we only see the source of light and what light hits. Now I begin to understand how we see the world around us. It is like a miracle! Now I can understand why space is totally black and we see only what objects are present in it, like a spaceship or an astronaut.*

Matter

- *For more than a week, I have been thinking about matter being mostly empty space. This is the weirdest thing I have ever heard. But it does make sense if you think about the distances between the nucleus and the orbiting electrons. Yet it is very strange.*
- *That matter in reality is 99% empty space is incredible. In reality my desk here is empty space and yet my hand cannot go through it.*
- *I knew that molecules are very very small. But it was after calculating the number of molecules contained in a glass of water that I really understood how tiny they really are. Although I knew all about Avogadro's number, now I can say that I understand its full meaning.*
- *I could never have imagined that there are more water molecules in a glass of water than there are glasses of water in the Mediterranean Sea. This I will never forget. It is perhaps the most impressive thing I have learned so far.*
- *Matter, what a strange concept! Everything around me looks different if I think about it, because we all touch, see, eat, and drink vacuum!*
- *That all subatomic particles contained in the bodies of all people on earth, if we could remove all empty space from their bodies, could pack easily into a ping-pong ball is astonishing. In reality we are all empty space!!*
- *Ever since I have learned about matter being 99% empty space, I see solids and liquids as empty space with some protons and electrons. If I think about it, every time I drink water, I drink nothing, except for a few protons and electrons.*

Forces and motion

- *That action is always equal to reaction, no matter what kind of objects we are talking about, is something out of my mind. The example regarding the head-on collision between a tiny car and a big truck makes me wonder about Newton. He must have been a genius to discover such a law!*
- *Although the law of action and reaction is perhaps the simplest law I have learned, if you really think about it, it is a strange law, perhaps the strangest law I have learned so far.*
- *The fact that there can be straight-line motion at extremely high constant speeds in the absence of a net force makes you see motion as mysterious a phenomenon as electricity and magnetism.*

The above comments do show that wonder can be associated with conscious learning, namely, with awareness of the meaning and the significance of science ideas. And, in this sense, the wonder experienced by students is similar in nature to the wonder experienced by leading scientists. The students' experience of wonder as awareness of some kind is indeed very similar to both Sagan's experience (as a young child in Brooklyn) regarding the nature of the Sun and Feynman's awareness that all trees are made of air and when they are burned this air goes back to the atmosphere (Feynman, 1969) and that the whole universe can be found in a glass of wine (Feynman, 1995).

The above comments, however, also show a change in students' perspective! Indeed, the comment that 'motion is as mysterious a phenomenon as electricity or magnetism' or that 'every time I drink water I drink nothing, except for a few protons and electrons' reveals that students experienced a shift in their perception of motion and water, respectively. Such change represents significant learning, according to various philosophers and educators.[14]

6.4 Implications for Science Education

The analysis of the notion of wonder and the evidence form research point to a number of implications, which can be seen as ideas behind a 'wonder pedagogy'. Such a pedagogy, while it does not imply that the classroom becomes a place where students are in a constant state of wonder, helps teachers evoke wonder whenever they (teachers) think it appropriate (e.g. as an introduction to a lesson in order to encourage conceptual engagement, when cognitive disequilibration becomes necessary). Wonder, of course, can be stimulated throughout the lesson (i.e. through questions) if the goal is the fostering of an inquiring attitude. However, in such a case, the stimulation of wonder can be carried to an extreme, and the result might very

---

[14] That significant learning should be directly related to a change of one's outlook, to the ability to perceive the world in a different, unhabitual way has been pointed out by Hirst (1972, p. 401); Jardine, Clifford, and Friesen (2003, p. 102); Peters (1967, p. 9); and Schank (2004, p. 37).

6.4 Implications for Science Education

well be a 'wonder overload' with most, if not all, students unable to concentrate on any specific idea and phenomenon.

6.4.1 Wonder Is Evoked

The first idea about wonder that needs to become very clear is that wonder is evoked. It cannot be experienced 'automatically', and this is the reason why the experience of wonder is not very common in the context of formal education (see Egan, 2014; Egan, Cant, & Judson, 2014). Perhaps, it is more common in the context of informal/free-choice education, even though such a claim is difficult to defend, unless one refers to the opportunities per se available in this specific context.

Given that the sources of wonder are multiple, one would expect that a sense of wonder can be experienced in any curriculum area and with any topic. Indeed, anything can be a source of wonder, from a stellar explosion and a total eclipse, to a drop of water and water crystal, from the motion of a comet and spaceship, to the motion of an ant and a wave swinger at an amusement park. However, while a spectacular, beautiful, and uncommon phenomenon can easily evoke a sense of wonder—as students feel astonished at the phenomenon and become aware of the beauty of it and also aware that phenomena like this can exist at all—there are cases in which one can experience wonder about everyday/ordinary and taken-for-granted entities and phenomena (e.g. air, light, a glass of water, the flight of a bird, the motion of child down a slide, an electric lamp, even a broken and useless styrofoam cup). And in these everyday and taken-for-granted cases, students, particularly the young ones, must first be encouraged to focus their attention on the object of wonder, before they are ready to imaginatively engage with questions about these cases.[15]

If one were to talk about a 'wonder pedagogy', then one could talk about a strategy that involved three steps, namely, direct students' attention; make them focus on an object, phenomenon, or idea; and then pose questions about them. It is crucially important that students be invited to direct their attention and focus on something, that is, their object of learning, as a potential source of wonder (e.g. a snail, a leaf, a rainbow, a sunset). The questions that follow this invitation can help reveal

---

[15] While at an amusement park, for example, a child may look at a helium-filled balloon or a wave swinger but may not notice what is really happening, even contrary to his or her own experience. Indeed a child seeing a balloon going up may not notice that that particular balloon is something that contradicts his or her own experience with balloons, which always fall down (unless there is strong wind to blow them away). The child, of course, if asked, will most likely say that the air takes it up. It is for this reason that the child must be invited to notice differences by seeing a regular balloon going gown and not up in the air, if he or she is to begin to reorganize and revise his or her ideas about the weight of objects and their motion. And in the case of a wave swinger, it is very unlikely that children, even adults, will notice that the angles the seats form to the vertical are all the same and independent of the weight they carry (i.e. both empty seats and loaded seats form the same angle to the vertical).

something strange, paradoxical, mysterious, surprising, amazing, and even astonishing about that object of learning.

A teacher can direct or catch the students' attention in a variety of ways: through a question, through a story—preferably one that reveals something strange and mysterious—through a poem, and through an image. However, the questions that follow are of crucial importance, as discussed later in this section. Of course, because there are various sources of wonder, as discussed in the next section (e.g. science ideas; everyday, ordinary objects and phenomena; mysterious situations and phenomena; amazing, incredible facts about natural entities and phenomena), one cannot have a 'wonder recipe' that can be applied across all instructional situations. However, as a general guideline, the teachers can implement the DAFA approach (i.e. Direct Attention, Focus, and Ask questions). Indeed, catching or directing the students' attention and asking them questions about the object of attention are at the very heart of a 'wonder pedagogy'.

The object of attention (i.e. the potential source of wonder) can be an entity (e.g. the Moon, molecule, rock, leaf), a phenomenon (e.g. water evaporation, rainbow, magnetic attraction and repulsion, sound refraction), or even an abstract idea (e.g. heat, light, energy transformation).

As two examples of 'wonder pedagogy', let us see how a teacher could introduce the concepts of sound and current electricity. In the case of sound, she or he could catch the students' attention with a short story about a strange and in fact mysterious event that took place on February 2, 1901 (i.e. the day the people of England mourned the death of Queen Victoria) and involved the sound of a cannon. This sound was heard by people in a town 150 km away but not by people living in the nearby area (where the cannon fired). After this short story, questions that make students wonder about sound, such as 'if a cannon fires in the air and there is no one around to hear the booming of the cannon, will there be a sound?' and 'is sound a sensation or a specific kind of motion taking place in the air?', can introduce students to the nature of sound and its sources and also to the difference between the physical event per se (i.e. the booming of the cannon and the travelling waves in the air) and the physiological event, that is, the sensation it causes (i.e. when one actually hears the sound from the cannon).

In the case of current electricity, a teacher could introduce the lesson through beautiful images and amazing or even 'awe-full' facts about lightning (e.g. about 100 every second, our planet is struck by lightning; each bolt can carry more than 3 billion KW; the air temperature in the channel surrounding the bolt is more than three times the temperature of the Sun's surface; due to the increase in temperature, the air expands and creates thunder that can be heard 25 km away) and then proceed to the task of asking questions that make students wonder about this spectacular phenomenon: Is lightning just bad because it can kill us? Is life on the planet possible without lightning (i.e. does lightning do anything useful for us? Why can't we use the electricity in the bolt?)? These questions can be a 'wonder-full' introduction to the concepts of voltage and electric current.

All these ideas I take up later in this section, but my point here was to illustrate, by way of example, what I mean by 'wonder pedagogy'. What follows will help make much clearer the central role of questions and the role of language in general in a 'wonder pedagogy'. Moreover, the role and value of phenomenological learning will also become clearer, given that a 'wonder pedagogy' differs from mainstream constructivist approaches to school science education.

6.4.2 Wonder vs Fun

Contrary to what is usually believed, wonder cannot, and should not, be confused and conflated with fun (see also Appelbaum & Clark, 2001) or even with the 'wow' factor (Feasey, 2005). Although the experience of wonder can be really fun, not all fun experiences with science are 'wonder-full'. A student may very well say that he or she had a wonderful experience in the classroom, a wonderful lesson, meaning though an interesting experience or lesson. Perhaps there should be a distinction between a 'wonderful' and a 'wonder-full' experience or science lesson, given that the word 'wonderful, more often than not, is used interchangeably with other words, which may not have the characteristics of wonder.[16] Moreover, the 'wow' factor in school science, although of crucial importance, should be accompanied by some kind of awareness.

No doubt both the fun of science and the surprise experienced by students are important and should be considered in science education, as they both refer to the emotional dimension of learning science. But without some kind of awareness on the students' part, they cannot be considered pedagogically appropriate sources of wonder that have the potential to contribute to students' understanding. Indeed, flashy, fun demonstrations, which leave students, and especially young children, dumbfounded, without awareness on their part of what really happens, or what these demonstrations mean, cannot be considered pedagogically appropriate science activities (unless, of course, the teacher's purpose is to make his/her students feel baffled and dumbfounded). The idea, of course, of surprise from something unexpected has been central to the process of 'cognitive conflict'—and which has been incorporated into the conceptual change teaching-learning model (Limon, 2001)—but it (surprise) has been considered the result of the awareness that one's beliefs are erroneous (and therefore in need of reconsideration).

---

[16] Kieran Egan believes that 'wonderful' is a 'word whose overuse has made it tired and empty' (Egan, 2014, p. 158).

6.4.3 The Centrality of Questions

Questions are at the very heart of wonder. They are also considered crucially important in the process of conceptual change.[17] Whether explicit or implicit, questions help students focus their attention on the learning object and help them notice those things that, more often than not, remain 'hidden' from the students, even though they see them all the time. For example, Richard Feynman's questions help one focus on the nature of sand and its relationship to rocks: 'Is the sand other than the rocks? That is, is the sand perhaps nothing but a great number of very tiny stones? Is the moon a great rock? If we understood rocks would we also understand the sand and the moon?' (Feynman 1995, pp. 23–24). The case of sound is another good example of a taken-for-granted phenomenon. Instead of asking students 'What exactly is sound?', a question that is more likely to evoke in students, a sense of wonder is the one posed in the eighteenth century, which 'set debates in the intellectual salons in Europe [...] If a tree falls in the forest [...] and no one is there to hear it, will there be a sound?' (Stevens, Warshofsky, & The Editors of Life, 1966, p. 9).

There are so many things that are taken for granted, but which can become sources of wonder once questions about them are asked. Such questions can help evoke wonder about anything. For example, questions can evoke wonder about what we see on our dinner table and also about what takes place at an amusement park. And speaking of dinner, two taken-for-granted things, such as water and chicken meat, can indeed become sources of wonder: Why does water dissolve more substances than any other liquid? Why does it behave so strangely compared with other liquids (e.g. why does it have such a high boiling point, climb up against gravity inside thin tubes)? And why do chickens (and turkeys for that matter) have both white and dark meat? Why, more specifically, is chickens' leg meat dark and breast meat white? What is so different about leg and breast muscles in a chicken? It is these questions that make students think about water as a special and 'wonder-full' substance and about the difference in colour between leg and breast muscles in the chicken they most likely eat all the time.

It should be noted that because wonder can be evoked from the perception of details—those things that students and all of us never notice—a simple question such

---

[17] Notwithstanding the various theoretical, methodological, and practical concerns and challenges that need to be seriously considered by the science education community, the conceptual change approach remains a powerful teaching/learning model (Duit & Treagust, 2003; Treagust & Duit, 2008). In discussing such concerns and challenges, Milne, Kirch, Jhumki Basu, Leou, and Pamela Fraser-Abder (2008, p. 433) recommend a question-based approach (which would also include the problem-based approach) and make the following comment:

> *A question-based approach would assist teachers because they would only need to focus on inquiry principles and would support the valuing of students' funds of knowledge [...] Also, a question-based approach would encourage learning on a needs basis engaging students in the integration of emotion and cognition as they search for answers that suggest further questions.*

6.4 Implications for Science Education

as 'Why does a natural entity look the way it does?' seems to do the trick: the natural entity, no matter how ordinary and familiar it is, becomes an object of wonder.

It should also be noted that the questions that are likely to evoke wonder are those that present the learning object (i.e. the content to be learned) as an 'unknown'. For example, in the case in which students are learning about tree leaves, after they are invited to observe closely a number of leaves, they can be asked questions such as the following: 'Why do these leaves have veins? Why do all tree leaves have veins? Do they have a function or do they serve a purpose? And why are there such a variety of leaf shapes?' By the same token, a teacher can ask his/her students questions like 'How come all tree trunks have rings? Do they mean anything? Are there any differences between the rings or are they all the same? How is it possible to tell the age of a tree by looking at its ring layers?' These questions differ from questions such as 'How can we tell the age of a tree by counting the rings in its trunk?' and 'What age is this tree in whose trunk we see 12 rings?' in the sense that in the first case the content to be learned is given as an 'unknown', while in the second case the content to be learned is given more like a 'known'—we know that trees have rings, so there must be a way to tell the age of any tree.

On the other hand, there are questions that help reveal the mysterious and the strange in nature (e.g. 'How come we can observe something here and now and see what it looked like millions of years ago?' How is it possible that any two bodies, whether near to each other or millions of km apart, always attract each other? How is it that such an attractive force exists between two planets and also between two people living in different continents?' 'How is it that huge masses—thousands of tons—of metal become airborne?' 'How do bats "see with their ears", how do whales and other echo-locating creatures communicate?'). Such questions capture the imagination and encourage one to wonder about various possibilities. And learning about such possibilities becomes an adventure into the mysteries of nature (see also, as an example, Huggins, 2010, for a challenging thought experiment involving the weighing of photons using bathroom scales!).

Questions, of course, that make students aware that their knowledge is either mistaken or incomplete deserve special attention. The reason in that such questions have always been considered crucial for learning science as they are the result of cognitive conflict:

- Why is it always cold on mountaintops (although we know that hot air always rises)?
- Why, during a head-on collision, does a very small car sustain more damage than a huge truck, even though the law of action and reaction says that the forces acting on both vehicles are of the same magnitude?
- How is it that our hands cannot go through a table even though we know that atoms are mostly empty space?
- How come both empty and loaded chairs in wave swingers form the same angle to the vertical?
- Why is the number of atoms or molecules contained in a certain volume, at a given pressure and temperature, independent of the nature of the gas? (i.e. How come equal volumes of all gases at the same temperature and pressure contain

the same number of atoms or molecules? Why such a number is independent of the weight of the atoms or molecules?)
- How is it possible for a basketball player to 'hang' in the air?
- Why is it that in a vertical jump, with both arms extended up in the air, lowering forcibly one arm just before the greatest height is reached raises the other arm several cm higher? Are the laws of motion violated?
- How is it that some balloons, in the absence of blowing wind, can go up even though we know that balloons go down if left to free fall?
- Why does the temperature of a certain mass of water remain constant although we keep offering it more and more heat?
- Why is it so cold in space? Is it so cold even if we face the Sun?
- Why do astronauts hop and not walk on the surface of the Moon?
- Why is it that one of the three friends who went to a concert complains that he did not hear anything, while the other two insist that they heard everything very clearly?
- How come two friends standing at opposite ends of a lake can hear each other in the morning, but not in the afternoon?
- How come birds, sitting on high-voltage power lines, do not get electrocuted?
- Why do flames come in different colours? Why is the flame of candle blue around the wick and when higher up yellow and red?
- Why can we boil water in a paper box?

With younger students questions that challenge them to think also deserve attention. These questions evoke wonder because they challenge students with something surprising and unexpected and make them aware that their knowledge is incomplete.

- How could we lift a pile of books with just a breath?
- How could we boil water without a fire?
- How could we make electricity from potatoes?
- How could we empty water from one glass into another without touching the glasses?
- How could we 'catch air' in/from our room?
- How could we see what is inside a room without being in the room?
- How could we use a light object to lift a heavy one (e.g. using a washer to lift five washers)?
- How could we make a volcano on our kitchen table?
- How could we make a soda pop can roll up an incline without pushing it?
- How can a piece of newspaper become stronger than a piece of wood?

It should be pointed out that the above questions are different from questions, such as 'How can we protect ourselves from lightning?' and 'How do sunglasses protect us from the sun?', in the sense that in these questions there is nothing unexpected and surprising. These are certainly important questions associated with important science knowledge. However, research shows that students, especially the young ones, become engaged with science after they experience something unexpected (e.g. Andersson & Gullberg, 2014; Hadzigeorgiou, 2001, 2012). For this reason, questions that present something surprising, paradoxical, or unexpected can be used to introduce a concept or idea. Two specific examples can help illustrate the point. The question 'How can we see what's inside a room without being in the room?' can be used to introduce students to the concept of light reflection, while the question 'Why does the temperature of a certain mass of water remain constant although we keep offering it more and more heat?' can be used to introduce students to the difference between heat and temperature.

However, it must be admitted that it is not always easy to predict which questions present something unexpected. It may very well be the case that a question evokes a sense of wonder only in some students. For example, the questions 'How do plants breathe?', 'Does the temperature of a magnet affect its strength?', and 'How can we increase the electrical resistance of our own body?' can evoke wonder only in students who sense something unexpected, that is, in students who have never thought of the possibility that plants breathe, that the temperature of a magnet may have an effect on its strength, or that we can deliberately increase the electrical resistance of our body (in order to reduce the electric current flowing through our body in case of an accident). For other students, the above questions may be taken simply as 'information questions'. The point therefore is how to increase the probabilities for students to experience wonder. Language can play a crucial role here.

But before I discuss the role of language in the process of evoking wonder, it must be stressed that wonder questions can help create anticipation, just like a story or narrative, if a comparison were to be made between the two, that is, between stories and wonderment questions. Challenging, for example, young students to make a volcano on a table using simple materials can be a potent stimulus for learning, perhaps just as potent as inviting them to listen to a Harry Potter story! Students' own questions can also create anticipation. These questions can be asked during 'wonder time', which can be a part of a 'wonder pedagogy'. Students can select a topic and ask any kind of question about that topic (e.g. electrical lamps, trees, bird flight, boats). Or, as Egan (2014) recommends, they can ask questions about anything. These questions can be totally unrelated to one another. Apparently, anticipation is created for the simple reason that students are curious or they feel a sense of wonder about the object of their questions (e.g. the different colours of flames, the tallest and heaviest trees, the size of the universe). Such anticipation can be fostered if there is a 'wonder box' in the classroom and all questions are put in there, let's say once a week. The teacher can then pick a certain number of questions, either every day or during 'wonder time' on a specific day of the week, in order to answer them.

6.4.4 The Role of Language

The crucially important role of language in teaching and learning science is well known (e.g. see Lemke, 1990; Saul, 2004; Sutton, 1996). With regard to fostering a sense of wonder, language plays a role in the way science ideas are presented and questions are asked. Consider, for example, a textbook definition for universal gravitation and compare this definition with the idea of 'an attractive force, acting in a freezing and empty space between any two objects, animate and inanimate, at any place on the Earth or in the universe'. Or consider an illustration of this same law presented through a question: 'How is it possible for any two students, one in one continent and the other in another continent, without even being aware of each other's existence, to be attracted to each other?' It is evident that in both these cases, the law of universal gravitation is contextualized in such a way as to reveal, through the use of language, the strangeness and 'surprisingness' inherent in this law. Such a presentation should precede the mathematical description of the law, if the science teacher aims to evoke a sense of wonder.

As another example, let us consider the concept of matter and compare a textbook definition of it with a short narrative focusing on 'the secrets and mysteries of matter' and/or on the idea that 'what we see is not what it really looks like'. It is this narrative and the ideas in them that capture the students' imagination and thus help evoke a sense of wonder. By the same token, an abstract concept, such as energy, can be introduced as an 'invisible agent who does things (work) for us' or as 'something unseen and untouched' (even though we see, hear, and sometimes even 'touch' its consequences). It can also be introduced through 'the joys of childhood' or the 'joys of life', which brings me to the role of metaphors in the process of evoking wonder.

Indeed, the role of metaphoric/poetic language should also be considered in the context of 'wonder-full' science. Here are some examples:

- Electrical forces as the glue of life
- Lightning as dazzling flashes that recharge a leaking planet
- Wind as the capricious bearer of mixed blessing
- The Earth as a huge battery
- The Earth's atmosphere as a giant heat engine
- Heat as the incorrigible thief of electric power
- Energy as the joy of life
- Energy transformation as a change in the guises of energy
- Nuclear transmutation as a change in the identity of elements

These examples illustrate the powerful role of language in capturing the students' imagination and evoking in them a sense of wonder through the unexpected possibilities that they (students) see. Such a powerful role can be illustrated by Descartes' idea: 'give me matter and motion and I will construct a universe'.

The point that is being made here is that language plays a central role in the process of evoking a sense of wonder. Traditional textbook definitions are, more often than not, unsuitable for such a task. A star may be defined as 'a huge ball of flaming gas' but, as C. S. Lewis wrote in his *The Voyage of the Dawn Treader*, 'that is not what a star is but only what it is made of' (cited in Schakel, 2002, p. 63).

What C. S. Lewis wrote should provide food for thought to all those involved in science education: just as a star is more than a huge ball of flaming gas, forces are more than just causes of changes in motion, flashes of lightning are more than sudden flows of electricity, and water is more than what its chemical formula says it is. Even an electric current is more than a flow of electrons. Indeed, definitions narrow the scope of an idea, by focusing upon the purely cognitive aspect of it. In other words, definitions cannot help evoke a sense of wonder. How can the concept of force, for example, evoke a sense of wonder if it is presented through such definitions as 'a force is that which causes a change in motion', or 'forces are vector quantities with the following characteristics: [...]'? It is quite obvious that such definitions dry up all the wonder inherent in science. It is for this reason that ideas need to be presented in 'non-definitional' forms. And the role of language can be quite crucial in the presentations of such forms.

What needs to be recognized is that language has the power to capture the students' attention by describing what is actually happening during a taken-for-granted phenomenon. For example, in the case of an airplane taking off, a question such as 'How is it that huge masses—thousands of tons—of metal become airborne?' can be used to make students focus their attention on the airplane and evoke wonder about both the phenomenon of flight and the airplane itself and thus introduce them to the physics of flight.

6.4.5 The Role of Content Knowledge

There is plenty of anecdotal evidence that supports the view that the more one knows about a topic (e.g. trees, airplanes, water), the more likely it is for one to experience a sense of wonder about it. The LiD (*Learning in Depth*) programme, developed by educational theorist Kieran Egan at the Simon Fraser University, in which students select and focus on a topic of personal interest for many years, can indeed facilitate the experience of wonder through the acquisition of 'increasingly intensive knowledge'. The central argument, supported by evidence from teachers who implemented LiD in their classrooms, is that 'what was once seemingly straightforward and known becomes increasingly strange and mysterious, and wonderful' (Egan, 2014, p. 158).

6.4.6 Emphasis on Phenomenological Approaches

If wonder presupposes students' attention to the details of physical reality, sense experience acquires a significance which, however, has been lost with the rise of the constructivist perspective on teaching and learning science. The reason is that the experience of wonder presupposes attention to the 'precognitive' aspects of the process of knowledge, that is, attention to the roles of sensing and feeling, and this (attention) can be seen as one of the differences between a phenomenological approach and a constructivist approach (Østergaard, Dahlin, & Hugo, 2008). Even though an emphasis on sensing and feeling can be criticized from the perspective of constructivism, from a phenomenological perspective, sensing and feeling are considered the foundations of knowledge.[18]

But if a phenomenological approach is to be implemented in the science classroom, then the teaching-learning process will not start with a hands-on activity, or with some technological applications, let alone a discussion aiming to elicit misconceptions. Instead, it will start from natural phenomena as the real sources of wonder. For example, instead of providing students with batteries, wires, and light bulbs in order to introduce concepts and ideas regarding current electricity, the teacher can evoke wonder and possibly awe by encouraging students, and especially young children, to observe spectacular natural phenomena (e.g. students can watch on a video lightning, the northern lights), by helping them become aware that phenomena such as these are significant since they are involved, in one way or another, in our own existence (e.g. our own body, the things we smell and touch). Thus the students are helped to see things in new light. As Witz (1996) pointed out, it would be more important to bring into focus the rich nature of a natural phenomenon than to focus merely on theoretical principles and/or technological applications associated with them.

> *The actual phenomenon or object – be it a manifestation of electricity, or the anatomy of the body, or the group behavior of chimpanzees – is the real source of wonder [....] Focusing on the actuality of the phenomenon, with its richness along many dimensions of human experience and its extraordinary wholeness, is not incompatible with a goal of 'cultural literacy' or 'understanding of mechanisms'. [...] For this purpose, a study of the more natural and less technological manifestations of electricity and magnetism which fascinated the early investigators might be preferable, with gradual extension to construction of simple prototypes of electrical apparatus only a few years later, and 'real' electricity and magnetism later.* (Witz, 1996, pp. 601–602)

It goes without saying that in order for students to experience a sense of wonder, they should be enabled to observe and listen carefully. Goethe's idea of 'delicate empiricism' is an important one to consider, as regards the phenomenology of natural phenomena (see Hadzigeorgiou & Schulz, 2014). To understand Goethe's approach to the study of nature, one has to understand Goethe's idea of 'active seeing'. Bortoft's (1996. p. 249) explains:

> *[...] when we see the leaves of a plant we just see the generality "leaf" and do not notice the particularity of any one leaf or the differences between the leaves. Attention does not go into sensory experience, but remains on the level of mental abstraction. This is the condition of automization, in which the particular is "tuned out" and only the general form of what*

---

[18] In distinguishing between 'private' and 'public' science, Elkana (2000) has argued that the methods of logic are insufficient for describing science as an *endeavour*, that is, 'science in the making'. His point is that 'logical tools are of limited use in understanding the development of science or, what is even more important, in the teaching of science' (p. 473). Private science is inevitably phenomenological, but the prevailing insistence on the 'logic' of science, when formulated in public language of 'final form science' as found in textbooks, does not give students the picture of science as a human activity, or even a historical activity, as pointed out by several previous researchers (e.g. Bauer, 1992; Matthews, 1994). Indeed, the intermediary nature of constructed scientific language, being posed between the phenomenon and cognition, can itself become alienating (e.g. Sutton, 1996).

things have in common is registered. This is our habitual state of passive awareness, which is reversed by the process of active seeing in Goethean science.

Apparently, the adoption of such an approach to school science has an important implication for science teacher education, namely, to focus on the development of the ability of observation. This, of course, entails 'training in careful observation', so that the observed phenomenon (e.g. a rainbow) and natural entity (e.g. a flower) can 'speak' to the observer in all the languages of nature, as Dahlin (2001) has phrased it.[19] Perhaps some of Goethe's writings need also to be included in the standard bibliography of science teacher education courses. If one considers how careful observation helped shape our world that we take for granted, we might as well attach more importance to this particular and often neglected skill. For example, it should be remembered that it was the careful and systematic observation of plant and animal flight that preceded the technological realization of flight. Indeed observing how seeds are seized and carried by the wind[20] can be a 'wonder-full' activity, but it could be objected that there is not enough time on the timetable and the curriculum, even if the students have already been trained for such kind of systematic observation. Nevertheless, changes in both the timetable and the curriculum could compensate for that constraint, once curriculum designers and teachers alike become aware that sense experience in general can directly link students to the natural phenomena and the world at large.

It is true, of course, that a phenomenological approach to school science gives primacy to students' 'lived experience'. One can indeed make a distinction between a 'lived experience' and a 'sensory experience'.[21] However, it is also true that a sensory experience can be a lived experience in the sense that it is not passive and has an impact on one's life. Observing a rainbow or a sunset can be a lived experience (if one perceives the beauty and the richness of these phenomena) comparable to an experience one can have through the perception of an artwork.

---

[19] The experiencing subject must pay attention to all the qualities of the perceived phenomenon. In the context of science education, this primacy means that students must observe, listen, and even taste, if necessary and if possible. Of course, a sense experience can be passive and simply result in a 'copy of reality', something that both Piaget (1970) and Dewey (1966/1916) had strongly criticized. In addition, because sense experience is associated, more often than not, with *observation*, there appears to be a serious problem in the sense that students cannot really observe without knowing what to observe in the first place. And this is because observation always presupposes some theory, some prior knowledge, or theoretical framework (see Driver, 1983). Yet, despite such criticisms, sense experience is indeed a pathway to learning, especially in the case of young students, for it is a source of scientific knowledge and understanding. Young students touch objects, make observations, smell substances, hear sounds, etc. This is the foundation of their knowledge, albeit intuitive and, by and large, not in line with standard scientific knowledge (Hadzigeorgiou, 2001; Saçkes, Trundle, Bell, & O'Connell, 2011).

[20] The best known examples of flyers (with rotating wheels) are the fruit of the maple and the lime tree and the seeds of coniferous trees.

[21] There is a difference, for example, between observing a rainbow and feeling action and reaction as there is between observing a tree leaf and feeling the heat of a room with and without humidity.

6.5 Sources of Wonder in School Science: Possibilities and Opportunities

Even though a sense of wonder can have multiple sources—indeed anything can be a source of wonder, as was said, provided that the latter is evoked—in the context of school science, it is its content knowledge (i.e. its ideas and the phenomena that these ideas describe and explain) that is the source of wonder. From an instructional point of view, this wonder can have specific sources, which relate, directly or indirectly, to science content knowledge. What follows is a brief description of these specific sources of wonder.

6.5.1 The Ideas of Science

It should be understood that wonder can have its source in the scientific ideas themselves. I mean abstract ideas like energy, matter, heat, and light. These abstract ideas are powerful, in the sense that they help us explain and predict a wide range of phenomena. Take matter, for example. It is an idea that is central to science. Richard Feynman, in fact, believed that perhaps the most powerful idea is the atomic hypothesis:

> *If in some cataclysm, all of scientific knowledge were to be destroyed, and only one sentence passed on to the next generations of creatures, what statement would contain the most information in the fewest words? I believe it is the atomic hypothesis that all things are made of atoms.* (Feynman, 1995, p. 4)

But how can such an idea become a source of wonder? Depending on students' grade level, a teacher can present the 'secrets of matter' (through a story about what matter 'really looks like'—if we could use an extremely strong microscope) or the evolution of ideas about matter (through narrative or storytelling) followed by a few illustrations of how such a relatively simple idea can explain a wide variety of phenomena (e.g. evaporation, contraction, melting).

In other cases, the ideas in and of themselves can be sources of wonder. Take, for example, the following 'symmetrical' idea: 'a changing magnetic field generates an electric field and a changing electric field generates a magnetic field', or the idea 'a charged particle generates an electric field while a moving charged particle generates both an electric and a magnetic field, which are perpendicular to each other', or the idea 'there can be motion at extremely high speeds in the absence of a net force'.

In considering what was said in the previous section in this chapter, the role of questions, the contextualization of the ideas, and the way these ideas are expressed determine the degree to which wonder can be evoked. It is very important that they are presented in a contextual way. The context can be presented verbally and/or through a demonstration (see Appendix E). As an example of verbal presentation, the idea 'Regardless of the velocity of the light source (i.e. whether a star is approaching us or moving away from us), the speed of light remains the same (something that does not happen with other objects like bullets)' can be used to introduce the idea that 'The speed of light is constant', while a demonstration involving 'A student pushing a broom stick against a bathroom scale and observing that his/her weight remains the same' can be used to introduce Newton's third law of action and reaction (see also Appendix E). The idea that was previously mentioned about 'motion at extremely high speeds in the absence of a net force' can be used to introduce (verbally) Newton's first law.

It deserves to be noted that with younger students the role of contextualization (e.g. through demonstrations, illustrations) is very crucial. Take, for example, the idea 'Mass does not affect motion in a gravitational field' (see Pendrill et al., 2014). While this abstract idea is in and of itself a source of wonder, the following three contexts can indeed help with the facilitation of the experience of wonder about such an abstract idea:

- Free fall of two objects of considerably different weights (i.e. both a pen and a bunch of ten identical pens fall simultaneously).
- Pendulum motion (i.e. the period of the pendulum does not depend upon the mass of the oscillating object).
- Wave swingers in amusement parks (i.e. both empty and loaded chairs form the same angle to the vertical).

It is important to stress here that while mathematical formulae (e.g. the formulae for free fall, the formula for the period of a pendulum) can help explain why motion is not affected by mass (e.g. the velocity of free-falling objects depends only on the acceleration due to gravity, while the time taken depends on this acceleration and the height from which the object is released, and in the case of a pendulum, the period is dependent only on the acceleration due to gravity and its length), the wonder experienced through a demonstration is 'still there'. In other words, the mathematical explanation of an abstract idea does not diminish the wonder experienced through the demonstration of this idea (Hadzigeorgiou, 2012; Hadzigeorgiou & Garganourakis, 2010).

As another example, let us consider the idea 'light is invisible'. As an idea, it captures the students' imagination, for it sounds strange, surprising, and may even astonish some students. However, a simple demonstration experiment can convince students that light cannot be seen when it travels. That is, they cannot see such a thing as beam of light. All they can see is the source of light and what light 'hits'. And this can be indeed a shock of awareness (Hadzigeorgiou, 2012). All that is required is a shoe box and a flashlight. The shoe box must have a hole on one of its sides and a small hole (the size of an eye) on its top side. The first hole is needed for the light of the flashlight to shine (travel) through, while the hole on the top side is used by students to look inside the dark box. As soon as the flashlight is placed and fitted in the hole and is turned on, the students can peer through the hole on the top side and see that there is no such thing as a light beam but only two bright spots, that is, the source of light and the small area on the opposite side of the box. No light can be seen in between these two bright spots.

The above demonstration is a successful one as it helps both upper primary and early secondary school students to understand how light works and, of course, the idea that light per se is indeed invisible (Hadzigeorgiou, 2007, 2012). It also points to the importance of context in the presentation of science ideas, especially if such context presents something surprising and unexpected. The following three demonstrations are examples of situations that were found to be sources of wonder that was in turn used by a teacher to introduce upper primary and early secondary school students to the concepts of density, light reflection and refraction, and air and atmospheric pressure, respectively (Hadzigeorgiou):

- An ice cube remains 'suspended' somewhere between the top and the bottom of a glass filled with water or a transparent liquid that looks like water (i.e. the cube neither floats nor sinks)!
- A coin disappears from a dish when some water is poured in!
- A tumbler immersed in a container filled with water leaves the piece of paper on the bottom of the tumbler completely dry!

All the aforementioned examples point to something crucial: ideas can be introduced through a context that evokes wonder. This context can be presented verbally (preferably through questions) or through a demonstration (see also Appendix E). Apparently all the questions that were used as examples in the previous section of this chapter can be used to introduce a science concept or idea. Even though the idea of 'contextualization' is not new, its value needs to be reclaimed in the context of 'wonder-full' science, which, as empirical evidence suggests, can foster both emotional and cognitive engagement with science content knowledge (Hadzigeorgiou, 2012; King et al., 2015). Have we ever considered the possibility of presenting the concept of angular momentum, not through its traditional or nontraditional definitions but through a photograph or, better, through a video showing a gymnast performing a twisting somersault (which appears to contradict the law of the conservation of angular momentum)? Have we ever considered the possibility of presenting something as ordinary as water, not through its chemical formula and its usual properties and utility but through a context that reveals its unusual, strange, and mysterious properties?

Regardless, however, of our answers to the above questions, in the light of recent evidence, the role of wonder for the learning of science becomes crucially important. Indeed, there is strong evidence from brain research that justifies the intentional teaching of science (especially in the early years) and the learning of unifying concepts or 'big ideas' (Brooks, 2011). The role, therefore, of wonder in learning about these 'big ideas' deserves particular attention.

6.5.2 Everyday Ordinary, Familiar Objects and Phenomena

The fact that wonder can have its source in everyday and taken-for-granted objects, situations, and phenomena gives science teachers a wide spectrum of choice. Their only concern is that what they choose must be connected to the target concepts and ideas. An incident at an amusement park, for example, can be used to evoke wonder at and about the phenomenon of motion, while a sunset can be used to evoke wonder about the concept of light. Even something so familiar and ordinary as water can become a source of wonder once two simple questions are asked: 'Why does water dissolve more substances than any other liquid?' 'Why does water behave so strangely compared with other liquids?'

However, wonder can also be evoked through storytelling, if certain images and associations are included in its plot. Egan (1992) gives as an example a useless object, namely, a broken styrofoam cup, and he describes how it can become a source of wonder, which can enlarge its significance for human life. Instead of seeing that cup as simply an environmentally damaging waste, Egan gives a 'romantic' alternative: he considers it a symbol of a heroic quality, that is, a symbol of 'an immense ingenuity'. Indeed we can hold burning liquids in it without our hand being burned! As he points out:

> We can enlarge its significance by considering it as a part of the heroic journey that is the human struggle to shape the world more closely to our desires, to find release from the constant toil, sickness, and pain that have been the lot of most people most of the time [...] The knowledge of chemical and physical processes that have gone into its design and making is prodigious. And we have learned the environmental costs entailed in applying this knowledge to create this convenience, and we are as a society recognizing that we must satisfy this particular desire in other ways that do not threaten our harmony with the natural world. One can flash such thoughts through the mind in less than half a second [...] They are associations that come with romantic image of the broken cup. (Egan, 1992, pp. 76–77)

It is quite clear that the unexplored relationships between ideas, objects, events, and human life can play a significant role in evoking a sense of wonder. Sometimes, even a single sentence can contain such an unexplored relationship. For example, Carl Sagan's motto 'we are star stuff', in which the idea is embedded that we are mostly composed of chemical elements first formed in stars, can help students see how we are connected with the cosmos.

6.5.3 Mysterious Situations and Phenomena

Strange and mysterious situations and phenomena can be used in order to introduce students to the concepts and ideas of science. In the literature one can find a large number of such situations and phenomena. The pilot, for example, who survived a free fall at the speed of 180 km/h, is a case in point. Because his airplane caught fire, the only way to save his self was to bail out. Unfortunately his parachute did not open, so he free-fell until he was stopped by tall trees (Arnold, 1997). This real incident can be used to introduce students—from high school or even upper elementary school—to the concepts of force and momentum. The difference between this particular story and the ones presented in Appendix G (i.e. stories/narratives for introducing students to scientific inquiry) is that the above incident is a real one. The 'realness' of the situation can certainly add more 'wonder' to students' thinking, as they are more interested in real events and phenomena, compared with students younger than 7–8 years (Egan, 1997, 2005). The following two short stories that refer to real incidents can be used to introduce students to the concept of sound refraction and the idea of the significance of trace elements for human health, respectively.

Example 1: On February 2, 1901, the people of England mourned the death of Queen Victoria. However, on that very day, a strange and mysterious phenomenon happened. The booming of the cannon that fired in the air to mark the death of the Queen was not heard by people in the surrounding countryside. That was something unexplained, given that the sound from that cannon shot was supposed to be very many decibels. And yet that sound was heard by people living in an area about 150 km away (Stevens et al., 1966, p. 21).

Example 2: In the 1920s a health mystery had puzzled public health officials in the US south. No one could explain a strange malnutrition pattern that was observed at that time. The reason was that even though the diet of both black and white sharecroppers was the same, the public health officials observed deficiency diseases among the whites but not among the black population. Indeed, the blacks were totally unaffected. The mystery was not solved until the public health officials learned that both whites and blacks boiled their staple foods for long periods to improve the palatability of their somewhat tasteless food. However, they also learned that the whites threw away the cooking water, as using the so-called potlikker was considered socially unacceptable. The blacks who were uninfluenced by the social stigma attached to potlikker, drank it all, or used it to soak corn bread[22] (Sebrell, Haggerty, & The Editors of Time-Life Books, 1967, p. 60).

6.5.4 Spectacular/Terrible Phenomena

Natural phenomena, such as rainbows, sunsets, flashes of lightning, waterfalls, geyser bubbles (appearing just before the water is ejected from the ground) and geysers themselves, volcanic eruptions, and solar eclipses, provide a context for teaching a wide variety of science concepts and ideas. These phenomena, with the use of technology (i.e. high-definition photographs), can become sources of both aesthetic pleasure and wonder. On the other hand, some very spectacular phenomena, such as aurora borealis and light pillars (forming around natural sources of light, such as a setting Sun and Moon, and caused by suspended ice crystals), or phenomena that are terrible, such as tsunamis, tornadoes, and fire tornados (created by vortexes that suck flames upwards) and volcanic lightning (when the volcanic ash gains electric charge), can literally catch the students' attention and evoke wonder and awe. All

---

[22] With the discarded liquid, all the essential nutrients, like vitamins and minerals, went away too. The blacks who drank the potlikker or used it to soak corn bread were able to recapture the essential nutrients.

the above phenomena make students aware of the beauty of the natural world and, at the same time, aware of the immensity and power of nature.

6.5.5 Unexpected Interconnections (of Phenomena, Entities, Ideas, and Human Life)

While all phenomena are interconnected, students, for instructional purposes, are encouraged to focus upon a single phenomenon, be it motion, increase in the temperature of an object, reflection of light, etc. However, if the instructional goal is to help students experience wonder, these interconnections become quite crucial. The interconnection of electric and magnetic phenomena, of motion and heat, and of light and heat is well known and can be used by teachers to show their students just that. However, there are some interconnections that are not so evident or so 'visible' to the students. Candle burning is a good example and I mention it because Faraday wrote a whole book entitled *The Chemical History of a Candle*, in which he shows that the burning of the candle taps all the known laws of physical science.

The interconnections of electricity can also be used as an example. Electricity, in addition to being a provider of energy that we use in our home and elsewhere, is connected in various ways to human life. In fact, it is electricity that makes life possible! And this is the way to start looking at electricity in a new light. First of all, electrical forces are the glue of life. Indeed, it is thanks to electric charges that we stand on our feet, take a step forward, or even kick a soccer ball. And our heart works because of electricity. On the other hand, it is electrical interactions that hold the molecules together. When we try to push through a solid surface, the molecules in our hand and those in the table interact and feel the solidity of the table. Such electrical forces are responsible for hydrogen bonding in water and also responsible for the fact that our hands or our full body can move easily through air but less easily through water or other liquids. Teachers must point out to students that if there were no electrical forces between water molecules (due to the polarity of each water molecule), the water would have other properties and the world, as we know it, would be a totally different place! It is all these connections of electricity with phenomena and human life that help evoke a sense of wonder about them and thus reveal their significance.

6.5.6 The Immensity/Vastness and Smallness of Physical Reality

The immensity and smallness of physical reality is of some concern to science teachers, given that students of all grade levels will have to think about how small or how big a physical entity or quantity really is. An elementary school student may know, for example, that atoms are very very small, while a high school student may know that the distance from the Earth to the Moon is about 384,000 km or that Avogadro's number (i.e. the number of atoms or molecules contained in a mol) is 6.023 multiplied by 10 to the power of 23. But are they really aware how small or big these numbers really are? Even in the case of a blue whale—which, as students learn, is the largest mammal on the planet—it is after students learn that its heart is as big as a small car that they begin to get an idea how big a blue whale really is.

It is a fact that there are so many things that students learn as regards the magnitude of physical entities and quantities. However, there is a question: are they really aware of this magnitude? A teacher can understand this problem if she or he begins to ask wonderment questions: How small is an oxygen atom? How small is a water molecule? How big is the speed of light? How far is the Moon or the Sun?

Girod (2007b), being interested in teaching his students about time scale, provided each of them with a page that had 10,000 dots on it. Each dot represented 1 year, and therefore each dotted page represented 10,000 years. The students, by laying out their dotted pages, created a tapestry of 100 pages (i.e. a tapestry with 1,000,000 dots) and thus began to get an idea about how long ago dinosaurs lived, how long ago the Ice Age was, etc. As Girod (2007b) points out, students were shocked to understand that mammoths and sabre-toothed tigers roamed the planet 10,000 years ago (i.e. the number of dots in one single page). In the case of dinosaurs, students understood that these creatures lived 65 times their 1,000,000 dot tapestry!

Calculations can also help raise awareness about the magnitude of distances, masses, volumes, etc. The following examples show that simple calculations made by students themselves can evoke wonder through their awareness of the real magnitude of physical reality, and most specifically of the smallness of water molecules, and the immensity of the speed of light and of the distance to the Moon.

- *The number of molecules contained in teaspoonful or a glass of water*: Using Avogadro's number and the value of mol for the water (18 g), high school students can actually calculate how many water molecules there are in a teaspoonful (about 10 g) or in glass of water (about 250 g). They can also compare this number with the number of glasses that a given body of water can contain (e.g. the Mediterranean Sea) and see for themselves how small molecules really are (Hadzigeorgiou, 2012).
- *The time it takes for electricity to reach a light bulb in Australia or on the Moon, if the switch is in London*: In considering that the distance between the Earth and the Moon is 384,000 km and the fact that electricity travels at the speed of light, students can divide it by 300,000 km/s (i.e. the speed of light) and see for themselves that the light bulb on the Moon will light in only 1.28 s! This is how fast electricity travels!
- *The time light takes to travel around the Earth*: Assuming that the circumference of the equatorial circle (i.e. the length of the equator) is about 40,000 km, light could go around the Earth seven times in just 1 s. Any elementary or high school student holding a tiny globe in one hand is unable to move the index of his/her

other hand around the equator (on the globe) more than one or two times in 1 s! This is how fast light travels!
- *The time it would take an aircraft flying at 1000 km/h to reach the Moon*: Assuming that the speed of an aircraft in an intercontinental flight is about 800 km/s, an airplane flying at 1000 km/h would need 16 days and nights to reach the Moon. Even at a speed greater than that of sound, for example, 1500 km/h, it would take us about 10 days! This is how far the Moon is!

6.5.7 Amazing, Surprising, Incredible Facts About Natural Entities and Phenomena

With the rise of the constructivist perspective on teaching and learning science, the role of facts has been downplayed, as more emphasis was placed on the way students learn. In short, the emphasis was not on 'what' but on 'how' one learns science. And more recently, with the emphasis on scientific inquiry, what matters most is how students think, investigate, and reach evidence-based conclusions (e.g. NRC, 2007, 2009). However, what has not been considered is that a scientific fact—whether it refers to a natural entity (e.g. a tree, a sea pebble, a river) or a natural phenomenon (e.g. lightning, photosynthesis)—can facilitate learning through conceptual engagement with the target concept. Indeed all natural entities and phenomena can be linked, in one way or another, to some incredible facts, which can be used as an introduction to almost all science concepts.

The fact, for example, that the ratio of the diameter of the Moon to that of the Sun is about the same as the ratio of the distances between the Earth and the Moon and the Earth and the Sun (i.e. the Moon's diameter is 1/400 of the Sun's, while the Moon's mean distance is about 1/400 of the Sun's), and therefore the Moon and the Sun appear to be nearly the same size when observed from the Earth, can be used to introduce students to the phenomenon of total solar eclipse (when the Moon completely blocks the Sun, thus casting a shadow several hundred km wide across the surface of the Earth). Even though the phenomenon of eclipse in general has fascinated people since the dawn of time, total solar eclipses are spectacular because the Sun's corona becomes visible, and therefore images of total solar eclipse can evoke wonder (see Hughes et al., 2014). But the role of the aforementioned fact (regarding the ratio of the diameter of the Moon to that of the Sun) is crucial in evoking wonder, as it is this fact that makes possible the observation of the phenomenon. Similarly, the fact that two mature trees produce as much as oxygen as two families of four need for a whole year can be used to evoke wonder about the role of trees in sustaining human and animal life and thus as an introduction to the phenomenon of photosynthesis. The following are some examples of incredible, amazing facts that can be used to introduce students to a variety of science concepts and ideas:

- Our body consists of atoms that are billions of years old. The hydrogen atoms in our bodies are as old as the universe itself, that is, about 14 billions years old,

while heavier atoms like carbon and oxygen are between 6 and 8 billion years old.
- The corn used to produce the ethanol consumed by an SUV in 1 day can feed as many as a hundred people for more than a week.
- Only 10 % of the electric energy consumed by a light bulb is transformed into light. The remaining 90 % is wasted as heat.
- About 100 lightning bolts strike the Earth every single second.
- Each bolt can carry more than 3 billion kilowatts of power, and the temperature in the bolt is about 20,000° (centigrade), which is about more than three times hotter than the Sun's surface.
- There are about 1000 organisms living in and on a rainforest tree.
- A 30-m-tall mature tree can absorb as much as 22 kg of carbon dioxide in year, which is about the same amount as that produced by a car driven for about 45,000 km.
- The same tree can produce more than 250 pounds of oxygen annually, which is enough to cover, over the same period of time, the oxygen needs for at least two people.
- Of all the water on the planet, only 1 % is not salt water and readily available for use (97 % is salt water in the oceans, and the rest 2 % is fresh water in the form of ice).
- The older a tree gets, the greater its capacity for photosynthesis.
- The average person walks in his/her lifetime the equivalent of 2–3 times around the world (along the equator).
- Coast redwood can be as tall as 120 m, while a giant sequoia tree can weigh up to 2000 tons.
- A blue whale's heart is as big as a small car.
- A glass in space boils rather than freezes (first becomes vapour which then turns into ice crystals).

6.6 Concluding Comments: The Need for Wonder in Science Education

If the problem in education is how to help students discover 'the imaginative mode of awareness' (Greene, 1978, p. 186), then the value of wonder in education needs to be reclaimed. Its role in countering 'the anaesthetic of familiarity', as Dawkins (1998) put it, its potential for self-initiated and self-directed inquiry, and its power to give significance to anything around us (Hadzigeorgiou, 2014; see also Duckworth, 2006) make it an indispensable teaching/learning tool. Egan (2014) sees wonder as that which can 'keep the mind awake, and not have it sink into an ossified slumber' (Egan, 2014, p. 160).

It is a misconception to believe that the passivity inherent in wonder is an obstacle to learning, and therefore the passive 'wonder at' state of mind cannot promote a 'wonder about' attitude, which is central to scientific inquiry. Having already pointed out that a 'wonder about' attitude cannot necessarily be identified with curiosity, it should be noted that a tight distinction between a passive and an active aspect of wonder (see Goodwin, 2001; Silverman, 1989, 2003) cannot be maintained. But even if one were to maintain such a distinction, and therefore a 'wonder about' attitude were identified with bewildered curiosity, the 'wonder at' attitude is important all the same.

It may, of course, be true that science teachers and science educators would be in favour of fostering a 'wonder about' attitude, while they should not bother with the passive, 'wonder at' aspect of wonder (i.e. wonder which has its source in the admiration of, and/or surprise by, an object or phenomenon, like a whale, a volcanic eruption, a rainbow, a sunset, a flash of lightning) on the grounds that it is the former that promotes scientific inquiry. Therefore the passive, 'wonder at' aspect should not be considered a component of scientific wonder and should not be encouraged and fostered in the context of science education. In fact, one may very well associate the passive aspect of wonder with religion (rather than with science) and thus associate or even identify the former with feelings of admiration, bewilderment, and even incomprehensibility, but not with bewildered curiosity, that is, the 'wonder about' attitude or state of mind that is central to science, for it is true that admiration leads one to accept what is being admired as something miraculous and therefore incomprehensible, rather than to become curious about the object of admiration.

And yet, even in this case of religious or 'passive' wonder, one can also speak of a particular kind of awareness that is developed, namely, the awareness of the beauty and immensity of the natural world. This kind of awareness, regardless of whether or not one is religious, is important, since it can encourage deeper involvement with, and also respect for, nature, which can be considered students' larger object of study (Witz, 1996). In the context of contemporary science education, respect for nature is an important educational goal, and the role of wonder in the achievement of this goal can be quite crucial (Hadzigeorgiou & Skoumios, 2013). In fact, such a role was pointed out by science educator Ann Howe more than four decades ago: 'The world needs people who can think and feel; people who know the earth and also love it; who know much about the forms of life and respect all life; who know what the stars are made of and can still look up at them and wonder' (Howe, 1971, cited in DeBoer, 1991, p. 179).

Therefore it should be understood that both aspects of wonder are crucially important in science education, and it is both aspects that make wonder not just useful but a very powerful tool in the hands of science teachers. Indeed, it is the experience of both aspects of wonder that makes students conscious of what they are learning and also helps them change their outlook on natural phenomena and generally on science itself. And such change of outlook is an important aim of education. As British educational philosopher R. S. Peters argued, 'To be educated is not to have arrived at a destination; it is to travel with a different view' (Peters, 1973, p. 20).

There is certainly a question with regard to today's educational reality and more specifically the teachers' ability to evoke in their students a sense of wonder, for it is true that such an ability presupposes passion as well as breadth and depth of knowledge. It is for this reason that I believe the first step towards making wonder an indispensable part of school science education is to help science teachers and science educators develop a sense of wonder themselves, by making *them* aware of the wonder in science. A retooling, therefore, of science teacher education programmes is imperative. On the other hand, the first step towards implementing a 'wonder pedagogy' in the context of science teacher education is to pay attention to those who have experienced wonder. Who, in reading Richard Feynman, cannot help but wonder at and about the physical laws governing our life in this cosmos?

> *A poet once said, "The whole universe is in a glass of wine." We will probably never know in what sense he meant that, for poets do not write to be understood. But it is true that if we look at a glass of wine closely enough we see the entire universe. There are the things of physics: the twisting liquid which evaporates depending on the wind and weather, the reflections in the glass, and our imagination adds the atoms. The glass is a distillation of the Earth's rocks, and in its composition we see the secrets of the universe's age, and the evolution of stars. What strange arrays of chemicals are in the wine? How did they come to be? There are the ferments, the enzymes, the substrates, and the products. There in wine is found the great generalization: all life is fermentation. Nobody can discover the chemistry of wine without discovering, as did Louis Pasteur, the cause of much disease. How vivid is the claret, pressing its existence into the consciousness that watches it! If our small minds, for some convenience, divide this glass of wine, this universe, into parts – physics, biology, geology, astronomy, psychology, and so on – hat Nature does not know it! So let us put it all back together, not forgetting ultimately what it is for. Let it give us one more final pleasure: drink it and forget it all!* (Feynman, 1995, p. 67)

Chapter 7
'Artistic' Science Education

> *Science is an art form and not a philosophical method. The great advances in science usually result from new tools rather than from new doctrines.*
>
> Freeman Dyson, in *The Scientist as Rebel*, pp. 17–18
>
> *Scientific experience is laden with aesthetic content of the beautiful, which is manifest both in the particulars of presenting and experiencing the phenomenon under investigation, and the broader theoretical formulation that binds the facts into unitary wholes.*
>
> Alfred Tauber, in *The Elusive Synthesis*, p. 1
>
> *Science excites poetry rather than extinguishes it […] The poet and the man of science can co-exist in equal activity […] During human progress every science is evolved out of its corresponding art.*
>
> Herbert Spencer, in *Education: Intellectual, Moral and Physical*, p. 71., p. 72, p. 139.

It is a fact that the perceived differences between the sciences and the humanities can be grounded in both their ontology and epistemology and, of course, in the way they are practised. Both Snow's (1959) famous book *The Two Cultures* and Hirst's (1972) work on the *Forms of Knowledge* provide a solid foundation for this difference. And because the humanities include the arts, any attempt to bring science and art together will more likely be doomed to failure. Is this really the case? Richard Feynman had made a comment that is worth quoting:

> I have a friend who's an artist, and he sometimes takes a view which I don't agree with. He'll hold up a flower and say, "Look how beautiful it is," and I'll agree. But then he'll say, "I, as an artist, can see how beautiful a flower is. But you, as a scientist, take it all apart and it becomes dull." I think he's kind of nutty […] There are all kinds of interesting questions that come from a knowledge of science, which only adds to the excitement and mystery of a flower. It only adds. (Feynman, 1989, p. 11)

Even though Feynman does not explicitly say anything about the relationship of art and science, his view does point to their complementary role in experiencing and understanding the world.

Scientific creativity, as was discussed in Chap. 5, is an imaginative endeavour and shares a number of similarities with artistic creativity. Certainly there are differences but a relationship between the two exists. Moreover, science and art share a common ground, or what is called a 'common aesthetic', that has been studied by a number of scholars (see Root-Bernstein, 2002; Tauber, 1996). Thus imagination and beauty can be considered the connecting ring between the two, with implications for school science education.

If art can indeed help us escape from the boredom of daily life, as Einstein (1949) himself had observed—because it has the power to quicken us from 'the slackness of routine' and to make us 'forget ourselves by finding ourselves in the delight of experiencing the world about us in its varied qualities and forms' (Dewey, 1934, p. 110)—then the role of art and the possibilities it opens for students and teachers should be seriously considered. It is a fact, and beyond any doubt, that mainstream science education—with its emphasis, on the one hand, on conceptual understanding (i.e. on students' alternative conceptions and their subsequent restructuring and change) and on standards and international comparisons, on the other—has failed to make science learning an engaging and, at the same time, rewarding activity that can make a difference to students' own lives. As Pugh and Girod (2007) have remarked, 'we often obsess over misconceptions but fail to ask whether students ever apply their 'correct' conceptions outside of school and use them to have aesthetic experiences in the world' (p. 10). It is indeed ironic that science curricula give attention to the teaching tools necessary for communicating science ideas and results of scientific investigations but not to the 'aesthetic tools' necessary to actually 'do science'.

Douglas Barnes in his *From Communication to Curriculum* raised a crucial issue concerning the difference between school knowledge (i.e. knowledge learned and used in the classroom in order to answer a teacher's question or complete an exam item) and knowledge that has scope beyond the classroom walls (i.e. knowledge that is applied and made meaningful in everyday life). The arts have the potential to make scientific knowledge come alive and take it outside the school classroom. Having discussed the notion of aesthetic experience (see Chap. 2), one may begin to perceive the potential of art and science connections to encourage both anticipation and deeper involvement with science. This chapter focuses on and discusses the relationship between art and science, the idea of beauty in science, and their implications for school science. However, a short historical background that explores this relationship can be helpful.

7.1 The Shift from a Positivist/Inductivist Conception of Science

The year 1687 is considered, according to Butterfield (1965b), the most important date in the history of science, since it represents the birth of Western science. Newton, with the publication of his *Principia* (*The Mathematical Principles of*

Natural Philosophy), laid the foundation for modern science.[1] More than that, Newtonian physics, having such power—since it could explain motion both on Earth and in the heavens and also predict the future motion of celestial bodies—became the foundation of a new worldview (see Appendix A for the story of the new science of motion). For three whole centuries, no one could even imagine to dispute the Newtonian worldview. However, at the turn of the last century, two theories, namely, quantum theory and the theory of relativity, began to seriously challenge the Newtonian worldview.

Although the formulation of electromagnetic theory had already started to shake the Newtonian foundation of science, it was the 'new physics', that is, the physics of quanta and relativity, that openly questioned the fundamental ideas of the Newtonian worldview. With the theory of relativity, the concepts of absolute time and space literally collapsed. With quantum theory precise measurement proved to be a dream, while recent chaos theory led us to abandon the idea of precise prediction. Also contemporary views on the nature of science dispute the idea of objectivity as well as the idea that the primary source of scientific knowledge is sensory experience (see Chalmers, 1990, 1999; Tauber, 1996).

We should be reminded, however, that it was with Romanticism (see Chap. 3) that the idea of 'cold objectivity' in science received severe criticism and, in fact, came under attack. For the romantic scientists aesthetic considerations played a crucial role in their study of nature. They imbued nature with aesthetic values and also sought unity between human being and nature. For them, science was not just an intellectual but also an aesthetic endeavour. Friedrich Schlegel's view, namely, that 'All art should become science and all science art', best captured the relationship between science and aesthetics (Hadzigeorgiou & Schulz, 2014).

Even though romantic science was placed in the background with the rise of logical positivism, changes that have taken place over the last three decades in the area of the philosophy of science have led to a shift from the positivist/inductivist/empirical view of scientific knowledge (Schulz, 2009, 2014). According to contemporary views, values, beliefs, and, generally, the theoretical/conceptual framework of the scientists determine the criteria by which scientific theories are judged and evaluated (Hadzigeorgiou & Fotinos, 2007; Trefil, 2003). Tomas Kuhn, in his seminal work *The Structure of Scientific Revolutions*, has argued that the notion of scientific truth is subject to conventions among the members of the scientific community (Kuhn, 1970; see also Kuhn, 1977). In short, and in other words, scientific truth is more a matter of a negotiation among the scientists than a search for an objective and absolute truth. Moreover, aesthetic factors have played, according to Kuhn, an important role in those conventions. This shift from the positivist/inductivist/empirical view of scientific knowledge appears to have important implications

---

[1] Newton had recognized Galileo's contribution to this new physics. Galileo was indeed the first to question the scientific status quo of his time. It is for this reason that he used to say 'I stand on the shoulders of giants'. In fact, he had also incorporated classical Greek metaphysical ideas about the structure of the universe (see Clavelin, 1974; Cohn, 1978; Klein, 1954).

for science education and leads to an exploration of the relationship between science and art.

7.2 The Relationship of Art and Science

The traditional distinction between science and art is well known: the aim of science is the search for truth about the world while the aim of art is aesthetic appreciation of the world. Even if one were to completely abandon the idea of truth as the aim of science—and to perceive science's main task the improvement of our existing knowledge—the distinction between science and art appears to be valid. In his *The Critique of Judgement,* Kant did state that there is neither a science of the beautiful, nor beautiful science, but only beautiful art. Beauty was not something to be empirically tested. Kant's view on aesthetic judgement (i.e. no aesthetic judgements can be made about science; in judging objects merely in terms of concepts, we lose all beauty) does not point to any relationship between art and science (Chevalley, 1996).

Paul Hirst, a British philosopher of education, made a distinction between sciences and fine arts, in his attempt to categorize human knowledge into various forms. According to his analysis, there are seven different/distinct ways in which one can experience and understand the world. He called these different ways 'forms of knowledge'. Each form has its own distinctive concepts and procedures (or methodologies), and therefore it makes sense to claim that each one of them represents a different way to understand the world (Hirst, 1972).

Of course, one might reasonably argue that all forms of knowledge should be considered complementary, on the grounds that they can all contribute to a fuller and richer experience and understanding of the world. So both art and science are needed for such fuller understanding. On the other hand, while complementarity is very evident between some forms of knowledge (i.e. mathematics and physical science or philosophy and human sciences), it is not so evident, or evident at all, between science and art. And this is not surprising given that science and art may be taken to represent the two extremes on our experiential continuum.

In looking at art and science, one may begin to perceive two characteristic differences. First, even though both art and science start from the real world, art proceeds to the exploration and invention of all possible worlds. This is a very crucial distinction between the two. The reason is that while an artist has an absolute freedom to create, a scientist can create only within the context and the limits of physical reality (of course, this observation does not entirely hold as is discussed later on in this section).

A second crucial distinction between art and science is that the former always projects the human presence in the universe. This is something that science is trying to avoid. For example, in their attempt to predict the behaviour of various systems, scientists develop models and equations, which do not include the human element.

Thus, it would not be an exaggeration to claim that science contributes to the 'dehumanization' of the physical world (Hadzigeorgiou & Fotinos, 2007).

However, there are also several similarities between the two. First, both science and art are human activities, taking place in a sociocultural context. The motive for both is to enrich human life and also, as Einstein (1949) had pointed out, the desire to escape from the boredom of everyday life. Second, they both, more or less, explore the natural world. For example, nineteenth-century painting was not just a means to represent nature but also a means to explore it. Third, artists in general, like scientists, strive for a deeper understanding of the world; and in this sense they are not just creators: they are thinkers and, in fact, conceptualizers. Fourth, in both science and art, creative imagination plays a central role. The creation, for example, of mental imagery and analogies, is common in the two fields. Fifth, in both science and art, observation skills are crucially important, as are exploration and experimentation. Sixth, both art and science are associated with the idea of 'wholeness', as scientific truth is not judged solely on the grounds that scientific ideas correspond to certain observable facts but also because they contribute to a sense of wholeness (Bohm, 1998). Such 'wholeness' can be seen, for example, in the case of a theory (e.g. the Newtonian theory which can explain all phenomena of motion at speeds lower than that of light) or even a specific product (e.g. the periodic table). It is the perceived unification of ideas in the above examples that makes scientists feel a sense of wholeness. Seventh, aesthetic factors—in addition to the idea of wholeness, which can also be associated with beauty—play a significant, and many times a decisive, role in scientific creativity (Kuhn, 1970; McAllister, 1996; Root-Bernstein, 2002). There is evidence that at the moment of creation, the boundaries between art and science cease to exist, and aesthetics play a central role (Miller, 2001). And last—even though one may identify some more similarities—both science and art are motivated by feelings of mystery, awe, and wonder (Hadzigeorgiou & Fotinos, 2007; Hadzigeorgiou & Schulz, 2014). It is well known that Einstein had pointed out that the experience of mystery is the most beautiful experience that we can have, and that this feeling is common in both art and science.

It should be understood that both scientific ideas and works of art are products of the imagination, which help us transform our perception of reality. For example, the scientific idea of 'time dilation' from the special theory of relativity (i.e. clocks tick more slowly if they move with a very high speed) may very well make us change our perception of time. By the same token, the idea 'solid matter is in reality mostly empty space' may very well lead us to change the way we see everyday objects, such as a table and a glass of water (see in Chap. 6 students' comments on how the wonder inherent in science ideas make them see things differently). The same, however, can happen through a work of art, be it Picasso's *Guernica* or Monet's *Water Lilies*. Thus, both science and art can help us transcend the boundaries of everyday reality.

What is also important to bear in mind is that with the advent of quantum physics and the theory of relativity, the distinction between art and science became blurred. The unification of time and space, the idea of simultaneity, and the idea that no two observers see exactly the same thing were common in art and science. The similarities between cubism and the theory of relativity are evidence of these common ideas. For example, with the theory of relativity, the concepts of absolute space and time collapsed, while cubism totally demolished the concept of perspective in art. Miller's (2001) analysis provides overwhelming evidence that both Einstein and Picasso were attempting to tackle the same problem, namely, the problem of simultaneity—Einstein was working on temporal simultaneity, while Picasso was working on spatial simultaneity.[2]

Miller's (2001) work does point to a fact: the origins of scientific ideas are not to be found only in science, and the origins of art are not to be found only in art. A characteristic example is how the discovery of X-rays made the distinction between the notions of 'inside' and 'outside' ambiguous, thus influencing artists, and more specifically the way they began to perceive space (Henderson, 1988). In looking at what transpired at the end of the nineteenth century—the discoveries of the X-rays in 1895, radioactivity in 1896, and the electron in 1897—one can see that the distinctions between the distant and near and the visible and the invisible become blurred. As Miller put it, 'what you see in not what you get' (p. 25).

A good example illustrating the intersection of art and science is Marcel Duchamp's *Nude Descending a Staircase*, which raised an issue regarding what actually that painting depicts. It was, according to Graig-Faxon (1996), a visual equivalent of Einstein's theory of relativity. The thing is that neither the nude figure nor the staircase is visible. However, this painting can be observed as existing simultaneously in the past, the present, and the future, and the only place in the whole universe that this observation would be possible would be aboard a beam of light.

It deserves to be noted that Kant's views on the difference between science and art—scientific knowledge becomes possible due to 'schematism', while aesthetic judgements are the result of symbolism (see Chevalley, 1996)—were found problematic at the beginning of the twentieth century by physicists. Thus Heisenberg had to break with Kant's theory of knowledge (i.e. how scientific knowledge becomes possible), since Kant's distinction between science and art (i.e. between schematic and symbolic knowledge) and between the 'noumena' (i.e. the things in themselves) and 'phenomena' (i.e. the things as they appear to us) perpetuated a Cartesian dualism, which, though, was inadequate when it came to describing the subatomic world, where sense experience cannot provide direct access to knowledge (Heisenberg, 1971). These reasons necessitated a radically different way of approaching nature. Thus phenomenological approaches and symbolic language (e.g. as in poetry) became useful and, in fact, dominant. Art played a crucial role in this new approach to physical reality and was found necessary to be included in a unified view of human knowledge (see Tauber, 1996).

The fact that art and science influence each other is no doubt important from a science education perspective. Art is indeed necessary for a holistic experience,

---

[2] Even though a direct link between cubism and the theory of relativity cannot be easily defended, the view that Poincare's writings (on the notion of simultaneity and on the non-Euclidean approach to geometry) influenced both Einstein and Picasso, as Miller (2001) says, seems to be plausible.

since it is necessary for a unified approach to the knowledge of the natural world (Hadzigeorgiou & Schulz, 2014). This means that art and aesthetic judgements are not useful and necessary only in cases, in which students must move beyond the phenomenal reality, beyond what they actually observe through their senses (e.g. when it comes to helping high school students understand ideas from relativity and quantum theory, both of which presuppose that we discard our traditional notions of time and space), but also in the context of school science in general. This role becomes clearer in the next section, which discusses the idea of aesthetics in art and science and, of course, in the section on the possibilities concerning the implementation of art/science activities.

7.3 Aesthetics and Beauty in Science

The deeper relationship between art and science concerns the idea of beauty, as they both share a common aesthetic element (Root-Bernstein, 1996, 2002; Tauber, 1996). Paul Dirac's famous aphorism 'It is more important to have beauty in one's equations than to have them fit the experiment' (Dirac, 1963, p. 47) best captures the aesthetic element common in art and science. Historically, the most representative example of this element is Leonardo da Vinci's work. But although Leonardo da Vinci stands out as a scientist and artist, whose life and work illustrates the unity of the two areas (see Deckert, 2001), there have been a number of scientists whose scientific work has been decisively influenced by the ideas of harmony and beauty (see Root-Bernstein, 1996).

Some scientists, of course, have explicitly talked about the beauty of science. Poincare, for example, had remarked that 'The scientist does not study nature because it is useful; he studies it because he delights in it, and he delights in it because it is beautiful' (cited by Silverman, 2003, p. 386), while Werner Heisenberg, in a discussion with Einstein, made reference to the aesthetic element of science, which he identified with the 'frightening simplicity and wholeness of the relationships which nature spreads before us' (cited in Chandrasekhar, 1987, p. 53). There are also scientists who have defended the centrality of beauty in science. Dawkins (1998), for example, argued that even scientific reductionism did not take away anything from the 'poetry of science'; it did not diminish the beauty of the natural phenomena. The rainbow, for example, did not lose its beauty when Newton reduced it to prismatic colours. Light, in general, does not lose its beauty when it is refracted, reflected, and digitized. Silverman (2003, p. 388) summarizes what beauty in science can mean:

> There is great beauty to science, whether in a bold and artful experimental solution to a seemingly insurmountable problem or in the aesthetic appeal and startling predictive power of a set of equations.

The idea of beauty in science has been linked to an 'artistic' approach to reality and scientific work. Root-Bernstein (1996) introduced the term 'synscientia' in order to explain scientific work in a variety of scientific areas.[3] This term refers to a way of knowing in a synthetic way, which enables one 'to conceive of objects or ideas interchangeably or concurrently in visual, verbal, mathematical, kinesthetic, or musical ways' (p. 66). There is no eminent scientist, according to Root-Bernstein's research:

> [...] who simply solves mathematical equations or pours chemicals into test tubes and analyzes the results or catalogues chromosomal abnormalities. Scientists, or at least scientists who are worth their salt, feel what the system they are studying does. They transform the equations into images; they sense the interactions of the individual atoms; they even claim to know the desires and propensities of the genes. (p. 66)

The centrality, however, of aesthetic beauty becomes very evident in creative moments. As Miller (2001) pointed out, during the creative moment, the borders between the different fields become blurred or even collapse and aesthetic considerations become dominant. Einstein's approach to space-time was not purely mathematical but also aesthetic. For Einstein, in fact, aesthetic factors were a given. In other words, aesthetics and laws of nature must, in some way, be connected. His reservations in regard to quantum theory are clearly seen in his claim that the light quantum leads to a deep typological distinction between particle and wave and also to a conflict between continuity and discontinuity—therefore to an asymmetry that does not exist in natural phenomena. This was not 'aesthetic', so he argued for a new principle of relativity, which could circumvent that asymmetry. This is the reason why Einstein gave a new and integrated definition of mass (instead of the two definitions given in Newtonian mechanics), which reveals the equivalence between gravity and acceleration (Miller).

The thinking of Niels Bohr was influenced by aesthetics too. He is described by Werner Heisenberg as an artist who, in using his brushes and various colours, tried to convey, just like an artist, his own images to other scientists. It is very likely that he was influenced by cubism in the formulation of the principle of complementarity. Just like in cubism (i.e. the way an object is observed determines what the object is), in physics an electron and photon acquire their reality from the way they are observed (Heisenberg, 1971). Bohr's aesthetics also included symbolic language, which became necessary when he became aware that ordinary, everyday language was totally inadequate to describe physical reality in the subatomic world. Indeed, he pointed out that 'language can be used as in poetry' (Tolstory, 1990, p. 16).

It is quite interesting to note that the history of science provides us with examples of famous scientists who produced literary works, even though the relationship between these works and their scientific work is underestimated. Erwin Schrodinger's, for example, literary imagination and his bilingualism (i.e. he was fluent in both English and his native German) must have played a role in his approach to physical reality (see Sofronieva, 2014).

---

[3] It is worth noting that Root-Bernstein (1987) has found about 400 cases of famous scientists, who had artistic abilities and also considered a career in the arts (see also Root-Bernstein, 1996; Root-Bernstein & Root-Bernstein, 2004).

7.4 Aesthetics and Beauty in Science Education

In the context of school science education, the aesthetic element of science can be directly linked to the notion of beauty, which can refer to such things as an experiment, an image provided by a microscope, an equation, and an idea. However, there is a quite straightforward question: what is a beautiful experiment, a beautiful formula, and a beautiful image? This is an important, in fact crucial, question if aesthetic beauty is to be judged by primary or secondary school students. The problems, however, which emerge, once we try to answer this question, are obvious.

First, in acknowledging the differences between scientists and students (e.g. different purposes, different motives, different conceptual frameworks), one can ask whether students, while engaged in school science, are capable of appreciating scientific beauty in the first place.[4] And second, there is the issue of the subjective nature of beauty. For a third grader, for example, beauty may lie in the colours perceived during a physical phenomenon or experiment, while for a tenth grader, beauty may lie in the perception of symmetry found in nature. For a scientist, on the other hand, beauty may be found in simplicity, in symmetry, or even in complexity. It may also be found in the interrelationships of various natural phenomena and ideas.

In looking at the world of science, an 'instructive' example regarding beauty can be taken from Michael Faraday's (1978/1861) *The Chemical History of a Candle*. In this book, Faraday makes an explicit reference to the idea of beauty, by saying that the phenomenon of candle burning is beautiful, not because of its flames (e.g. the prettiness of its colours or its shapes) but because of the fact that it taps all known laws of the universe: not the best-looking but the best-acting thing. In other words, for Faraday, the aesthetic dimension of science, and specifically the beauty of the phenomenon of candle burning, is identified with his experience of wonder (i.e. his admiration towards the phenomenon of candle burning, which includes the awareness that all known physical laws are involved in this phenomenon). This kind of beauty, perceived only through one's awareness that all known laws are in some way interconnected, requires 'work' on the part of science teachers. It is certainly easy for a scientist like Faraday or Feynman, who too finds beauty in the interrelationships of various phenomena and ideas (Feynman, 1969), but it is not so easy for school students.

It has been argued that, in the context of science education, the 'aesthetic dimension' of scientific knowledge cannot, and should not, be reduced, or linked only, to symmetry and abstract mathematical concepts, as is often done in the physical sciences today. It needs to be linked to one's personal experience of doing science and hence linked to such notions as mystery, awe, wonder, imagination, and inspiration. Indeed, it was these elements that were present in a truly aesthetic experience and

---

[4] Students can make aesthetic judgements, as was discussed in Chap. 2, about beautiful and ugly things, but such judgements are not the same as scientific beauty, usually associated with symmetry or simplicity.

had been considered central to the study of nature by the romantic scientists (Hadzigeorgiou & Schulz, 2014).[5] Therefore, even a theatrical play, performed or watched by the students, can address the aesthetic dimension of scientific knowledge if students perceive it as a 'wonder-full' and inspirational activity.

Moreover, and as a consequence of what has been just pointed out, the 'aesthetic dimension' of scientific knowledge cannot, and should not, be linked only to contemporary physics (i.e. relativity and quantum theory where mathematical abstraction and poetic language are required to describe physical reality). School students too can appreciate the beauty of science if they are provided with opportunities to do so. They can appreciate it in familiar natural phenomena and entities. A sunset or a tree, for example, can both provide such an opportunity. Not just the beautiful colours of a sunset and not just the variety of symmetrical shapes of tree leaves but also the dancing of electrons and photons, and the exchange of chemical molecules, like carbon dioxide and water, taking place during a sunset and the process of photosynthesis, respectively, can reveal an aesthetic beauty that does not emerge in traditional science classrooms.

Notwithstanding the potential of the ideas discussed in the previous chapters (e.g. storytelling, wonder, romantic understanding) to reveal to the students the aesthetic dimension of science, the integration of the arts into the science curriculum can be helpful in making science learning a truly 'artistic' activity, during which students can perceive and appreciate beauty.

7.5 The Pedagogical Importance of Art

Having discussed the art-science relationship, the question regarding the importance of art in the context of school science becomes quite clear, notwithstanding the caution that was voiced in the previous section about the notion of 'scientific beauty'. However, art, in addition to helping students become aware of the creativity and beauty in science, can help with the achievement of significant goals in the context of science education and education in general. Thus, there are also pedagogical reasons, beyond the purely common aesthetic element in art and science, which justify the use of art in science education. These reasons, according to the literature (e.g. Caine & Caine, 2001; Gardner, 1993b, 1997, 2010; Hardiman, 2012; Rinne, Gregory, Yarmolinskaya, & Hardiman, 2011; Waterman, 1992, 2004), point to the potential of the arts to encourage the having of holistic experiences, and they are as follows:

- *Engagement*: Art activities encourage deeper involvement because they provide opportunities for anticipation and also for immersion experiences.
- *Creativity*: Art activities offer opportunities for 'prolonged creativity'.

---

[5] Nature, as has been already discussed in Chap. 3, was considered a poem and a work of art by the romantic scientists (Hadzigeorgiou & Schulz, 2014).

- *Beauty*: Art activities offer opportunities for students to appreciate the aesthetic element.
- *Identity building*: Art activities, through immersion and creative expression, provide students with opportunities for exploring their 'unexplored selves'.
- *Self–actualization*: Art activities, through self-directed learning and self-expression, provide opportunities for self-actualization.
- *Cognitive skills*: Art activities offer excellent opportunities for observation and other scientific skills (e.g. classification).
- *Risk taking*: Art activities encourage improvisation and experimentation in a nonthreatening environment.
- *Retention*: Arts integration improves long-term memory.
- *Brain growth*: Art activities facilitate the creation of neural pathways and the development of stronger synapses.
- *Fine motor skills*: Art activities help very young children build the same skills that they use in writing.
- *Learning styles*: Art activities can offer opportunities for kinaesthetic, auditory, and visual learners.
- *Multiple intelligences*: Art activities offer opportunities for stimulating and exercising a variety of intelligences.
- *Self-expression*: Art activities can help students express not only their thoughts but also their imagination and their feelings.
- *Change of outlook/perspective*: Art activities, by encouraging a more focused way of seeing reality (i.e. through focusing on detail), can facilitate a change in students' perspective/outlook on the world in general.

Perhaps the most important pedagogical characteristic of art—inherent in most of the aforementioned reasons—lies in its power to charge emotionally that which it depicts. Art, indeed, is much more likely to do this than mere data (i.e. numbers and graphs). An excellent example is the use of satellite visualizations, which show both the beauty and frailty of the Earth. Scientific facts, especially those regarding the state of the planet, such as global warming and its consequences, can be drawn, illustrated in photos and collages, dramatized, etc. (see Caddy, 2015). Moreover, art encourages change in perception. Given that aesthetic perception presupposes a focus on detail, on nuances, it helps develop awareness of the significance of the object of perception, be it a tree or a tree leaf, a piece of rock or a mountain, or a water crystal or a waterfall. As Jackson (1998) pointed out, 'Not only must we perceive art objects in order to appreciate their worth, but doing so is at least one means by which we come to better perceive other objects and events, including ourselves and others' (p. 113).

Of course, one could ask a practical question: can all science concepts be linked to or presented through some art form? This question can be answered in the affirmative, provided that we select situations in which the target science concepts are embedded (e.g. the motion of clouds in the case of force, the motion of sea waves in the case of energy). In other words, we need to select phenomena and situations (e.g. a volcano, a tornado, the water cycle, a flash of lightning, the motion of a wave swinger, the twisting somersault or the balancing act of a gymnast), which 'aesthetically' exemplify the target concepts. What follows in the next section are teaching/learning possibilities that the arts can offer to school science education.

7.6 Teaching/Learning Possibilities

The aesthetic element of science, which reflects the very nature of science, as a form of inquiry, can be certainly considered by science educators, in designing curriculum and instruction. In fact, art and aesthetics appear to offer an alternative or even a complementary way to science teaching and learning, in line with recent changes in science and science education (NRC, 2007, 2009).

Photography, holography, sculpture, dance, drawing and painting, theatre and drama, and poetry represent ways for introducing scientific ideas. They can all foster anticipation and they can all lead to 'consummation' (see Chap. 2) and hence to an aesthetic experience. Computer technology can be an important tool for such an approach to science learning. Students, for example, can work on a project requiring the photographic recording of various technological advances and explore the scientific ideas behind them. They can write a poem containing science ideas, draw models, and study the principles of mechanics through dance. They can also study the life of the great scientists (e.g. Curie, Galileo, Newton, Planck, Maxwell, Bohr, Einstein), in order to write a script for a theatrical play (i.e. dramatizing important events from the scientists' lives).

It may be argued that while students could learn ideas from the nature of science through their participation in drama/role-play activities (i.e. the dramatization of historical events implicitly or explicitly presents such ideas), science content ideas are certainly much more difficult to be learned. True, one cannot learn about photosynthesis by looking at tree leaves, any more than one can learn about electricity by looking at or drawing a light bulb, no matter how bright the latter is! But one can use the arts—all forms of art—as a medium to represent ideas. Knowledge representations, as Bruner (1966) in his *Toward a Theory of Instruction* had stressed, are crucially important for effective learning. These representations can be enactive, iconic, and symbolic, and therefore the role of art—because the latter is based upon such representations—in teaching science should be given more serious thought in mainstream science education (Taber, 2013). Moreover, all art forms, be it visual arts, dramatic arts, music, or poetry, tap students' creative imagination. What follows is a brief description of such possibilities.

7.6.1 Visual Arts

Visual arts like photography, painting, and drawing, even sculpture[6] and filmmaking, have the potential, especially nowadays with the educational uses of sophisticated technologies, to raise students' awareness of the beauty of natural entities and phenomena. Images, for example, of water drops and water crystals, snowflakes, tree leaves, mineral rocks, flashes of lightning, rainbows, and light interference patterns, can be used to raise an aesthetic awareness. Such visual art forms/products can be used as teaching tools (i.e. both as instructional and assessment tools). With young children, the role of visual art to help them understand such concepts as line, space, shape, colour, and form, and also to help them bridge the gap between what they actually observe and their preconceived ideas about the observed object(s), should be recognized (Yokoi & Yee, 2011).

Depending upon grade level and the target concept, as well as the teaching model/strategy, the classroom context, and the available materials, the teacher may decide on the most appropriate visual art form, even though all forms can be used with all concepts in all areas of science (see Allen, 2012; Alrutz, 2004; Buczynski, Ireland, Reed, & Lacanienta, 2012; Liguori, 2014). Here are some possibilities[7] for science teaching:

- *Making artistic products based on scientific principles* (e.g. containers with layers of various liquids; shadows of various shapes and sizes, using both the students' own body and objects and both natural and artificial light; magnetic lines of force, using a variety of magnets and iron filings).
- *Making artistic products to represent a natural phenomenon* (e.g. a rainbow, using bottles with coloured water or painted beans; a flash of lightning; a volcanic eruption).
- *Using any visual art form to represent an abstract science concept or idea* (e.g. wave, light, heat energy, momentum, time dilation, the uncertainty principle).
- *Using any visual art form to represent a natural phenomenon* (e.g. a rainbow, a mirage, the swinging motion of a pendulum, colour mixing, the greenhouse effect, flashes of lightning, motion of charged particles in electric and magnetic fields, weightlessness).
- *Taking/using photos to represent a natural phenomenon* (e.g. a rainbow, a mirage, flashes of lightning).
- *Drawing unobservable processes/phenomena* (e.g. sound waves in the air, in liquids, and in solids; changes in intermolecular distances in solids as a result of contraction and expansion).

---

[6] Battles and Rhoades (2005), in the case of an interdisciplinary course in art and geology, use metalwork and jewellery in order to introduce the properties of minerals and the medium of sculpture in order to give students the opportunity to investigate these properties, while they are making a sculpted piece.

[7] These possibilities refer to both primary and secondary science education. The teacher can select those that match his/her subject matter and also his/her curricular and instructional objectives.

- *Drawing the results of an investigation* (e.g. the effects of mass distribution on rotational motion; the effects of electric current on the behaviour of a compass; the effect of heat on the strength of a magnet).
- *Creating a sequence of drawings to describe the motion of a physical entity in a given context* (e.g. an atom in a solid, a water molecule in a heated container with water, an electron in a copper wire carrying an electric current, a human being walking in the street, an astronaut walking on the Moon, a gymnast performing a simple or twisting somersault).
- *Creating a collage of photos or drawings to represent an idea in a variety of contexts* (e.g. motion of athletes, dancers, astronauts, cars, and animals; light reflection by mirrors, water, and other surfaces; conservation of energy in the case of the motion of a ball in the air, up and down an incline).
- *Creating a sequence of photos or drawings to illustrate events in between the generation of something and its final use* (e.g. in the case of electricity, several events take place between its generation at a power plant and its final use/consumption, such as generator friction, insufficient insulation, heat loss on the wires and transformers, short circuits, and wind and rain damage).
- *Drawing or creating a photocollage to illustrate a sequence of phases/states or physical events* (e.g. water turning into ice and vice versa, water turning into steam, the concentration of electrons on the lower part of a cloud and the lightning bolt that is created between the cloud and the ground, the phases of the Moon).
- *Creating a collage with various physical entities* (e.g. dried leaves, dried flowers, pulses).
- *Creating a collage to represent the absence of something* (e.g. phenomena on the surface of the Earth without an atmosphere, without gravity).
- *Creating a collage or a web of images to represent abstract concepts* (e.g. energy) *or more 'concrete' concepts* (e.g. motion).
- *Using visual art to make scientific tables appear more 'artistic'* (e.g. the periodic table, tables showing physical characteristics/properties of various substances, planets).
- *Drawing or using any visual art form to represent objects and phenomena in which a variety of colours is involved* (e.g. a light bulb showing the various types of radiation emitted by it, a rainbow, a sunset, aurora borealis).
- *Drawing various objects that call for a decision regarding their position in the context of the drawing* (e.g. fruit in an aquarium that will either float or sink, balloons filled with various gases that will be up in the air or on the ground, an astronaut in a state of weightlessness in a spaceship, a ball moving horizontally and vertically inside a moving train).
- *Creating a collage that deliberately depicts misconceptions* (e.g. about the laws of motion, the flow of heat, the colour of an object illuminated by light of various colours, the physical properties of elements).
- *Making an innovative artistic graphic representation of experimental data* (e.g. heights to which super balls bounce off the floor, the time taken to suffocate a candle by using glasses of different sizes and volumes).

7.6 Teaching/Learning Possibilities

- *Using at least two forms of visual art to illustrate/represent a natural phenomenon* (e.g. lightning, rainbow, magnetic attraction, planetary motion) and *natural entities* (e.g. renewable energy sources, water crystals, tree leaves).
- *Using modelling clay to make models* (e.g. models of atoms, molecules, molecular structure, wave motion).
- *Using fruits, candies, and vegetables to model atoms and molecules* (e.g. hydrogen atoms, oxygen atoms, the carbon dioxide molecule, the water molecule).
- *Using fruits, candies, and vegetables to construct 3-D models of various objects/entities* (e.g. a car, an airplane, a boat, a tree, a cell).
- *Using recycled materials to create/represent physical entities* (e.g. leaves, clouds, models of the planets).

It is evident that all the aforementioned activities require divergent, creative thinking and provide opportunities for students to be creative, collaborate, and even participate in a classroom contest for the most "beautiful" creation. What, of course, must be clear is that any art product, from any of the above activities, needs to be accompanied by a rationale/interpretation of what it means and how it relates to the science idea that it supposedly represents. While for the teacher such a rationale/interpretation may be easier in some cases (e.g. a photocollage referring to the various states of water, a collage of leaves) and more difficult in others (e.g. an abstract piece referring to the concept of energy, to life without gravity, or even to the uncertainty principle), it must be the student himself/herself who will provide the rationale/interpretation behind his/her artistic creation.

7.6.2 Drama/Role Play

Dramatic arts include theatre and dance, and in science education the former has been used mainly in the form of role play. Perhaps the best known example of dramatic performance in the history of science is 'The Blegdamsvej Faust', which was successfully performed in 1932 by Bohr's students (Pantidos, Spathi, & Vitoratos, 2001). What is now called 'dramatic science' has received, in comparison with other artistic forms, more attention (McGregor & Precious, 2015; Ødegaard, 2003).[8] There are, of course, some concerns regarding the theoretical framework underpinning dramatization, given that studies that explored its role and effect on student

---

[8] The rationale for including the drama activities in the science curriculum is that drama activities foster empowerment, inclusion, engagement, embodied thinking, social interaction, argumentation, opportunities for multiple modalities, and deeper understanding of science (e.g. Aubusson et al., 1997; Darlington, 2010; Dorion, 2009; McGregor, 2012; McGregor & Precious, 2015; Varelas et al., 2010). Ødegaard (2003)) pointed out that 'Though making scientific concepts come to life through the use of a dramatic model of them is not uncommon, it is particularly in addressing the nature of science and science in a societal context that drama has a lot to offer science education. It is in this area that, in general, drama seems to be an untapped resource in the science classroom' (Ødegaard, 2003, p. 79).

learning have been based upon a variety of ideas, such as social constructivism, embodied thinking, and play (see also Braund, 2015, who is concerned about the fact that the theory behind drama in education does not draw on ideas about theatre in education).[9] However, there is now evidence, as discussed later in this chapter, that drama helps with the acquisition of affective, cognitive, and procedural knowledge, as it offers a more lively way to learn science and involves all children in a science classroom.

Perhaps the most important reason for including 'dramatic' activities for teaching science is student's motivation and their enjoyment of science (Aubusson, Fogwill, Barr, & Percovic, 1997; Darlington, 2010; McGregor, 2012). And the most important reason why such activities might fail to foster science learning is that teachers do not fully incorporate them into the lessons and use them simply as an 'add-on' (Darlington). That is why workshops aiming to familiarize teachers with the potential of drama in science education are quite crucial (see McGregor, 2012; Precious & McGregor, 2014). Such workshops can also help teachers overcome their lack of confidence in using drama in their classrooms.

Dramatization involves what Egan (2005) calls 'change of context'. This change gives students the opportunity to engage differently with knowledge (i.e. in ways that they do not experience it in the traditional curriculum), for the simple reason that the classroom is transformed into another place, and the students are taken to another time and take on roles relevant to that place and time. Thus dramatization is a very imaginative activity. Imagination is involved even in the case in which young children use their body to represent physical reality (e.g. a tree, a water molecule, wave motion). Indeed, the mental image created by the imagination from the various elements of reality is embodied and then realized again in reality (i.e. the reality of the stage). Thus dramatization, as Vygotsky (2004) pointed out, expresses the 'full circle of imagination'.

Drama can be used with children in the early grades of primary school (McGregor, 2012; Varelas et al., 2010), as well as with older students from both primary school (Cakici & Bayir, 2012; Dennis, Duggan, & McGregor, 2014) and secondary school (Darlington, 2010; Dorion, 2009; Pongsophon, Yutakom, & Boujaoude, 2010). It has also been used with undergraduate students (Simons, 2013). It can also be used in a variety of ways or 'dramatic activities', with or without improvisation, in the context of the school classroom, or even outside the classroom, which range from large ongoing projects to smaller 5-min mimes and can easily be adapted for the delivery of a variety of science topics to a wide range of students.

In 'dramatic' activities the roles that students can play are mainly three: (a) enactment of physical entities (e.g. students play the role of electrons in molecules in their attempt to illustrate physical changes and natural phenomena); (b) role play in the context of the history of science (e.g. students play the role of scientists, by dramatizing historical events, like the discovery of magnetic induction by Michael

---

[9] According to Braund (2015) 'theorisation of drama in learning, at least in science, has been lacking and no attempt has been made to integrate drama theory in science education with that of theatre' (p. 202).

Faraday, the trial of Galileo, the Galvani-Volta controversy); and (c) role play in the context of contemporary society (e.g. students play the role of citizens participating in public debates about controversial socio-scientific issues, like the use of GMF and alternative sources of energy).

Regardless of the role that students play, or even regardless of what form the drama activity takes (e.g. mime, improvisation, structured or spontaneous role play), imagination is always present. As Dorion (2009) found through his empirical research, 'drama-based activities may be viewed as a potentially rich classroom resource for interactive and imaginative learning' (p. 2268). Whenever students take on the role of a physical entity or of another person, imagination must be called into play.[10,11] Moreover, when students take on the role of another person (e.g. a scientist of the past or present), they can develop empathy and thus better understand the perspective of another human being. This is an important reason to use drama in environmental education.

Drama activities, for the case of teaching/learning science content knowledge, can be used to illustrate and also assess the learning of:

- *Science concepts* (e.g. moment of force, angular momentum, electric resistance, electric current, electric circuit, heat, conductivity, entropy).
- *Chemical compounds, ions, free radicals, chemical bonding* (e.g. water, carbon dioxide, hydroxyl radical, carbon chains with various types of bonds).
- *Changes in molecular structure and molecular motion* (e.g. change of water to ice, change of water to vapour, the expansion of an iron rod)
- *Chemical reactions* (e.g. the formation of water, oxidation, photosynthesis)
- *Natural phenomena* (e.g. earthquakes, sound and wave travelling, resonance, winds of various types and intensities, such as breeze, squalls, gale force, winds, tornados)
- *Physical process/changes* (e.g. evaporation, condensation, melting, wave propagation, change of electrical resistance with temperature)

The inclusion of dance in some of the above activities can also be used (e.g. dance can be used to represent weather patterns, atmospheric conditions, and even seasons, to illustrate the concept of angular momentum).

For the case of teaching/learning ideas about the nature of science, drama activities can illustrate:

---

[10] Indeed students imagine that they are, for example, water molecules in a drop of water or iron atoms in a rod. And in illustrating the travels of a water drop or its transformation to ice, and the expansion of the iron rod, they must imagine what that travel may be or the image of ice and the rod with an increased length, respectively.

[11] Even though the dramatic activities, in which students enact physical entities and processes, are primarily sensorimotor (Hadzigeorgiou & Savage, 2001), the role of the imagination cannot be downplayed. It is true that all conceptual knowledge is embodied, that is, 'mapped within our sensory-motor system […] the sensory-motor system not only provides structure to conceptual content, but also characterizes the semantic content of concepts in terms of the way that we function with our bodies in the world' (Lakoff and Nunez, 2000, pp. 455–456), but imagination plays always a role in the human sensorimotor system (Johnson, 1987; see also Lakoff & Johnson, 1980).

- *The life and work of a scientist* (e.g. Newton, Curie, Einstein, Feynman)
- *Historical debates and controversies* (e.g. Lord Kelvin and 'the age of the Earth' debate, the Tesla-Edison and Galvani-Volta controversies)
- *Historical experiments* (e.g. the Magdeburg experiment, Galileo's experiments)
- *Contemporary social controversies* (e.g. the use of glass as a substitute for sand on a beach, the use of nuclear energy for the production of electricity)

With regard to the use of dramatization in the form of debates for teaching ideas about the nature of science, the potential of the former to facilitate conceptual change needs to be recognized. From a historical perspective, the most famous dramatization is that created by Galileo (in his *Dialogues Concerning Two New Sciences*) in order to present the arguments between Salviato (representing a Galilean) and Simplicio (representing an Aristotelian). The debate ended with Salviato disproving Simplicio's Aristotelian idea that heavier objects fall faster than lighter ones. This debate can be enacted in a science classroom for the purpose of helping young students learn that both heavy and light objects, if dropped from the same height above the ground, fall down simultaneously.

Another very good example with regard to the nature of science concerns the debate between Lord Kelvin and Huxley about the age of the Earth. Even though the historical events could be presented through storytelling, or generally in a narrative form, it is dramatization that helps bring forth the human emotions associated with Lord Kelvin's arrogance during the debate.[12] Such dramatization can also show how the established field of physics had to collaborate with the emerging fields of geology and biology in order to answer the question regarding the age of the Earth!

It must be pointed out that drama in science education is an imaginative activity and an excellent way to illustrate science ideas and assess the understanding of these ideas, in addition to the use of graphic and symbolic representations. An activity, for example, during which a class will have to 'perform' or dramatize an electric circuit—with one student playing the role of a battery, fully charged with energy (e.g. in the form of a basket full of apples), another one playing the role of a resistor (i.e. an electric appliance like car toy or flashlight, which consumes energy), and the rest of the students playing the electrons carrying from the battery to the resistor electrical energy (e.g. in the form of an apple)—will determine the extent to which students understand the concept of the electrical circuit with all its necessary components and their specific role in that circuit. The battery runs out at some point, after its energy (i.e. a bag full of apples) has been consumed, after the students have jogged several times around the classroom, from the battery to the resistor.

Given that embodied thinking plays an important role in role-play activities (see Braund, 1999; Hadzigeorgiou, 2001, 2002a; McGregor, 2012; Metcalfe, Abbot, Bray, Exley, & Wisnia, 1984; Varelas et al., 2010), it is crucially important that Johnson's (1987) thesis on the bodily basis of meaning, imagination, and reason be

---

[12] Indeed it is a dramatic performance that best presents and reveals the arrogance with which physicist Lord Kelvin talked to the biologist T. H. Huxley. Looking straight in Huxley's eyes, Lord Kelvin said that he knew everything about the physical laws of the universe (Stinner & Teichmann, 2003).

7.6 Teaching/Learning Possibilities 203

considered more seriously in the context of education, particularly science education.

7.6.3 Music

Music, as is well known, can be linked to the physics of sound. Musical sounds from any instrument, with the aid of a computer, can be used to illustrate the characteristics of sound waves, like wavelength, frequency (pitch), and intensity. In this way, these concepts are directly linked to a musical experience and thus become more concrete.

Students can produce sounds by using musical instruments or everyday objects that produce sounds, which can be musical sounds too. Giving students, for example, bottles of various shapes and sizes, or identical bottles with different amounts of water in them, and asking them to blow across their bottle's top can create beautiful sounds with various pitches. In this context, the possibility for students to form a band that plays music created by their own bottles can also be considered. The production of sounds by these different bottles—each with a different pitch—can help students link sound, music, and physics.

There is also a possibility for students to make a variety of simple musical instruments (e.g. a drum, a simple xylophone, a simple pipe with a few holes), in order to explore the physics of sound. Such a creative experience involves 'artistic design' too and can therefore be quite rewarding. However, music can also be used in 'artistic/creative' activities that link music and science. The following are some possibilities that tap students' creative imagination:

- *Reviewing musical albums and selecting pieces to match various physical phenomena* (e.g. motion/rest, magnetic attraction/repulsion, lighting, rainbows, photosynthesis, electrostatic/electromagnetic forces)
- *Composing a tune that best describes the above phenomena*
- *Creating a rap tune that may help students remember scientific facts* (e.g. what happens during the process of photosynthesis; the effects of electric current; numbers referring to plant and animal life; names, such as those of chemical elements; types of energy; experimental results, such as how water and vinegar react with various white powders)
- *Inserting sound effects into the students' experiments or inquiries in order to represent the sequence of steps they took*
- *Selecting or creating music that can be used as an aid to the introduction of such skills as counting, patterning, and sequencing*
- *Combining music and dance to represent physical phenomena* (e.g. the flow of electricity in a circuit, when it is off and on, under low or high voltage; changes in the states of matter; chemical reactions; an increase in the entropy of a system)

It is apparent that while linking music with the physics of sound can result in engaging activities, and even lead to an aesthetic experience (at least for some students), it is the aforementioned activities that offer possibilities for creative thinking and, therefore, can be considered truly 'artistic' activities. It is interesting to note that there is evidence to suggest that using 'science songs' can enhance learning, especially if these songs are combined with visuals and movement (see in Monk & Poston, 1999, how music can inform changes to the national curriculum with regard to science). According to Crowther (2012), 'science songs' enhance the students' memory, reduce stress, and offer opportunities for multimodal learning and also opportunities for an in-depth exploration of science content.

7.6.4 Poetry

Poetry can be used for a variety of reasons, such as to spark interest in the beauty and mystery of science (Herrick and Cording (2013), to integrate writing into the science lessons (LaBonty & Danielson, 2005), to familiarize students with the nature of science (Frazier & Murray, 2009), and generally to tap students' emotions and imagination (Eastwell, 2002; Watts, 2001). If the affective domain is important in education, because feelings and emotions shape attitudes, tastes, and motivations for learning (see Chap. 1), then the value of poetry in the context of school science should be recognized and seriously considered by science teachers and science educators (Watts, 2001; see also Alsop, 2005).

It should be noted that integrating poetry into school science helps to address two neglected domains in school science education, namely, the affective domain and that of creativity (see Chap. 5). In considering the theory of multiple intelligences (Gardner, 1983, 1993a) and the fact that there are many female students with both high mathematical and high verbal abilities (Wang, Eccles, & Kenny, 2013), poetry can tap the potential of those females, so that they are attracted to science.

However, poetry, in addition to enhancing students' creativity and affective skills, helps foster the ability for imagery, metaphor, and analogy and also communication. All these skills are crucially important in science and science education. Images, and especially metaphors and analogies, can help one clarify the meaning of scientific ideas.

It should also be noted that by writing a poem, one shows what one has learned, for it is a fact that in the process of writing a poem, apart from literary skills, science content knowledge also comes into play (Frazier & Murray, 2009). Indeed, students can illustrate concepts and phenomena (e.g. forces and motion, light and rainbows, electricity, energy transformations) in the poems they write or even attempt to write. They can also write poems about their own learning experience in the context of science—an excellent way to demonstrate their feelings and attitudes towards the subject of science.

Poetry in science can be used at the beginning of the lesson, in order to introduce students to the idea(s) of the lesson, and to motivate those with verbal and linguistic ability, but also as a culminating activity, or even as an assignment, in which students will try to show what they know about science. It can also be used in the context of the compulsory curriculum or as an enrichment activity or as part of an enrichment programme. Apparently, poems can be written by individual students or in collaboration with their peers, and their creations can be shared in the classroom, followed by judgements about whether these creations demonstrate science understanding. And they may be entered in *The International School Science Poetry Competition* (see www.sceinceeducationreview.com/poetcomp.html).

Whether children read poetry or have poetry read to them, the centrality of metaphoric language needs to be recognized. Metaphoric language that describes, for example, heat as 'the incorrigible thief of electrical power', or atoms as 'the most sociable entities on the universe', is crucially important in poetry. Richard Feynman's (2015) description of rushing sea waves as 'mountains of molecules' and of himself as 'an atom in the universe' and 'a universe of atoms' is a characteristic example of metaphoric/poetic language.

Feynman's poetic language is, of course, admirable, but young students can write poems too (see Eastwell, 2002; Watts, 2001), even though their poems lack the literary sophistication one can find in poetry written by skilful poets and scientists. While not as popular as drama activities, poetry can make science learning be both a scientific and a literary experience. The following two examples show how science content knowledge can be presented in a poetic form and illustrate the combination/integration of scientific knowledge with verbal and literary abilities:

On forces

Forces around us push and pull,
They are not scalar or just vector quantities-
they are really cool.

They make my bike move fast
They help my memories – although metaphorically –
move past.

They make birds and airplanes fly
They help me lift my spirit- although metaphorically –
up in the sky

They make a rocket go to the moon
They make lovers walk fast
and thus go home soon.

Forces in nature are of many kinds and magnitude
And for this very reason
I can express my real gratitude

They are gravitational, nuclear and electromagnetic
Which keep our world going
So let us learn about them-don't be pathetic

They always act on two separate bodies, in motion or at rest
And we call them Action and Reaction,
so let us learn about them and forget all the rest.

(Written by two Greek 9th grade students, who wanted to integrate physics, poetry, and english as an assignment on a unit on forces)

On acid rain

There once was an oxide called SOx
Who, along with another named NOx
Unleashed acid rain
As this lesson will explain,
Causing "eco-illogical" shocks

From smokestacks and auto exhaust
Exacting a terrible cost
Acid rain's killing lakes
Do we have what it takes-
To make certain that no more will be lost?

The tough problems passed by pollution
Are crying out for a solution
To bring NOx and SOx down
Let us meet in our town
For we each have a key contribution.

(Based on the poem by Lincoln Bergan, published in the GEMS guide *Acid Rain* and adapted by Linda McCarter who submitted it as part of an assignment for a summer (2002) course on Inclusive Science in the College of Education, at the University of Northern Iowa)

7.6.5 Integrating the Arts

It should be pointed out that, while the most common integration of art forms refers to drama, dance, and music, all forms of art can be integrated in a single activity. On the other hand, any concept/idea or phenomenon (e.g. tree, electricity, sunset) can help integrate all forms of art. Here are some possibilities:

- *Combining art forms to represent a science concept/idea* (e.g. electricity, molecular motion).
- *Combining art forms to represent a phenomenon/process* (e.g. the consequences of global warming both in the physical environment, like glacier melting, water level rising, and change of landscape, and the human environment, such as change of human geography).
- *Composing sound effects that will accompany the visual presentation of a phenomenon* (e.g. an earthquake, water movement/transport, such as sea waves, rain, rivers).
- *Illustrate the effects of heat or electricity using at least two art forms* (i.e. students may choose between visual arts, poetry, music and drama).

On the other hand, storytelling is an activity that helps integrate all art forms. In the 'Story of Force', for example, students can listen to a story about the evolution of ideas from Aristotle, to Galileo, to Newton (see Appendix A), can role play to dramatize the scientists, can use photos and drawing representing the concepts of force and motion, can compose poems and music relating to the laws of motion, and can prepare a performance in front of an audience, with the objective to make their knowledge of the laws of motion understood by that audience. A story about lighthouses (see Appendix H) can also help integrate all forms of art while students are introduced to a unit on light, with such concepts as light reflection and refraction.

7.7 What the Research Shows

Of all art forms drama is the one that has been researched the most. There are, however, some studies on the effect of visual arts on the learning process, but their number is still very limited compared with the number of theoretical or position articles on the value and role of visual art in science education (e.g. Alberts, 2010; Allen, 2012; Alrutz, 2004; Buczynski et al., 2012; Merten, 2011). Notwithstanding the dearth of research papers on the impact of integrating art and science on learning science, the available evidence is quite positive.

Mills' (2013) study, which explored the nature of a creative learning experience in the context of an art project, provided evidence for positive learning benefits. The study was carried out in a US urban high school with a sample of eight students in an advanced placement art class. The students were asked to research and create a unique art piece for a pollinator garden that could withstand the climate of the region. The time frame for this project was 6 weeks. Qualitative data for the study (e.g. field notes, personal journal, student journals, interviews) revealed a number of leaning benefits. More specifically, the students became more collegial and collaborative, while they also learned to work independently in order to plan and complete their art project. They also recognized the link between science and art and became engaged with the natural environment.

Dhanapal, Kanapathy, and Mastan (2014) also found positive effects of visual arts on the learning of science. Their purpose was to study the perceptions of grade 3 students and teachers about the role of visual arts in the context of school science. Their data were obtained from an international school in Malaysia through a survey distributed to both teachers and students. Their main findings were as follows: (a) Art did not simply enhance and stimulate the learning of science; it also had an impact on children's mental and physical development. (b) The young students 'enjoyed the freedom of choosing their preferred form of art to express their learning of science'.

However, positive learning benefits have been reported by Alberts (2010), whose study—at The Lab School of Washington, DC, with students in grades 1 through 12—investigated how art-based activities impact students with learning disabilities. It was found that art-based activities can help this particular group of students comprehend abstract scientific theories and improve their critical thinking skills.

Studies with undergraduate students have also found evidence of the positive role of visual arts. For example, Steele and Ashworth (2013) studied student teachers' ('elementary generalist teacher-candidates') views about integrating art and science. Using grounded theory, they analysed reflection papers written by the student teachers. They identified emergent topics, all of which point to the value of integration of art and science. These topics referred to different opportunities for learning, the possibility for student teachers to articulate a pedagogy of integration, the opportunities for cognitive 'disruption', and the positive impact on practice.

Ghanbari (2015) investigated the role of art integration in the construction of knowledge, in the contexts of two established US university programmes that integrated art with a STEM (science, technology, engineering, mathematics) discipline. Using sociocultural theory and experiential learning theory, as a theoretical framework for the study, Ghanbari found that the arts have the potential to open up new ways of seeing, thinking, and learning and that they can be successfully integrated into STEM disciplines. It is suggested that it is reasonable to make the STEM acronym STEAM.

Another two studies with university students are those conducted by Bruna (2013) and Wells and Haaf (2013), respectively. The study by Bruna used as sample first-year veterinary medicine undergraduate students—who often feel discouraged by courses requiring complex thinking. For this reason Bruna incorporated three art and biochemistry sessions into the students' biochemistry course. Working in groups, the students were asked to represent a biochemical concept or process of their own choice through any art form and then to present it to their classmates and to a panel of professors, both of whom would evaluate their artistic representation using a global perspective rubric. The assessment of such an activity in the context of the biochemistry course over three consecutive academic years suggested that the integration of the arts was successful, as it both motivated students and helped them to understand and learn biochemistry.

The study conducted by Wells and Haaf (2013) investigated the role of art objects in an undergraduate chemistry (attended primarily by nonscience majors) and an art course at a US university. This course was team-taught by an art historian and a chemist, for the purpose of helping students become aware of the complementary nature of the two disciplines. The students, in their exploration of art objects, used such techniques as visible and infrared spectroscopy, mass spectrometry, X-ray fluorescence spectroscopy, gas chromatography, and microscopy. At the end of the course, they participated in a culminating collaborative project in which they had to select an art object, generate a list of questions about the object, and subsequently use available scientific tools in an attempt to answer these questions. Two case studies from the students' research projects reveal the positive effects of such a project on thinking and collaborative skills as well as knowledge acquisition. The project also showed students the importance of technology used in chemical research in the study of a work of art.

With regard to drama, even though the literature is not extensive—the number of studies is still limited, compared, for example, with such areas as student misconceptions and conceptual change—over the past 20 years, there has been enough evidence to suggest that drama in science education has a positive impact on learning. The literature on drama in science education can fall into two broad categories, namely, (a) teachers' views on the role of drama in learning science and (b) intervention studies with primary and secondary school students. The second category includes studies that explored the learning of science concepts and attitudes/emotional responses to learning and also studies that focused on learning ideas about the nature of science.

With regard to the first category, Alrutz's (2004) case study explored the possibilities for using drama to teach science to students of a fourth grade elementary school class. Despite some realistic limitations to its implementation in the classroom,[13] she found strong support for using drama to teach science. Active learning, motivation to participate, and enjoyment were found to be the most important reasons for including it in the curriculum. Dorion (2009), on the other hand, studied teachers' own drama activities in five science lessons taught across schools in three cities in England and more specifically the drama forms along with the teaching objectives and the characteristics by which those forms were perceived to enable science learning. The age range of the students in those schools was 12–16, and the science subjects were chemistry, biology, and physics. According to the findings, teachers used drama activities for teaching a variety of topics—even topics that have not yet been reported in the related academic literature. Mime and role play were found to be the most dominant forms of drama activities used for the teaching of abstract scientific concepts. These activities were thought to encourage children to develop their visualization skill and possibly thought experiment skills and also to use a variety of modalities along with anthropomorphic metaphors and embodied thinking.

McGregor (2012), working on an intervention pilot project with 20 primary school teachers from ten schools in a city in England, reported positive effects of drama on teachers' knowledge. The teachers experimented with the dramatic approach and kept reflective journals. Data based on journal extracts, interviews, and classroom observations indicated that the drama activities[14] increased teachers' scientific understanding. Even though these were the findings of a pilot study, and therefore some further developments were still needed with regard to the design and

---

[13] There is limited time; as drama activities, by their nature, can take more time than a teacher has planned, some teachers may lack confidence in implementing dramatic activities and managing the classroom.

[14] McGregor (2012) used eight strategies: (a) mini historical play (providing a narrative of a scientist's life that in turn provides opportunities for young children to act out), (b) spontaneous role play (children are placed in a variety of contexts and take a position or stance on an issue and argue), (c) on the table (examining objects—oftentimes unusual—that the scientist might use or create), (d) hot seating (children play the role of an expert in order to answer questions asked by their peers), (e) miming movement (children use body movements or actions to mime/represent scientific ideas), (f) freeze frame (children pause their action in order to represent a specific thing, like a skill that a scientist used), (g) modelling (children physically (re)create a concept or phenomenon), and (h) mind movies (children use audio and/or visual stimuli to imagine a different context/location and to consider consequences of happenings and events).

the implementation of the activities, a survey based on several focus group discussions with over 200 children (ages 5–7) also revealed that children had positive views and feelings about the dramatic approach. Most children thought that (a) using drama to learn science is more fun, (b) drama helps understanding of more difficult ideas, (c) acting out helps learning, and (d) discussion helps learning.

Notwithstanding the effects of a Hawthorne effect—as children may have been motivated by the novelty of the approach—the positive outcomes cannot be ignored, especially the development of the teachers' confidence in teaching science. As McGregor (2012) pointed out:

The drama provided more than just an art form for learning science; it had also proffered insights into children's perceptions about science and of science processes. It enabled teachers to 'see' what children think. This is particularly useful for new entrants to teaching who may make assumptions about what their learners already know. (p. 1161)

The same dramatic activities were also used by Precious and McGregor (2014) in a project carried out in the form of a workshop. The aim of the project was to introduce teachers of the upper elementary school grades, who pretended to be famous scientists form the past, to the use of mini-speeches and monologues (on how they lived, how they made their discoveries and/or inventions, how they developed their ideas), in order to help children appreciate the human dimension of science (i.e. to think about these scientists as everyday people). They reported on one case study, which provided evidence of the positive effect of the use of mini-speeches and monologues coupled with dramatic activities that students participated in.

The evidence provided by the studies conducted by McGregor (2012) and by Precious and McGregor (2014), with regard to the teachers' efficacy to implement drama activities/strategies, agrees with the evidence from a study with student teachers, carried out by Tanriseven (2013). This study investigated specifically the effect of student teachers' sense of efficacy with regard to planning, implementing, and evaluating drama activities in the context of preschool education. The study was conducted with 52 undergraduate (second year) students at Mersin University. A quasi-experimental design model (with control and treatment groups and pretest and posttest conditions) was used. The student teachers of the experimental group implemented and evaluated drama activities that they themselves design, while those in the control group simply used drama activities as classroom applications. Tanriseven used the 'Sense of Efficacy Scale Relating to Use of Drama in Education' scale, developed for this study. The results showed that the student teachers from the treatment group improved their efficacy relating to the use of drama in education, which, in turn, is evidence that the development and evaluation of drama activities by the teaches themselves can have a positive effect on their sense of efficacy.

With regard to intervention studies per se, Hendrix, Eick, and Shannon (2012) conducted action research to explore the effect of integrating creative drama in an inquiry-based elementary science programme. The participants were 38 children in an upper elementary enrichment programme, who were taught by the teacher-researcher the Full Option Science System (FOSS) modules of sound (fourth grade) and solar energy (fifth grade) with the integration of creative drama activities in the treatment classes. In order to assess the effect of the integration of drama on learning outcomes, they used a mixed method approach. In regard to the qualitative data of their study, Hendrix et al. (2012) looked into the treatment group students' perceptions of how the use of creative drama helped them to learn science. The analysis of those date showed that children enjoyed participating in the drama activities as they thought them to be a fun way to learn science and that creative drama activities helped them to remember and also think about science. As regards the quantitative data of the study, their analysis revealed that the students who were taught science with the inclusion of creative drama showed greater understanding of the science content than the students in the control group taught without the inclusion of creative drama. In other words, the combination of inquiry science and creative drama was found to be more effective than inquiry science alone.

Even though the data from this study do not provide evidence that the inclusion of creative drama in a science inquiry programme increases the students' attitudes towards learning science (compared with an inquiry-based instruction without the use of creative drama)—indeed it was found that the students of both groups showed a small decline in attitude towards science—the positive perceptions expressed by the students of the treatment group and their greater understanding compared to that of students' from the control group need to be acknowledged. This is the reason why Hendrix et al. (2012) recommend the inclusion of creative drama as an extension to inquiry science.

Another intervention study by Varelas et al. (2010) also supports the positive benefits of learning science through drama. Their study was quite interesting as its primary aim was to explore the role of drama as a method that fosters multimodal learning. More specifically, they explored how young children (grades 1–3, from six urban elementary schools) construct meaning of scientific concepts during the teaching of a unit on matter and a unit on forests, through the use of gestures, talk, and movement.

Basing their theoretical framework on the three metafunctions of communicative activity (i.e. ideational, interpersonal, and textual), Varelas et al. (2010) were able to study the process of meaning construction as children gestured, talked, moved, and positioned themselves in space. They found that embodied meaning was central to their attempt to engage with science and that children's whole bodies became central tools used to represent the 'imaginary scientific world'. They also found that children's bodies operated on three levels: as material objects that moved through space, as social objects that negotiated classroom relationships and rules, and as metaphorical objects that stood for various physical entities, such as plants, animals, or water molecules in different states of matter.

The effect, however, of drama activities on the learning process has also been explored in the case of older students. Pongsophon et al. (2010) investigated how process drama can help secondary students from a school in Bangkok, Thailand, to develop scientific literacy in the context of global warming. The 31 students of their sample participated in a 7-day workshop, during which the students were actively engaged in a series of activities, including lab exercises, review of global warming issues presented in selected printed media, a documentary film, as well as instruction on acting, elements of drama, and storytelling. The students developed a script and performed the drama to the public. The results (based on data from informal interviews, questionnaires, participant observations, and student journals) showed that the students developed an understanding of the basic ideas related to global warming as well as conceptual understanding of the causes, processes, and consequences of global warming. Pongsophon et al. also report that the students' views about the solution of the problem of global warming were both creative and critical.

Of particular interest may be studies that provide evidence for the effect of drama/role play on learning biological and ecological concepts. Dennis et al. (2014) explored whether evolution and inheritance, which traditionally are considered secondary science topics, can be introduced to grade 6 students (ages 10–11).[15] Despite the complexity inherent in the above ideas, and the fact that evolution describes changes occurring over long periods of time (which is difficult for children to understand), it was found that drama in the form of role play provided a sequential learning experience which helped the children to progress in their thinking about evolution. And Bailey and Watson (1998), who investigated the advantages of using drama/role play with elementary school students (aged 7–11), found that structured role play (in the form of the 'eco game') was an effective way to help students understand ecological concepts, which more likely are not easily understood through first-hand exploration of habitats.

Interesting though are studies that investigate the effect of drama/role play on students' understanding of the nature of science. Cakici and Bayir's (2012) study is worth noting. This study investigated the effect of role play (i.e. portraying a scientist's life story) on young students' views of the nature of science.[16] The study, which was conducted at the Children's University of Trakya in Turkey, used as a sample 18 children of the age range 10–11 years. The intervention, which lasted

---

[15] The new National Curriculum in England expects grade 6 students to (a) recognize that living things have changed over time; (b) recognize that living things produce offspring of the same kind, but normally offspring vary and are not identical to their parents; and (c) identify how animals and plants are adapted to suit their environment in different ways and that adaptation may lead to evolution.

[16] The study focused specifically on the following six aspects which were deliberately included in the role play: science is tentative (e.g. Newton's different ideas about light and colour, Marie Curie's new ideas about atoms), science is empirical (e.g. Newton's construction of a reflecting telescope after many trials, Marie Curie's experiments with atoms), science uses multiple methods of investigation (e.g. Newton's insights within maths and astronomy by questioning and reasoning on the farm, Marie Curie's investigation of several substances by observing them in the lab), science is subjective-theory laden (e.g. Newton's different ideas from previous scientists' work, like Galileo and Kepler), science is creative-imaginative (e.g. Newton's revelation of the gravity law after watching an apple fall in the orchard, Marie Curie's development of the technique of investigating samples by electrometer), science is embedded in sociocultural context (e.g. Marie Curie's situation in which women were not allowed to study at universities in Poland, the dispersal of the University of Cambridge students due to plague and the return of Newton to the farm, the financial problems in Marie Curies' life).

10 days, consisted in role-play activities, regarding the life stories of Isaac Newton and Marie Curie. During the intervention the children both practised and performed. Using pretest and posttest assessment (the children had to respond to 16 open-ended questions), Cakici and Bayir found that children had more informed views about aspects relating to the nature of science in comparison with their views prior to the role-play activities. Even though the majority of children (about 80 %) had naive ideas about the nature of science, after the intervention there was a 70 % change in their views concerning the aspect of scientific method. However, there was also a 40 % positive change in children's views of the tentative, empirical, and creative/imaginative aspect of the nature of science and also a 50–60 % positive change in their views regarding the subjective as well as sociocultural dimension of science. Such results, despite the limitations that a 10-day intervention may have, are encouraging.

These findings are in line with a study by BouJaoude, Sowwan, and Abd-El-Khalick (2005), who explored the effect of drama on grade 10 and 11 students' conceptions of the nature of science. The time frame of this two-group design was 12 weeks, and the sample consisted of 14 students (experimental group) who participated in extracurricular drama activities aimed at developing of the concept of light by the enacting the lives of four scientists. At the end of the study, it was found that the students who participated in the drama activities possessed more informed views about the nature of science than the control group (consisting of 18 students who participated in other 'nondramatic' activities over the same period).

However, the effect of watching theatrical plays on science learning has also been investigated. Casey (2010) reports on his experience as an impersonator, who brings scientists to life for the purpose of making science an interesting and engaging subject. During his visits to schools (primary and secondary) around the UK, he plays various roles as Newton, Galileo, and Darwin and then interacts with the students, who can also ask him questions. This theatrical performance and the interactive experience that follows it appear to be engaging and conducive to learning.

Peleg and Baram-Tsabari (2011), on the other hand, studied the effect of the play 'atom surprise', that is, a play portraying various concepts relating to the topic of matter. They used a mixed method approach and investigated both the knowledge and attitudes of children (grades 1–6) who watched the play. The study was conducted in two elementary schools—one public and one private—in Israel. In both cases the play was presented as an enrichment activity to the entire student body. The analysis of data from questionnaires and in-depth interviews suggested that in both schools, children's knowledge on the topic of matter increased after the play with younger children gaining more conceptual knowledge in comparison with older students. It is also interesting to note that public school girls showed greater gains in conceptual knowledge than boys. Even though no significant changes in students' general attitudes towards science were found, students did demonstrate positive changes towards science learning.

Even though in Peleg and Baram-Tsabari's (2011) study, the children did not participate in any role-play activities, the effect of their watching something that they found interesting cannot be downplayed because there was no peer interaction.

As far as the development of conceptual knowledge was concerned, the use of that particular theatrical play was found to be an effective teaching/learning medium. Indeed, there is also evidence that movie watching can also have such an effect. For example, Cajigal, Chamrat, Tippins, Mueller, and Thomson (2011) report on how the movie *Eight Below*, which included action and thrilling scenes, captured high school students' attention and helped them develop an understanding of basic chemistry concepts, in addition to their engagement with thinking about the ethical issues and dilemmas embedded in the movie. Hadzigeorgiou and Garganourakis (2010) also report on the learning benefits from watching the movie *The Prestige*. The high school students who watched this film, as was pointed out in Chap. 6, developed a sense of wonder at and about Nikola Tesla's experiments on electricity, participated in self-directed inquiry about the concepts of DC and AC electricity, and began to view physics differently as a result of their investigations.

Interesting, and different from what has been reviewed so far, is a study at the university level, which also serves as evidence of the positive impact of drama/role play on student learning. Sloman and Thompson (2010) did their study with a sample of 86 undergraduate marine biology students at a university in England. The students in the sample were to participate in a large-group cross-year drama activity (all 3 years in order to explore the topic of funding in the context of scientific research as well as the links between science and society. In the drama, first year students role-played the general public, who had to decide which environmental research areas should be given priority for funding, while second year students played the scientists, who had to prepare their research proposals, and third year students were the research panel, who were going to decide which proposals would receive funding. Using questionnaires that were distributed at the end of the dramatic activity, Sloman and Thompson were able to gather the students' perceptions on the cross-year nature of the exercise and on the use of peer assessment and also their views on the dramatic activity itself. They reported that the majority of students thought that their knowledge of the topic of research funding increased, as a result of the drama they participated in and also that they enhanced such skills as critical thinking and communication. In addition, the students felt that their participation in the dramatic activity increased their confidence in presenting work to others and that they were comfortable with the use of cross-year peer assessment.

7.8 Concluding Comments: The Need for Integrating Art and Science

In light of what was discussed and the evidence from the research studies, 'artistic' and especially 'dramatic' activities represent one of the best approaches to integrating creativity, imagination, and science (both its content and its nature). In considering also the empirical evidence that inter-/multidisciplinary approaches foster

7.8 Concluding Comments: The Need for Integrating Art and Science

student learning (Park Rogers & Abell, 2007a, 2007b), the integration of art with school science becomes imperative.

It should be pointed out that art can help 'bring out' such elements as feelings, intuition, and beauty, which, as was discussed in Chap. 5, are core elements of extracognition. John Dewey's (1934) view that art has the power to 'quicken' students from the slackness of their routine provides food for thought to all those involved with the educational process, from education secretaries and stakeholders to science teachers and parents. Given also the importance of multimodal representations for understanding (see Moje, 2007; Tytler, Prain, Hubber, & Waldrip, 2013), art-based science activities deserve particular attention as they can provide students with opportunities for both visual representations (e.g. in the form of photos, drawings, models) and enactive representations (e.g. in the form of drama/role play).

The crucial importance though of 'artistic' science activities can be seen in the opportunities they give students, not simply for imaginative engagement but, as Dewey (1934) argued, opportunities for 'seeing and feeling things as they compose an integral whole' and especially opportunities for a 'large and generous blending of interests at the point where the mind comes in contact with the world [...] when old and familiar things are made new in experience' (Dewey, 1934, p. 267). In short, art-based activities are worthwhile educative activities that can foster a change of outlook on science and the world. And this is a significant instructional goal that could also be considered an important aim of education itself (Peters, 1973).

ERRATUM

Imaginative Science Education

The Central Role of Imagination in Science Education

Yannis Hadzigeorgiou

© Springer International Publishing Switzerland 2016
Y. Hadzigeorgiou, *Imaginative Science Education*,
DOI 10.1007/978-3-319-29526-8

DOI 10.1007/978-3-319-29526-8_8

In "Reference" section of the back matter, the author reference is correct but name of the author is to be updated. The name should be listed as Gilbert, A.

In Chapter 6 titled "'Wonder-Full' Science Education", the in-text citations referring to "(Andrew, 2013)" on page 143 and 155 is also changed to "(Gilbert, 2013)".

The updated original online version for this book can be found at
DOI 10.1007/978-3-319-29526-8

Conclusion: Towards an 'Imaginative Future' of Science Education

Ever since Dewey (1938, 1966) became concerned with the notion of educative experience, there have been many scholars and educators who believe that schools are not achieving their purposes (see Egan, 2010, for an excellent discussion of the contradictory educational goals that schools strive to achieve). Jamshed Bharucha, professor of psychology at Tufts University, in considering new research findings from the field of human cognition, has voiced a similar concern about what we have been doing in our schools:

> *The more we discover about cognition and the brain, the more we realize that education as we know it does not accomplish what we believe it does [...] How much of the learning persists beyond the time at which acquisition is certified? How does learning impact the lives of our students?* (Bharucha, 2006, pp. 1–2)

The imaginative approaches discussed in this book, although not revolutionary, can reform science teaching at all levels of education. They have the potential to make science learning an activity that encourages deeper involvement with content knowledge and also an appreciation of its cultural value. It is this deeper involvement with, and appreciation of, science that can have an impact upon the lives of students. But in order for this to happen, we should all change the way we view learning and education in general. We should be reminded once again what the educational philosopher R. S. Peters pointed out, namely, 'To be educated is not to have arrived at a destination; it is to travel with a different view' (Peters, 1973, p. 20). It is high time we thought about such a possibility and therefore seriously considered the crucial role of imaginative engagement in the teaching/learning process.

Imaginative engagement, as discussed and illustrated in the preceding chapters, can indeed be a potent stimulus for learning. In considering the problems that beset science education, one is tempted to propose imaginative approaches as alternatives and/or complements to traditional/mainstream science education. Given the shortcomings of the conceptual change approach and the limitations of the sociocultural perspective (Schulz, 2009, 2014), imaginative approaches need to be more seriously considered by the science education community. The reason behind such a consideration is that all imaginative approaches (e.g. learning science through storytelling, through the experience of wonder, through dramatization, poetry, the development of romantic understanding) are not in opposition to the ideas behind both the conceptual change approach and the sociocultural perspective on science teaching and learning. However, what must also become clear is that it is imaginative engagement—a common feature of all imaginative approaches—that increases the possibilities for students to have aesthetic experiences and thus to become involved with science, especially with its ideas. The empirical evidence, although limited at the time of writing, is quite encouraging. Further research, of course, on the effectiveness of imaginative approaches is imperative, but it is important to point out that, at least, according to the evidence cited in the previous chapters, imaginative approaches do tap students' 'imaginative mode of awareness' (Greene, 1978, p. 186). In considering, in fact, that 'epicazation' (i.e. the creation of epics and grand narratives in the history of the field of science education) and 'novelization' (i.e. the renewal of narratives) are 'at the core of the imagination of science in science education' (van Eijck & Roth, 2013, p. xiv), one can justifiably recommend imaginative approaches to teaching and learning school science.

It is well known that over the last two decades, the idea of science as another world (Costa, 1995), another culture (or subculture), has been specifically stressed from the cultural perspective on science education (Aikenhead, 1996, 2002). Such a perspective has been very helpful. Indeed, the appropriation of concepts from cultural anthropology has helped science educators understand several problems associated with the teaching and learning of science as a culture. However, there are some questions to be raised: Is science a culture that is beyond most students' grasp? Are border crossings extremely difficult, if not impossible, for some students (see Hadzigeorgiou, 2005b; see also Giroux, 1992 for a better understanding of the idea of 'border crossing')? How about providing students with opportunities for deeper involvement, through storytelling and creative drama? How about fostering an aesthetic experience? And how about helping students understand science 'romantically' by evoking in them a feeling of a sense of mystery and wonder? Such feelings, in fact, of mystery and wonder, can help them view science as a 'grand adventure'—to use Feynman's (1964) own words. It is this feeling of 'grand adventure' that has the potential to facilitate 'border crossings', and that is why the curriculum should include such elements as mystery and wonder, and if possible awe, although in the context of education this would be more difficult.

The value therefore of imaginative approaches in the process of attracting students to science needs to be seriously considered. Even students from marginalized groups can be attracted to science and thus helped to expand their cognitive horizons (Hadzigeorgiou, 2015), for it is crucially important to bring those 'outsiders' in (Brickhouse, 1994, 2001, 2003). Given the recent evidence that the ability to study science is not gender related (Wang et al., 2013), the potential of female students to study not just life sciences but also math-based sciences should not be overlooked. On the other hand, if 'Science for All' is a vision and not just a slogan, then it should provide guidance for science education for *all* students. Therefore, the achievement of an important goal, that is, attracting the next generation of scientists and engineers, and also science educators and science teachers, should not be overlooked either.

In a recent study in the USA, Wang et al. (2013) found that individuals with both mathematical and high verbal abilities were less likely to pursue STEM careers in comparison with individuals with high mathematical abilities but moderate verbal abilities. And they found that those with the highest scores on both mathematical ability and verbal ability (according to their SAT scores) were women. But if gender differences in attitudes towards science are not due to lack of ability on the part of women (Wang et al.) and if women still lag behind in math-based science fields, such as engineering, geophysics, and physical science (Ceci, Ginther, Kahn, & Williams, 2014), then it is high time that attention was paid to the task of attracting women and all the 'outsiders' to science. Female students with both high verbal abilities and high mathematical abilities can find science more attractive if imaginative approaches like storytelling, dramatization, and poetry, which require verbal skills, are implemented in the science classroom. It is these approaches that can be really emotionally engaging for those students. That is why imaginative approaches need particular attention by both curriculum designers and science teachers.

Even though imaginative/creative approaches to school science have been gaining some momentum (e.g. Begoray & Stinner, 2005; Dorion, 2009; Hadzigeorgiou, 2012; Klassen & Froese-Klassen, 2014a, 2014b; Pantidos et al., 2001, 2014), the notion of imagination is not *explicitly stated* or acknowledged. Over the past decades, we have mainly focused on the paradigmatic (or logico-mathematical) mode of thinking. It is time that more attention was paid to the narrative mode. Even in the context of early childhood science education (see Cabe Trundle & Sackes, 2015), the role of the narrative mode needs to be recognized and the value of the imagination be reclaimed. Educational theorist Kieran Egan's view should be considered:

> *A feature of young children's mental life that is commonly asserted as an implication of research on their logico-mathematical thinking is that their thought is perception-dominated. If we focus instead on their imaginative lives we can see rather an enormously energetic realm of intellectual activity that is conception-driven.* (Egan, 1999, p. 9)

There are, of course, obstacles to providing learning opportunities for imaginative engagement with school science, especially opportunities for the experience of wonder and aesthetic experiences in general. Ministerial interest in accountability and 'performativity' is perhaps the most difficult obstacles to be overcome. But it would be worth the effort, as learning science can become both a fun and a truly educative experience. It is true that oftentimes, students, and especially young children, experience wonder and have aesthetic/transformative experiences but outside and independent of schools. Indeed, a simple walk through a forest, during which they listen, observe, touch, and smell, can be an aesthetic/transformative experience, as can an experience at a playground and science museum. The question, therefore, is how to remove the obstacles so that chance discovery, surprise, or even astonishment are experienced in the classroom as well. It is for this reason that imaginative approaches to school science deserve particular attention. This book, while it does not say anything about how to remove the aforementioned obstacles to the implementation of imaginative approaches, provides a repertoire of possibilities that can help us reimagine school science education.

We need, anyway, to pay more attention to the workings of the imagination, given that science, at all levels, is an imaginative activity. It would be indeed an injustice not only to the nature of scientific inquiry in general but also to the discipline of school science itself, to identify imagination only with certain open inquiry activities that explicitly call for a divergent kind of thinking and/or with science fiction. Imagination, as the sine qua non of science, is required for effective learning in all areas and branches of science. The fact that science deals with unobservable entities and also with unobservable, and, many times, not easily observed processes, makes imagination a central mental ability that should be tapped and further developed in the context of school science education, at all levels.

The history of science, as was discussed in Chap. 1, has provided ample evidence that scientific discovery is indeed an imaginary endeavour. It is for this reason that creative imagination is considered a defining element of the nature of science and not just of science fiction. Thus it would be a great misconception to link creative imagination solely to science fiction. Science fiction, of course, has a special value (see Allday, 2003; Brake & Thornton, 2003), not only because it can help students develop reading and writing skills but also because it can provide excellent opportunities for discussions, that is, opportunities for 'talking science' (see Lemke, 1990, 2001). In addition, getting students to discuss hypothetical principles which, however, do not contradict accepted principles is an important step towards scientific discovery. True, it might be argued, it is naïve to even talk about scientific discovery in the context of science education. But one should also be reminded that the problems that humankind has to solve, such as the depletion of energy sources and overpopulation, will require people with imaginative powers who, at the time of writing, are probably in elementary and high school, in various parts of the world. And if science fiction, as Carl Sagan himself confessed, led him to science (Sagan, 1980, p. 3), then science fiction should begin to be taken seriously in science education. It would be really hard to argue that this world would be a worse place with no Sagans, Asimovs, and Clarkes.

It would be really hard to argue that the great physicists, who come up with new ideas, and propound new theories and change our perspective on the world, do not engage imaginatively with physical reality. Did Einstein himself not approach the world of atoms and of planets and outer space imaginatively since, as he commented, 'The most beautiful experience that we can have is the mysterious' (Einstein, 1949)? Did Feynman not view all natural phenomena in an imaginative way? Certainly, the comparison between those two scientists and students may be unrealistic, but, I would argue, it is not.

Of course, imaginative approaches to science do not, and should not, downplay the crucial role of dialogue, argumentation, and generally discussion. Indeed science understanding requires an ongoing discussion. It would be entirely naïve on anyone's part to believe that capturing and engaging the students' imagination is sufficient to enable them to understand science. For example, it would be naive to believe that just the presentation of an idea in a way that it conflicts with everyday experience is in and of itself sufficient to develop conceptual understanding. Although conflict with everyday experience or even conflict with accepted beliefs is necessary and in fact crucial in capturing the students' imagination, it is what comes next that determines whether students will develop an understanding of science. The way questions are posed, the opportunities given to the students to respond, and the discussion that ensues are all crucial for understanding science. Even an exciting and indeed 'wonder-full' experience in the classroom or at a science museum, which has the potential to become an 'aesthetic experience', will remain simply an ordinary experience if it is not followed by a discussion. The historical development of ideas, as these are presented through the various exhibits, should be followed by a discussion of the possibilities these ideas have opened, and can open, for humankind. It is indeed this discussion that has the potential to expand the students' intellectual horizons.

And speaking of intellectual horizons, we have to seriously consider the possibility that capturing the students' imagination can make the content of science (i.e. its abstract ideas) more meaningful. Even remote and apparently irrelevant content, such as stellar explosions and black holes, can also become meaningful (Egan, 2005). In actual fact, some of the most abstract ideas of contemporary science (e.g. annihilation, pair production, black holes) can be approached and become understood even by high school and elementary school students mainly through narratives, and through the engagement of their imagination in general (Hadzigeorgiou, 1999, 2001, 2002b). Whether or not one accepts Lewis Thomas' view that science education, *at all grade levels*, should start from cosmological mysteries (see Chap. 1)—something that necessitates a restructuring of science curricula worldwide—imaginative engagement could be given a more prominent role in the current science curricula. Topics that foster imaginative engagement, even cosmological, and other mysteries, could be featured in these curricula alongside the 'standard' science content. Moreover, imaginative engagement, in any form, within the current 'standard' curriculum, can be a good starter (see an example in Appendix I) and can be incorporated in all teaching/learning models (Stefanich & Hadzigeorgiou, 2001).

Giving imagination, on the other hand, its proper place in science education can also help reaffirm the value of liberal education. Recent reform efforts in science education have placed an emphasis on the notion of application and problem-solving. Although applications and problem-solving presuppose the workings of the imagination, an emphasis on presenting science mainly through applications (technological and practical) and problems, because they are relevant to students' lives, may help consolidate an instrumentalist view of science. And, perhaps, worse, it may give students the impression that everything can be reduced to a problem, and all that matters is problem-solving. It is of note that Osborne, Simon, and Tytler (2009), in discussing the topic of attitudes towards school science, point out that the only defensible position from which we might justify the study of science, as a compulsory subject for all students, is a liberal one, that is, one that views science as a subject that can help students develop the skills of thinking insightfully, critically, and creatively and generally as a subject that can enrich their life (see Hadzigeorgiou, 2015).

Notwithstanding the importance of helping students to become aware of the relevance of science to everyday life, and also the importance of problem-solving, we have to bear in mind that such an instrumentalist approach to science education may undermine or limit the possibility for students to come to appreciate science as one of the various ways to know the world: to appreciate the fact that it has something to say about some matters of significance (Jenkins, 1996, 2007; see also Hadzigeorgiou & Schulz, 2014; Matthews, 2015). Moreover, an imaginative approach to school science that attempts to capture the students' imagination, through mystery, wonder, and awe, may help students develop intellectual humility, that is, an intellectual attitude that is indeed crucial in today's world (Paul & Elder, 2013; Piersol, 2014).

Although there are no recipes about how a student learns science, certain ideas—as research over the past three decades has shown—such as preconceptions, misconceptions, conceptual change, social context, and dialogue should be considered in planning curriculum and instruction. These, no doubt, are important ideas, but they have represented a limited view of how students learn science, especially physical science (Hadzigeorgiou & Schulz, 2014). They should certainly be considered in developing more holistic models of learning, but they should also allow for a place for the imagination. For example, the social dimension of science, in fact, the idea that 'science is itself constitutively social' (Woolgar, 1993, p. 13) should neither 'obscure' nor downplay the imaginative element inherent in scientific inquiry. The discursive approach to science learning and generally the recent emphasis on the social nature of science learning should provide opportunities for imaginative thinking. Egan's (1999, 2005) idea, that the neglect of the imagination, especially in the early years, might be a reason for the students' later failure in science, should be given more serious thought. Most likely imagination was a missing or downplayed factor in the process of conceptual change, even though it should have been a crucially important element in their education. Mary Midgley (1992, p. 24) has eloquently argued for this crucial importance of imagination:

> *Attending to the workings of imagination is not a soft option, and it is not a mere gossip [...] a harmless, licensed argument. It plays a part in shaping the world-pictures that determine our standards of thought – the standards by which we judge what is possible and plausible.*

Appendices

Appendix A

Two Stories for Current Electricity and One for Mechanics: The Main Historical Events, the Main Ideas

The Galvani-Volta Controversy[1]

The discovery of current electricity is quite interesting for a number of reasons. Perhaps the most obvious, and perhaps the most important from a societal point of view, is the production of electricity by mechanical means. A not so obvious reason is that the contribution of research came from at least three different countries: Italy, Denmark, and England. Although it was in England, the first week of September 1821, that electromagnetic induction was discovered, through a series of experiments conducted by Michael Faraday in the basement of the Royal Institution, the history of current electricity has its beginnings in Italy in the late 1700s. At that time, two Italian scientists, Luigi Galvani, a medical doctor, and Alessandro Giuseppe Volta, a physicist, became involved in a great dispute over the nature of current electricity.

As the story goes, Galvani made his discovery by observing that the legs of a frog twitched while he was dissecting it. Galvani was startled since the frog was dead: what made the frog's legs twitch? Quite accidentally Galvani also observed that the twitching took place both during electrical storms and when the frog was placed next to a static electricity machine that gave off sparks. This was a significant discovery since Galvani made the following inference: the twitching must be connected in some way to electricity.

Being a scientist, Galvani embarked upon a set of experiments and he was able to get the legs twitch under good weather conditions and away from the static

---

[1] Based on Burns (2003), Googindg (1985), Hamilton (2002), Meyer (1971), and Pera (1993).

electricity generator. So he concluded that the source of the twitch must be something inside the frog, and he put forward his theory: the source of the twitch is a form of electricity, which is a kind of 'animal electricity'.

Volta, however, challenged Galvani by proposing a different theory: the source of the twitch must be outside the frog! He performed experiments that showed the twitch could be produced only when two dissimilar metals were used in the process of dissection. Therefore, the twitch did not come from the legs but from the two metals that were touching the two legs. It must be a kind of 'contact electricity'.

Galvani, following this challenge, undertook more experimental work and did show that the legs could twitch even when two similar metals were used for the dissection of the frog. But Volta was insistent that a dissimilarity in those obviously similar metals had to be involved. Galvani did more experimental work and produced startling evidence: the frog's legs could twitch in the absence of any metals (by just touching two end nerves)! It was a clever counterattack on the part of Galvani to experiment without the metals, since his theory was based on the idea that the source of the twitch was inside the frog and not outside of it. Who could object such evidence? Apparently only Volta himself who made a cleverer counterattack. If Galvani got rid of the metal, he got rid of the whole frog!

Volta did experiments by using two different metals in a liquid. The simplest combination that could be used in those experiments was a piece of copper and a piece of zinc, both the size of a coin and a cotton pad (which had been previously immersed in a solution of sodium chloride) in between them. Although the electricity that could be generated by such an element was small, a series of much bigger elements (e.g. a box filled with sulphuric acid in which a number of pairs of bigger pieces of metal were immersed) could produce a larger amount. So Volta managed to produce a much larger amount than the amount produced through the twitching of the frog's legs. Moreover, scientists and technicians could reproduce quite easily Volta's experimental work (without the surgical skills that Galvani's technique might have required).

Of course, one could think of many applications of such discovery. But there were some problems. The apparatus consisting of Volta's elements was quite heavy (e.g. if it had to use a source of energy for a bicycle, where would it be strapped?) and even dangerous, since the acid could cause accidents and even prove fatal.

In the meantime, Galvani had to admit his scientific defeat and he felt humiliated, disgraced. In less than a year, he died. In looking back on this controversy, one can see its positive side, that is, the competitive force that drove scientific research in Italy. But in looking into the present and even into future, one can also see that Galvani's death was an unfair consequence of his work. Today the idea that animal cells in general have their own electricity is well established and its applications, namely, the electrocardiogram and electroencephalogram, are widely used in the medical profession. The irony of it all, of course, was that both Volta and Galvani were right.

But while this debate with its sad ending was taking place in Italy and in France Charles Coulomb was discovering the repulsion between electric charges, in England, another scientist was about to discover the production of current electricity using moving magnets. His name was Michael Faraday. Originally a bookbinder, he was self-trained, and at the age of 21, he was appointed assistant to Humphrey Davy in the laboratory of the Royal Institution in London. Between October 1813 and April 1815, Faraday accompanied Davy, as his assistant, on a scientific tour of Europe. In Italy they met with aged Volta.

However, it was a discovery made in 1820 by the Danish natural philosopher Hans Christian Oersted that must have played a crucial role in the thinking of Faraday and his subsequent experimental work. Oersted, while giving a lecture to his students at the university, noticed that the needle of a compass changed direction when an electric current passed through a nearby wire. Oersted had made a great discovery: an electric current produces a magnetic field. Faraday, in reading the paper that Oersted had published on his discovery, commented as follows: 'I have very little to say on M. Oersted's theory, for I must confess I do not quite understand it'. However, on September 3 and 4, 1821, Faraday did try to duplicate Oersted's experiment in his basement laboratory at the Royal Institution.

In the course of his experiments, Faraday discovered that a suspended magnet would revolve around a current bearing wire, leading him to propose that magnetism was a circular force. He undertook a set of experiments which culminated in his discovery of electromagnetic rotation—the principle behind the electric motor. But Faraday went even further. Since an electric current could cause a magnetic field—something that Oersted had discovered—a magnetic field should also be able to produce an electric current. This was the principle of induction which Faraday demonstrated about 10 years later, that is, in 1831. It was a landmark in applied science since it made possible the generator and the transformer, thus turning the production and the transference of current electricity from a mystery to a reality. Current electricity became a commodity at the service of humanity.

The Genius Who Lit and Transformed Our World: His Life and Work

Nikola Tesla was born in the Croatian village of Smiljan, then part of Austro-Hungarian empire, in 1856.[2] The strange thing is that he was born at the stroke of midnight between July 9 and 10. His father, Milutin Tesla, was an Orthodox Christian priest, whose desire was that Nikola become a clergyman. Tesla, however, had completely different plans. During his later years at high school, Tesla became increasingly fascinated with the science of electricity, so he expressed his desire to become an electrical engineer. Although his father insisted that Nikola become a clergyman, the fact that Nikola had contracted a very powerful form of cholera shortly after his graduation from high school and that he had been bedridden for about a year with no hope to survive death made his father relent after Nikola said to him: 'Perhaps I will recover if you let me be an electrical engineer'.

---

[2] Based on Cheney (1981), Cheney and Uth (1999), Johnston (1983), Jonnes (2003), Lomas (2000), O'Neil (1992), Seifer (1998), and Tesla (1982).

Then Tesla began his college education at the Graz Polytechnic Institute, pursuing the study of the topic that fascinated him the most: electricity. He studied daily (most likely to impress his parents) from 3 am to 11 pm every single day. It was while he was studying at Graz that he first began thinking about AC. Tesla was able, even at that early age as a student, to 'perceive' that AC holds the key to the future in comparison with the electric current of the time known as direct current (DC). There was of course theoretical talk about AC, but no one could think how to use it in practice and how to make it useful in everyday life. But even Tesla's mere mention of AC in the lecture rooms brought scorn and he even enraged his professors by pursuing the idea of AC. Yet Tesla's passion for AC and his unbelievable persistence made him never abandon this idea. And if persistence is a characteristic of creativity, then Tesla's ingenuity and creativity, as we'll see, is due to his extraordinary persistence and willpower.

Tesla had realized that the challenge to make AC work in practice is due to the nature of AC itself. Because the direction of the AC changes many times per second (e.g. 50, 60, or much more), in sharp contrast to that of DC, which remains always constant, there is a problem: the constant change of the direction of current inevitably produces violent vibrations, which will affect the efficiency of every appliance, of every application. This was the reason why Tesla's professors poured scorn on the idea of AC. It was totally impractical. But Tesla had realized, through his persistence and his ability to visualize, that the nature of AC itself holds the key to the problem, because AC can be transformed. If a current can have that 'wavy' form, its characteristics (e.g. frequency, amplitude) may be changed. So Tesla continued to think about the problem of AC throughout his time at Graz Polytechnic.

Unfortunately, due to a stroke and subsequent death of his father, Nikola, without any funds for his tuition, had to abandon his study at the Graz Polytechnic. But his passion for electricity made him travel to the capital, Budapest, to secure a position as an engineer in company that was planning to install a telephone system. Unfortunately, once again, the telephone system was in its initial stage, so he was forced to take, instead, another job—a very poorly paid job—at the government telegraph office. But little by little, the stresses of his life led to a nervous breakdown, which manifested itself in a form of extreme hypersensitivity. (The ticking of a watch even several rooms away was painfully deafening, the footsteps of passersby sounded like earthquakes to him, so rubber cushions had to be inserted underneath the legs of his bed to absorb the vibrations, and light was painful both to his eyes and to his skin.) Tesla, however, did recover because of his extraordinary willpower and desire to live and continue the work on AC.

As part of his recovery process, Tesla took up walking in the Budapest Park, where, in fact, he tackled and solved the problem of AC. Unfortunately no one in Budapest was interested in his new, AC power system. Through an acquaintance and due to his incredible persistence, Tesla managed to get a job with the Continental Edison Company, a French company in Paris, which manufactured and sold motors, dynamos, and lighting systems under licence from Thomas Edison. This job was the beginning of a brilliant career and Tesla's first real opportunity to demonstrate his amazing intellectual abilities.

It was during his employment at the Continental Edison Company that Tesla tackled one of the most difficult problems concerning the lighting system and powerhouse of the Strasbourg railway station. Tesla, without making a single engineering drawing, managed to build his first AC motor and dynamo. This work was the foundation for the production of AC. The idea, of course, behind the production of AC is quite simple, as Michael Faraday had demonstrated: AC is produced whenever a coil (a series of loops) of wire is moved inside a magnetic field or whenever a magnetic field rotates around a set of stationary wire coils. We can visualize this, just as Tesla did. In both cases, there is motion (mechanical energy), which is transformed into electricity (electrical energy). The reverse principle is used by the electric motor. An AC running through a coil creates a magnetic field that changes direction many times a second, thus pushing periodically against the magnetic field of the magnet. This way the coil rotates.

So in order to produce electricity, we just need coils, magnets, and motion. It is as simple as that, but it took Tesla's genius and his extraordinary power of visualization to turn ideas into mental images and, therefore, to build the technology that is required to successfully run AC motors and dynamos.

Tesla, 1 year later, that is, 1884, having realized that Europe was not very interested in his idea of AC, and under the advice of Continental Edison's chairman, decided to travel to America and work for Thomas Edison. His adolescence dream, as he used to say, to harness the power of Niagara Falls, may after all be realized. So Tesla bought himself a ticket to New York and left on the first available boat. Upon his arrival, he immediately began working for Thomas Edison, who was really impressed at Tesla's skill as a troubleshooter and also at his incredible work rate. Usually Tesla started work at 10.30 am and finished at 5.30 am the following morning. Eventually Edison presented Tesla with a challenge: if the efficiency of the DC dynamos were increased by 25%, Edison would present him with a $50,000 bonus. Tesla was given a 2-month deadline. The amazing thing is not that Tesla kept to that deadline, but that he increased efficiency of some dynamos up to 50%.

However, when he asked for the bonus, Edison said that that was a joke—an American one—that Tesla did not understand. Unfortunately, once again, Tesla had believed in Edison's offer. Moreover he did not know that Edison was notoriously tightfisted. Tesla was infuriated by that kind of betrayal—he had felt that he had been betrayed by his own idol—so he immediately quit in disgust.

This was the beginning of a controversy between the two men that took the form of a battle, known as the 'War of Currents'. In an attempt to demonstrate the dangers of AC power, Edison sponsored an electrical engineer to travel the country electrocuting animals with both DC and AC. Because the frequency of AC confuses the heart, animals that were electrocuted by AC died, whereas animals that were electrocuted by DC victims were stunned but survived. Edison used these 'experiments' to contrast the danger of AC with the relative safety of DC. Although the effect of any type of electric current on a human being is very difficult to predict as it depends on a number of factors (i.e. condition of the skin, amount of fluids in the body, point of contact), Tesla had been experimenting with very high frequency currents, which, as Tesla showed, did no harm. With his theatrical flair, Tesla could draw sparks to his own fingers and even walk through sparks without being hurt. He had realized that the high frequency of the current kept it on his skin. It was this strange effect, known as skin effect, which made Tesla famous. He even sent sparks to the audiences, making people realize that there was safety, at least in the AC current used by Tesla.

However, Edison was bent on showing the bad effects of AC and smearing Tesla's name. Whether Tesla was aware that Edison was a supporter of DC because of the investments his company had made and the economic benefits that he was enjoying is pointless, since what really mattered to Tesla was his work on AC. So Tesla, while Edison was campaigning in order to show the superiority of DC, went into partnership with some investors who were interested in making arc lights. However, several business errors left Tesla penniless. The result? One of the greatest geniuses and best engineers in the world was reduced to digging ditches for about 2 years and for a dollar a day, suffering, at the same time, the indignities of the immigrants of that era.

But Tesla was not discouraged a bit. His passion for, and belief in the ultimate victory of, AC made him give lectures, and the result was that Westinghouse, who owned the Westinghouse electric company, heard about Tesla and invited him to his laboratory in Pittsburgh. So Tesla, about 4 years after he arrived in New York as an immigrant, began to work for Westinghouse. The year was 1888. Westinghouse's belief in Tesla's work on AC, on one hand, and Tesla's adolescence dream to harness the power of Niagara Falls, on the other, made Tesla focus on large-scale production of AC power. Having already the experience with building dynamos and motors, Tesla was feverishly working on generators, which were, in a way, big dynamos and worked on the same principle. Both Tesla and Westinghouse had realized that in order to generate big amounts of electricity by utilizing the energy of the falling water at Niagara, they would need gigantic electricity generators. The idea that a copper coil can turn around repeatedly inside a permanent magnet, if there is falling water on the coil, thus generating a current that flows through the coil, is quite simple. Yet it took Tesla's genius to build the technology that would make this simple idea work.

Tesla's, of course, passion and/or obsession with AC had their source in his awareness that DC was inefficient and incapable of transmitting power over long distances, simply because there is a progressive loss of power due to electrical resistance. (The further one gets from the source, the greater the loss.) Owing to the motion of electrons, which do not travel in straight lines inside the wire, and their constant collisions with the atoms in the wire, the resistance increases considerably, and the electrical energy is dissipated as heat (which is proportional to the square of the intensity of the electric current). Therefore, the longer the wires, the more electrical energy would be dissipated as heat. This is the reason why power stations had to be installed every couple of kilometres.

Tesla knew that AC could solve this problem since AC, due its wavy character, could be transformed. This meant that its voltage could be decreased or increased, according to our needs or the problem at hand. There were two advantages of this process, as Tesla thought: First, because for electricity to be efficiently used, it needs to be generated with low voltage and high current (to prevent sparking) and then transmitted at high voltage and low current (so there is little loss as heat) and converted back to low voltage for use in the home (to prevent electrocution). This cannot be done with DC. Second, since the direction of AC changes many times a second, there is a fascinating side effect, known as the 'skin effect', taking place. This means that the current only flows through the surface of the wire. Therefore, losses due to resistance are greatly reduced and AC power stations have a much greater range. Tesla had been already experimenting with high frequency currents, which, due to their high frequency, were not doing any harm. So what Tesla really needed at that point was the technology, the electrical device that would allow him to transform AC. Luckily enough, the Westinghouse company had that technology and Tesla worked to improve it. That electrical device was the transformer.

A transformer consists of two coils of wire wrapped around a central iron bar (core). The first coil consists of a few dozen turns of copper wire, while the second coil consists of a very large number (even many thousands) of turns of much thinner copper wire. When Tesla fed a current in the first coil, a powerful magnetic field was formed. When he cut off abruptly the current, the magnetic field ceased to exist. Tesla had realized that since the energy is conserved, the energy of the current in the first coil cannot disappear. So the released energy from the first electric current in the first coil will appear as energy in the second coil. The amazing thing, however, is that because the second coil has many more turns, the release causes an increase in the voltage. Tesla could get an output of several thousands, even millions of volts, a record that has not been broken ever since. In other words, the voltage output in the second coil is voltage input in the first coil times the ratio of the number of turns in the second and first coil: a simple mathematical formula, one would say, which, however, held the key to the future.

Although Tesla had already improved the technology behind dynamos and motors—it is said by some engineers that Tesla's AC motor is one of the greatest inventions of all time—and although his work with the Westinghouse also led to the improvement of transformers, Tesla's ingenuity, persistence, and his exceedingly long hours of work resulted in a device known as the Tesla coil. This was like a regular transformer, but instead of iron core, there was an air core. Tesla realized that this would allow him to produce extremely high frequency currents and extremely high voltages. This device was meant to be the basis of Tesla's idea of wireless transmission of electrical power.

In the meantime—and while Edison was still campaigning against Tesla and Westinghouse, electrocuting animals—Tesla astonished the world by demonstrating the wonders of alternating current electricity at the World Exposition in Chicago in 1893. AC was used to light the whole area of the exposition, and Tesla demonstrated the safety of high frequency AC by sending sparks to the audience that was left totally unharmed. He was using the Tesla coil.

Then came the final victory of AC over DC. Two years after those demonstrations in Chicago, that is, 1985, the Westinghouse company was successful in generating and transmitting power from Niagara Falls to Buffalo 34 km away. The water was used to turn big coils that were inside huge magnets, and then, through the use of transformers, the voltage was stepped up, so that the current that would flow in the wires and would transmit electrical power would decrease, and thus the losses would decrease. This was a heroic achievement at the time, it received world coverage, and Tesla was praised as a world hero.

However, a mysterious fire, right after the success at Niagara, destroyed completely Tesla's laboratory in New York, which he had set up from the money that he had received from Westinghouse. The mystery becomes bigger since the fire broke out on the eve of an experiment to demonstrate a radio-controlled boat on the Hudson river.

Tesla was not insured, so he lost everything, except for the ideas. Owing to his amazing mental abilities, all the ideas and models had remained intact in his mind. But now was the time to go west. Managing to secure funding, and free electrical power, Tesla set up a laboratory at Colorado Springs. After his justification regarding the superiority of AC, it was high time he pursued his humanitarian and global vision: sending free energy to all people, no matter where they lived on the planet. It was a big, impossible idea, scientists would say, but Tesla's ingenuity had many ideas in store. Tesla felt he was now ready to experiment with his idea of wireless transmission of electrical power.

The idea of wireless transmission emerged when Tesla realized that such an idea is indispensable when interconnecting wires are impossible or simply inconvenient, or dangerous. Although air could be used as a medium to transmit energy, Tesla had observed that the Tesla coil could make light bulbs light from many metres away. This meant he was transmitting electricity through the air. But Tesla thought that, since the transmission of power over long distances would result in significant losses, instead of using air as a medium, he could send electrical energy through the ground. By experimenting, Tesla found that the ground, if he charged it highly enough, could become a conductor of electricity. So, the planet—the entire planet—could be transformed into an imminence transmitter of electrical energy.

That is why at Colorado Springs, Tesla constructed an enormous power transmitter. This was a massive Tesla coil and on top of it an antenna about 70 m tall. This transmitter could generate voltages up to 100,000,000 V. The people of Colorado Springs, of course, were naturally curious about what Tesla was doing in his lab, since no access was allowed. The 'KEEP OUT – GREAT DANGER!' sign made them more curious. But they sensed that something extraordinary was taking place, since they could see sparks leaping from the ground as they walked the streets, the grass around Tesla's lab glowing with a bluish light, and metal objects near fire hydrants drawing small lightning bolts.

One night, in one of his experiments, Tesla pumped electricity into the earth's surface, and he was able to light 200 lamps without the use of any wires from a distance of about 40 km!! But this was just the beginning. Late one night in the fall of 1899, Tesla's tower pumped 10 million volts into the earth's surface. The current raced through the surface of the earth at the speed of light, and, as it said, it bounced back and returned to the lab. On its return, the current was weakened a lot, but Tesla was happy enough that his idea was working. At that night, Tesla was able to make the longest man-made lightning—that was an arc of lightning. The thunder could be heard at a distance of 30 km, and the whole experiment resulted in a blackout. After that, the company that was supplying free power stopped doing so.

Tesla, after he could not receive free energy, headed back to New York to continue his work on the wireless transmission. He managed to set up another lab, with another tower on Long Island, which was bigger and better than the one in Colorado Springs. J. P. Morgan, the wealthiest person in America, agreed, after Westinghouse passed on Tesla's idea, to sponsor Tesla's work, supposedly on wireless communication and not on wireless transmission of electrical energy.

He was experimenting there, but a number of accidents made Tesla run out money. Morgan, after those accidents, and after hearing from Tesla that the experiments were on wireless transmission of energy, refused to give him more funds. That was Tesla's last opportunity to secure funding for research on 'free energy to the whole world'. Thus, little by little, he declined into obscurity. He stopped his public appearances, he had very few friends, and several times he was evicted from hotels because of his habit of keeping pigeons in his room. Nikola Tesla died of heart failure in his hotel room, in New York City, on January 5, 1943.

The questions why such a genius went into decline, why his humanitarian/idealistic vision was not realized, and why he did not receive the credit, let alone the distinction he deserved, can have a couple of answers. Perhaps, the most reasonable one may be that because he did not have a university degree, he could not be accepted by the academia. One may also say that he was not a good businessman. However, one fact remains. He was a genius who worked all his life with one passion and one vision: electricity to all people on Earth. We should remember him as a genius, with over 700 registered patents—even the US Supreme Court, right after his death, hold Marconi's patent for the radio invalid and recognized Tesla's contribution to the invention of the radio technology—as a person with heroic qualities, all of which helped divert the course of human history and thus change our world.

The Birth of the 'New Science' of Motion: The Contributions of Galileo and Newton

Galileo Galilei, the son of a musician, was born in Florence in 1564.[3] This was also the year William Shakespeare was born and Michelangelo died. As a young child, he had many interests and the scope of his activities was quite broad. As he was growing up, he developed a keen interest in mechanical devices and also kept a workshop at his home. His mind was always busy with ideas; he never missed an opportunity to think about what was going on around him. A characteristic example is when Galileo, as a student at the University of Pisa, which he entered at the age of 17 to study medicine, went to attend mass at the cathedral. He must have felt a bit bored, so he started to observe things around him. So he noticed that the time of a full swing of the lamp that hung from the cathedral's high ceiling was independent

---

[3] Based on Arnold (1997), Ball (1983), Clavelin (1974), Goudsmit, Clairborne, and The Editors of Life (1967), Klein (1954), Mandelbrote (2001), McAllister (1996), and White (1997).

of the arc of the swing (i.e. both narrow and wide arcs required the same amount of time to be completed). Galileo, measuring the time of a full swing with what he had readily available, that is, his own pulse beat—let us not forget that at the time he was a student at the University of Pisa—discovered an important physical law concerning pendulum motion: that the period of a pendulum does not depend on the amplitude of the swing (but only on its length). And being really interested in mechanical devises, he had an idea to apply the idea of pendulum to time keeping and thus use pendulums in clocks in order to make them more accurate.

Galileo, however, while a student the University of Pisa, developed an interest in mathematics, especially geometry. So he received private lessons to improve his knowledge about mathematics. He made such a great progress that in 1589, that is, at the age of 25, he became a lecturer of mathematics at the University of Pisa, while in 1592 he was awarded the Chair of Mathematics at the University of Padua. Galileo was lucky—not that he did not deserve that position—for the reason that he had applied for that position because of some financial difficulties. It was there, however, that he did his revolutionary scientific work and his famous experiments on the motion of objects.

Contrary to Aristotle's method, namely, to explain the causes of phenomena, such as the fall of an object to the ground, Galileo focused on the mathematical variables involved in a phenomenon and their relationship. Such a relationship described what took place during a phenomenon (e.g. free fall), but not why it took place. Galileo's new mathematical approach to the study of motion made him focus on certain parts of the phenomenon and totally ignore some others.

His ingenious experiments showed that gravity was a characteristic property of the Earth. By using a heavy plank of wood, in which he had cut a completely straight groove, he was able to calculate the time taken by a bronze ball to cover different distances along the groove. He then extrapolated to the vertical plane, and he concluded that free fall was an accelerated motion in which equal increments in velocity take place in equal time intervals. He defined accelerated motion as a motion in which velocity is proportional to time. Any other possibility (e.g. velocity being proportional to the distance, proportional to the square of time) would lead to a logical inconsistence.

In those experiments with the rolling of bronze balls down inclined planes of various lengths, Galileo also showed his ingenuity when it came to measuring time. What he actually did was to measure time by measuring the weights of the water in a cup as the watch and chronometer had not yet been invented. So he measured time by weighing the water dripping from a clepsydra into a cup. By comparing the weights of water in that cup, he was able to compare the times taken by a body to roll down inclines of various lengths.

He realized from these experiments with the inclined planes that velocity was not enough to describe motion. So he found it necessary to consider the change in velocity with time, that is, acceleration. But this idea was really revolutionary in the seventeenth century. However, a genius from England was to incorporate it in his theory of motion some years after Galileo's death.

In his book *Discourses and Mathematical Demonstrations Concerning Two New Sciences*, which took him 30 years to complete, Galileo made it clear that when he considered the free fall of a stone, the increments of speed would take place through a simple rule. Thus, he suspected that there was a universal law—a constant—that was later identified by Newton as the force of gravity. And in this way Galileo also proved that the distance of a free-falling body varies with the square of the time, and it does not depend upon any property of the falling body (e.g. heavy, light). Those two mathematical proofs led to one conclusion: both heavy and light bodies fall with same speed and reach the ground simultaneously if dropped from the dame height (and if the air resistance is negligible). According to a story—about which there is no historical evidence—Galileo also dropped balls of various sizes and weights from the tower of Pisa in order to test his ideas about free-falling objects. Thus, Galileo disproved Aristotle's idea about free fall. Moreover, he had evidence that also in another experiment, Galileo placed two inclined planes in such a way that they formed an angle (i.e. in the form of a V) and began to observe the behaviour of a rolling ball that was released from rest from the first plane. Using his imagination, he concluded that in the ideal case in which there was no friction during the motion of the ball (i.e. if the planes were very smooth and there was no air resistance), the points of release on the first plane and instantaneous rest on the second plane would be at exactly the same height above the horizontal plane. And using again his creative imagination, he visualized the ideal case in which a ball rolls down a frictionless inclined plane and then onto a frictionless horizontal surface. He concludes that this ball will never stop and will continue to move on the horizontal surface forever.

In this way, Galileo discovered an important law, namely, it is the continuous application of a force that makes a body gain or lose speed, and therefore, if no forces act on it, the body will either remain at rest or it will keep moving in straight line with constant speed forever. This last part was later incorporated by Newton in his famous laws of motion and is now known as Newton's first law. Thus he disproved another Aristotelian idea, namely, that objects move only when they are in contact with something else, that is, an agent that supplies the force (e.g. Aristotle believed that an arrow moves even after the force that sets in motion ceases to be applied, because the air pushes them forward). Indeed his experiments—both real and imaginary—led to a logical conclusion, namely, objects can move in the absence of contact with anything.

Galileo's ingenuity, however, can also be seen in his experiments with complex motion. Those experiments led him to consider complex motion as two separate components at right angles to each other. He had realized, while studying the motion of projectiles, that a body could have a uniform horizontal component and an accelerated vertical component. Further he had observed that as the body moves along a horizontal but tilted surface, the horizontal (inertial) motion of the body becomes coupled with an accelerated downward motion. Thus he was able to discover the principle of simultaneous and independent motions, which could explain why objects on Earth are not left behind by its rotation and its revolution around the Sun. And nowadays we can understand why an object that is dropped from a horizontally flying plane will reach the ground in a time that is independent of the horizontal velocity of the plane. In this way, he discovered that the path of a projectile is not any curve but, instead, a well defined one, namely, a parabola. In his *Dialogue Concerning The Two World Systems*, he pointed out that a stone dropped from the mast of a ship always lands at the foot of the mast, regardless of whether or not the ship is moving in straight line with constant speed.

Of course, Galileo's interests were so diverse that he also began his exploration of the sky. While Galileo was busy studying the motion of bodies near the surface of the Earth, his contemporary, Kepler, was spending long hours in Poland studying the motion of planets. These two scientific areas, that is, earthly motion and heavenly motion, were considered at the time of Galileo independent. In 1609 he invented the telescope, which he set up in the garden of his home, and every night he was observing the starry sky. With his telescope, he was able to observe the mountains and valleys of the Moon. So he discovered the Moons around Saturn and Jupiter. Those observations were indeed crucial, as they provided evidence that heavenly bodies had properties that were not different from those of the Earth and therefore a clue as to the validity of the heliocentric theory. Thus, his interest to Copernican theory finally became a deep-seated belief which was the cause of his trial in Rome in 1633. Thus it was considered a blasphemy to claim that the Earth was a planetary body, just like any other planetary body, that goes around the Sun.

Galileo was aware of the problems that his ideas created, and therefore it was very difficult for him to write a book in which he could explain his ideas. So he found a way to write and publish a book and with the Pope's permission! Luckily enough, in 1623 a friend of his, who became the Pope of the Catholic Church, suggested that Galileo could write a book, but without analysing and explaining his own views. So Galileo came up with an ingenious way to present his ideas, namely, to present an argument in a dramatic form, that is, as a dialogue/debate between two characters, Salviati and Simplicio, the former representing an individual who believed in Galilean ideas, while the latter representing an Aristotelian. This book was entitled the *Dialogue Concerning the Two Chief World Systems*.

Nevertheless the Catholic Church was not fooled, so in 1633 Galileo was summoned to Rome to stand trial. Even though he admitted that he was right in claiming that the Earth moves, because of his age, he was not sentenced to death, but, instead, to remain closed in his home for the rest of his life and under strict surveillance. He lived his last 9 years there, isolated. Hopefully his work bore fruit. His contribution to science was immense and can be summarized as follows:

- He took motion out of the domain of philosophical speculation and carried it into domain of mathematics (i.e. arithmetic and geometry).
- He introduced the idea of the quantitative description of phenomena independently of any explanations—physical or otherwise—as to why they occur. He had realized that speculations about the reasons behind the phenomena had not helped advance science.
- He brought together two separate fields at his time, namely, mathematics and physics.

- He did not inquire what space and time were, or why they existed. In fact, he completely ignored the metaphysics involved in these to fundamental concepts. He simply used them in his mathematical equations.
- He introduced thought experiments. His ingenuity can be seen in the fact that he was able not just to observe phenomena but also to imagine them.

All these ideas became the foundation for further exploration which was meant to be carried out by another scientist in England. Indeed, by a striking historical coincidence, the year Galileo died, in Woolsthorpe, an hamlet in Lincolnshire, on Christmas Day, a widowed woman gave birth to a premature child that was meant to make perhaps the most important contribution to the history of western science, some even say western civilization. That frail baby was Isaac Newton who lived, however, a full life—he was 85 when he died—and enjoyed a great fame, even though he always described himself as a young boy playing on a seashore standing and diverting himself in finding pretty and smooth pebbles while the great ocean of knowledge lay undiscovered before him.

Newton's childhood was not a happy period. When he was barely 3 years old, his mother left him with his grandmother, because she wanted to remarry. She returned to Woolsthorpe, several years later, when her second husband died. Those childhood events—without his mother's attention and care but with a growing hatred for his stepfather—were crucial for young Isaac Newton's emotional development. And they must have played a role in the emotional and nervous breakdowns that he suffered several times in his lifetime. Those childhood events may also explain his quite complex character and the attacks he made against colleagues and even friends on several occasions.

With regard to his intellectual promise, Isaac Newton, as a young child, was not considered gifted either. However, he did show a keen interest, just like Galileo did, in mechanical contrivances. In actual fact, young Isaac had designed a toy, which used a mouse on a turning wheel!

At any rate, upon his mother's return, Newton was taken from school in order to fulfil his mother's wishes to become a farmer. But Newton was not the least interested in farming. So his mother, seeing that her son's future in the farming was bleak, decided to sent him off to Cambridge. That was in the summer of 1661, when Newton 19 years old. Four years later, that is, in 1665, Newton took his bachelor's degree. During those 4 years, Newton devoted his time to the private study of the works of René Descartes, Thomas Hobbes, and other scholars of the time, all of whom were to play a significant role in the scientific revolution that was coming.

Due to the spread of plague, however, the University of Cambridge closed (and remained closed for the next 2 years) and Newton returned to his native Woolsthorpe. There, he became literally a recluse for 18 months and spent about 20 h a day thinking and writing. Those 18 months were, as Newton himself admitted, the most productive and fruitful years of his career. Indeed during those 18 months, he conceived the notion of calculus and laid the foundation for his theory of light and his theory of motion and gravitation. According to Newton's own recollections, it was in 1666 that he began to think of gravity extending to the orbit of the Moon.

In April 1667, right after his return to Cambridge, Newton was elected a minor fellow at Trinity College, and a year later he earned his master's degree and became a senior fellow. The circumstances were such that a year later, that is, in 1669, at the age of 27, Newton became Professor of Mathematics. Three years later, that is, in 1672, he was elected to the Royal Society and gave his first public paper on the nature of colour.

However, Newton's time and intellectual energy, after his return to Cambridge, and for the next 20 years, were taken up with his work on the study of motion and the mathematical treatment of gravitation and planetary motion. Thus, in 1687 Newton published a work (consisting of three books) that was meant to be, according to the historians of science, an intellectual masterpiece. That book had the Latin title *Philosophiae Naturalis Principia Mathematica*, or *The Mathematical Principles of Natural Philosophy*, and in it Newton explained how the whole universe worked. It included his theory on gravity and three laws of motion. This book was an immense intellectual achievement, as it could explain, using the same laws, the motion of everything: from planets and comets to squids and snails and from falling stones and horses to birds and human beings.

The ideas found in Newton's *Principia* were based upon those of his predecessors. Even though Aristotle's ideas regarding motion had already been disputed, disproved, and rejected by Galileo, Newton did incorporate Classical Greek beliefs about the universe, such as that the world is governed by order and simplicity, there exist universal causes, number relationships are the key to the universe, the world is subject to eternal irreversible change, and space, time, and matter are components of human experience.

Galileo's contribution to Newton's thinking was also significant. Newton took advantage of four Galilean ideas: (a) the law of inertia, (b) the ideas of relativity (i.e. the selection of a stationary frame of reference is a matter of convenience), (c) the new approach to view the heavens, and (d) the geometrization of motion. Using Galileo's law of inertia, and imagining the Moon as a stone in a sling, he inferred that there must be a force keeping it a circular motion; otherwise it would keep moving in a straight line forever. So some force must be acting on the Moon so as to deflect it from the straight line path. The same should hold for the planets. Thus, both the Earth and the Sun must attract bodies towards them. Newton then extended the gravitational law to include the motion of objects on or near the surface of the Earth.

At this point, Newton made the masterstroke and showed that a single universal force keeps the planets and satellites in their orbits, causes objects to fall, and holds objects on Earth. And in this way, Newton managed to unify the motion of the Earth and the Sun with the motion of objects on the surface of the Earth. In addition, he was able to quantify that action.

Just like Galileo, he did not inquired into the causes of gravity—indeed it always remained a mystery for him—but instead used mathematics to work out the force of gravity and come up with the famous law of universal attraction, with one mathematical formula that relates the masses of any two objects in the universe, that is, between the Earth and an apple, between the Sun and the Earth, between the

Moon and the Sun, and so forth. This law also showed an important fact, namely, that the mass of an object is not the same as its weight. While its mass is always constant, its weight varies according to its distance from the surface of the Earth.

Being interested, just like Galileo, in the motion of bodies on the Earth's surface, Newton found that the force acting an any body produces changes in its velocity, which are proportional to the applied force but inversely proportional to its mass. In combining this law with his law of universal gravitation, he was able to show with mathematical formulae in just a few minutes Kepler's third law (i.e. the square of the time of revolution of any planet is proportional to the cube of its average distance from the Sun), which took the latter years of careful observation and a painstaking process of trial and error.

However, there was a dispute between Newton and Robert Hooke—then president of the Royal Society in London—over the latter's contribution to the formulation of the idea of universal gravitation. Indeed, it seems that this idea, even though an insight on Newton's part during the plague years in Woolsthorpe, took about 20 years to be fully articulated, and Robert Hooke must have played a role in this process. Hooke believed that he helped Newton to clarify, reconsider, and articulate his ideas, through the exchange of correspondence in late 1679/early 1680.

Historians of science have found evidence that during the plague years (mid-1660s), Newton's thinking was focused on the Moon's centrifugal tendency and not on the Moon's attraction towards the Earth. He was thinking that such a centrifugal force had to be balanced by some unknown force. It was Hooke who proposed the idea that planetary motion is possible due to an attractive force towards a central body and also that this attracting force varies with the square of the distance.

The degree to which Hooke's letters to Newton helped the latter with the formulation of the theory of universal gravitation and planetary motion is open to speculation, although historians of science believe that it was Newton's mathematical skills and ingenuity that played the crucial, catalytic role in the formulation of the ideas propounded in his *Principia*. At that time, however, Newton was really infuriated with Hooke, so much so that he was initially unwilling to publish the third part of the *Principia*. But in the end, he decided to publish the whole work. Edmund Halley played a major role in this publication process, as he himself both financed the *Principia* and made sure that they got published. Halley's visit to Cambridge in 1684, during which he was amazed at Newton's work on the relation between an inverse square attraction and elliptical orbits, must have played a role in making him deeply appreciate Newton's work and ingenuity.

However, Newton's interest in alchemy must have also played a role in his theory of gravitation. Due to an emotional breakdown in 1678 and his mother's death in the following year, Newton isolated himself from all people and began to study alchemy. He believed that alchemy could help him penetrate the hidden workings of nature. Despite the criticisms on the part of his colleagues—especially by Cartesian philosophers such as Christian Huygens and Gottfried Leibnitz—Newton's interest in alchemy became stronger and stronger, as he saw that alchemy could open for him new avenues for investigation, which he could not find in natural philosophy. It is believed that it was alchemy that made Newton consider the ideas of attraction and action-at-a-distance. It has been said that Newton reintroduced into physical theory concepts what would have been familiar to Renaissance practitioners of astrology, alchemy, and magic.

After the publication of the *Principia*, Newton began to show an interest in social and public affairs, and in 1689 he became a representative for Cambridge University in Parliament. However, in 1693, another nervous breakdown (some speculate that it was mercury poisoning due to his alchemical experiments, while others believe it was stress due to his acrimonious dispute with Hooke and/or the long working hours) made him consider leaving Cambridge.

So in 1696 he moved to London and was successful in finding a new job as the Master of the Mint. During his time in London, Newton acquired fame and power. In 1703, after Robert Hooke died, he was elected president of the Royal Society, and he was reelected on an annual basis until his death in 1727. It was during that time, as president of the Royal society, that he published his second major work the *Opticks* (1704) and was knighted (1705).

Notwithstanding his bitter dispute with Robert Hooke, and the fact that he avoided to make any reference to him with regard to gravitational attraction, Newton was a genius and *The Principia* has been perhaps the most influential scientific work—two more editions appeared during Newton's lifetime with the last 1 year before he died. He was successful in combining the ideas of Galileo, Descartes, Copernicus, and Kepler into a new and, at the same time, powerful synthesis. And his study of alchemy and theology can be considered as a search for unity, and hence for the ultimate synthesis, in which the study of the natural world could be combined with the study of the Scriptures.

Appendix B

Bringing the Stories into the Classroom: Three Planning Frameworks

A Planning Framework for the 'Galvani: Volta Controversy' Story

- *Central idea to evoke a sense of wonder*: a chance discovery concerning the legs of a frog that changed the face of the whole world.
- *Human values*: ingenuity, insistence, persistence, curiosity, patience, commitment.
- *Protagonists*: Luigi Galvani, Alessandro Giuseppe Volta, and Michael Faraday. Featuring also Hans Christian Oersted.
- *Mental images*: (a) Galvani and Volta in their labs to experiment with patience and persistence and to defend their ideas. (b) The happiness on Faraday's face when he was almost certain he had discovered a safer and easier way to produce electricity from magnets.

Appendices

- *Plot of the story*: (a) The conflict and scientific debate between Galvani and Volta and (b) the significant discovery by Faraday about the relationship between electricity and magnetism (after turning on his head, an accidental discovery by Oersted).
- *Ideas to be learned by the students*: (1) *Physics content*: (a) production of current electricity from animal cells, (b) from different metals immersed in an electrolyte, and (c) from magnets and (d) the relationship between electricity and magnetism. (2) *Nature of science*: (a) science is a social activity; the ideas put forward by one scientist open new ways for other scientists, and scientists think further what other scientists have thought before them. (b) Science is a human endeavour (i.e. it is tied to the scientists' struggle, to the anxieties, frustrations, hopes). (c) The accidental nature of scientific discovery. (d) The difficulties scientists experience sometimes in understanding what other scientists are saying. (e) The existence of conflict in the scientific community.
- *Moral*: (a) Galvani's discovery, despite the humiliating defeat that he suffered, was very significant since nowadays that discovery can save human lives (e.g. use of electrocardiography). (b) The scientific community should not ridicule and reject scientific ideas however strange or even crazy they may seem and sound at first (Source: Hadzigeorgiou, 2006b).

A Planning Framework for the 'Tesla and His Work on AC Electricity' Story

- *Introduction to create anticipation*: The following Saturday we are going to listen to a story about something that we take for granted, but which literally transformed our lives. In actual fact, ever since, our world has not been the same. This 'something' is the electric current, which makes almost all of our home appliances work, which is used in industry and, generally, which changed completely the world about a century ago. No one would dare to imagine what our life and the world at large would be like without the use of electric current. Moreover, many of us are unaware of the fact that useable electric current was an idea of a man; a product of his legendary, extraordinary, even heroic qualities; and the victor of a long controversy and battle. This battle, known as the 'War of the Currents', was finally won by that man, in 1895, when, for the first time, he was successful in transmitting electrical power from Niagara Falls to Buffalo City, about 34 km away. That feat was something unthinkable, even unimaginable, at that time, and that's why it was covered by the world press, and that man was praised as a hero worldwide. That man was Nikola Tesla, who, according to Life Magazine's special issue (September, 1997), is considered among the 100 most famous people of the millennium. Although current electricity was not a new idea at the time of Tesla, the transmission of current electricity over such a distance, at that time, was a heroic achievement, which is directly associated with Tesla's extraordinary mind and especially his passion for, and even obsession with, a particular kind of current that is called alternating current (AC).

- *Central ideas to evoke a sense of wonder*: (a) One of the pioneers of radio, who also built the first AC power system that resulted in changing the face of the whole world, Tesla's contribution to science has been marginalized by history. (b) Tesla's imaginative vision to transform the entire planet into a colossal electrical transmitter so that energy could reach the most distant places on the planet. (c) Through Tesla's theatrical flair, science became magic that kept audiences spellbound.
- *Human values*: ingenuity, imaginativeness, insistence, persistence, curiosity, patience, commitment.
- *Protagonists*: Nikola Tesla and Thomas Alva Edison. Also featuring Guglielmo Marconi.
- *Mental images*: (a) Tesla, bedridden for months after contracting a very powerful form of cholera, pleading with his father and saying: 'Perhaps I will recover if you let me be an electrical engineer'. (b) Tesla walking in a park with some friends, becoming entranced with the sight of a beautiful sunset, having a vision of a vortex whirling eternally in the Sun and blurting out 'See my motor here... watch me reverse it'. (c) Tesla working nonstop for 2 or 3 days and nights in his New York laboratory until he solved the mysteries of electricity. (d) Tesla's commitment, while working as a ditch digger and suffering all the indignities of the immigrants of his time (circa 1880), to the idea of AC polyphase current and his insistence to attract the attention of new financial supporters. (e) The happiness on Tesla's face when he discovered that high frequency currents were harmless for the human body. (f) The happiness on Tesla's face when he was successful at generating electricity at Niagara Falls and sending AC power to Buffalo, 20 miles away.
- *Plot of the story*: (a) Tesla's strong desire to live and work on AC current and his humanitarian purpose to supply free energy to all people made him immigrate to America. (b) The conflict and scientific/business debate between Tesla and Edison over the safety and effective transmission/utility of DC and AC currents. (c) The significant discovery (reinvention) by Tesla of AC current in the form polyphase current. (d) The first successful transmission of electrical power from Niagara Falls to Buffalo, NY.
- *Ideas to be learned by the students*: (1) *Physics content*: (a) the idea of AC current, (b) the advantages of AC current when it comes to transmitting power to long distances, (c) the physiological effects of DC and AC currents, and (d) the idea of wireless transmission of energy. (2) *Nature of science*: (a) science is a social activity; the ideas put forward by one scientist open new ways for other scientists, and scientists think further what other scientists have thought before them. (b) Science is a human endeavour (i.e. it is tied to the scientists' struggle, to their ambitions, anxieties, frustrations, hopes). (c) The accidental nature of scientific discovery. (d) The difficulties scientists experience sometimes in understanding what other scientists are saying. (e) The existence of conflict in the scientific community. (f) Financial considerations always play a role in scientific research.

- *Moral*: (a) Never pour scorn on scientific ideas however strange or even crazy they may seem and sound at first. (b) Trying to make a career outside the academic world and the security it offers and attempting to cooperate with businessmen are not considered, at least historically, appropriate ways to do science (Source: Hadzigeorgiou & Garganourakis, 2010).

A Planning Framework for the 'Birth of the New Science of Motion' Story

- *Introduction to create anticipation*: Today we are going to listen to a story about the life and work of two scientists who changed the way we view everyday phenomena of motion. The two scientists—one known by his first name, that is, Galileo, and the other one by his surname, that is, Newton—are considered two of the most important scientists in our Western civilization, and their names are associated with the scientific revolution that took place in the seventeenth century. While Galileo was the first to understand that Aristotle's ideas about motion were wrong, Newton was the first to provide a wonderful synthesis of ideas, which explained how the whole universe worked. We can now use these ideas to study the motion of everything: from animals, cars, and athletes to planets, satellites, and rockets. We used these ideas to send man to the Moon and bring him safely back to Earth.
- *Central ideas to evoke a sense of wonder*: (a) unification of the laws of motion—the motion of all bodies on Earth and in the sky is governed by the same laws. (b) All kinds of motion involve the same metaphysics of space and time: the flight of a mosquito, the kicking of a soccer ball, and the motion of a comet are described and explained by the same laws. (c) Motion does not require a force (it is change in motion, that is, acceleration, that requires a force).
- *Human values*: scepticism/doubt, inquisitiveness, ingenuity, imaginativeness, insistence, persistence, patience, and commitment.
- *Protagonists*: Galileo Galilei and Isaac Newton. Also featuring Robert Hooke and Edmund Halley.
- *Mental images*: (a) Galileo experimenting with falling and rolling objects in order to study accelerated motion. (b) The happiness on his face when he had scientific evidence that the Aristotelian theory of motion was wrong. (c) Newton walking and thinking deeply about motion (upon his return to his hometown after the closure of Cambridge, due to the spread of Plague). (d) Robert Hooke and Edmund Halley, in the coffeehouses of London, struggling unsuccessfully with the problem of planetary motion. (e) Halley's amazement at Newton's ingenuity and mathematical skill, as the latter had already worked out the mathematical formula for planetary motion (e.g. the planets' elliptical paths).
- *Plot of the story*: (a) Galileo's dissatisfaction with the Aristotelian theory of motion, and his experiments at the University of Pisa and later at the University of Padua. (b) Galileo's discovery of the law of inertia and the laws governing free-falling objects. (c) Newton's birth in the year Galileo died. (d) Newton's return from Cambridge, after he received his bachelor's degree, to his hometown

(due to the spread of plague and the closure of Cambridge) where he worked for 18 months on his ideas about motion and other things. (e) Newton's return to Cambridge and his work for the next 20 years on the mathematical treatment of gravitation, planetary motion, and the unification of the laws of motion. (f) The publication of Newton' intellectual masterpiece *The Mathematical Principles of Natural Philosophy*.
- *Ideas to be learned by the students*: (1) *Physics content*: (a) a single universal force keeps the planets and satellites in their orbits, causes objects to fall, and holds objects on Earth. (b) Galileo's law of inertia, (c) Galileo's laws of free fall (i.e. distance of a free-falling body varies with the square of the time, and it does not depend upon any property of the falling body (e.g. heavy, light)), (d) Galileo's principle of simultaneous and independent motions. (e) Newton's laws of motion. (f) Newton's law of universal gravitation (i.e. any two bodies in the universe attract each other with a force that is proportional to the product of the masses of the two bodies and inversely proportional to the square of distance between them). (g) The mass of an object is not the same as its weight (whereas its mass is always constant, its weight varies according to its distance from the surface of the Earth). (2) *Nature of science*: (a) science is a social activity; the ideas put forward by one scientist open new ways for other scientists, and scientists think further what other scientists have thought before them. (b) There is conflict between the scientists. (c) Some scientists deliberately avoid to mention the contribution of other scientists.
- *Moral*: (a) All ideas, no matter their source, can be subjected to scientific scrutiny, which may result in their reconsideration or even their rejection. (b) Commitment and persistence on the part of scientists bear fruits.

Appendix C

The Romantic Elements of the Tesla Story

Humanization of Meaning

The story of Nikola Tesla, which is based upon real events concerning his life and his work, makes it quite apparent that scientific knowledge (i.e. about alternating current) was the product of a human pursuit motivated by personal ambitions, as well as humanistic ideals and a sense of social responsibility, and, at the same time, frustrated by the establishment and especially the human greed of his rivals. The final victory of the idea of alternating current over that of direct current, which was the accepted technology of his time, is directly linked to Tesla's character, his ambitions, and his humanistic ideals (i.e. provide free electricity to all people on the planet).

Heroic Qualities

According to the September 1997 special issue of *Life Magazine*, Tesla is among the 100 most famous people of the millennium that have helped to change the course of human history. In reading Tesla's biography, one cannot help but admire and, in fact, be astonished at his qualities, talents, and mental powers, exemplifying four distinct heroic qualities—ingenuity, imaginative vision, willpower, and the power of visualization.

Tesla's innovation began with the realization that direct current was inefficient for transmitting electrical power over long distances. Although there was theoretical talk about alternating current, no one had been able to determine how to make it work. Tesla's heroic feat in transmitting electricity from Niagara Falls to Buffalo City, using alternating current, resulted in him being the first to send electricity to people's homes. This achievement was widely covered in the press, and Tesla, at 39 years of age, was praised as a hero around the world.

While some consider Tesla's alternating current induction motor as one of the ten greatest discoveries of all time (Tesla Society, 2011), his genius was also demonstrated in his invention of the Tesla coil, which is widely used, even today, in radios and television sets and in other electronic equipment and wireless communication.

A scientist of vision, Tesla conceived of the virtually instantaneous transmission of wireless energy—energy that was free, convenient, and hazard-free—to all the corners of the planet, demonstrating a notable, humanitarian ideal. Knowing that the wireless transmission of power over long distances would result in significant losses of electrical power, Tesla conceived of the idea of transmitting electrical energy through the ground, instead of the air. By experimenting, Tesla found that the ground, if it were charged highly enough, could become a conductor of electricity, potentially allowing the entire planet to be transformed into a transmitter of electrical energy. This type of extraordinary ingenuity was realized in his registration of over 700 patents, worldwide, and his contribution to the invention of radio technology was duly recognized by the United States Supreme Court in 1943, the year of Tesla's death, rendering Marconi's most important patent invalid.

Vision, to be translated into action, must be accompanied with perseverance and willpower. Tesla unrelentingly pursued his adolescent dream of wireless electrical power transmission through alternating current for six decades, from when he arrived in the USA in 1884 until his death, despite the discouraging hindrances and undesirable circumstances. Not only was he forced to dig ditches for 2 years, after Edison failed to keep his word to pay Tesla the promised $50,000 for his work on improving DC motors, but Edison also waged a war against Tesla's idea of alternating current. Even the complete loss of his New York laboratory in a mysterious fire, in 1895, did not cause Tesla to give up his dream.

Alongside his personal character traits, his mental, conceptual ability—the ability to visualize diagrams, working out the details in his mind—also played a role in Tesla's ultimate success, as demonstrated in his solving the inherent problem in direct current motors. In this instance, he visualized two coils positioned at right angles, supplied with alternating current 90° out of phase, causing a rotating magnetic field and thus inducing the AC motor. His conception of all forms of energy being cyclical in nature led him to defend alternating current.

In discussing, of course, the heroic element in relation to the Tesla story, one could argue that helping students to learn science through the stories about single 'heroes' might make them think that science is done only by very 'special' people and not by themselves. Having previously discussed the association with heroic qualities as a cognitive tool of romantic understanding in some depth, it becomes apparent that Tesla becomes a hero only for those students who choose to make him their hero. Tesla cannot be imposed as a hero. Moreover, empirical evidence from students' journals reveals that Tesla was associated with certain human qualities and did inspire science learning. And in this case we can say that students' understanding was romantic. However, there is a possibility for some students to be discouraged by Tesla and think that science can be done only by special people, but in that case we cannot talk about a development of romantic understanding of science. Case studies with few students can certainly shed light into such possibility. However, some students' comments (see last section) provide evidence that Tesla did make them think and feel that they can do science despite the fact that Tesla was considered an incredible human being possessing superhuman mental and physiological abilities.

Extremes of Reality and Experience

Extremes of physiology and life events can also manifest as heroic qualities, whether innate or trained, such as Tesla's highly acute sense of sight and hearing and his ability to work unceasingly without sleep for up to 3 days and nights. His remarkable memory made it possible for him to recall all of the details of the laboratory assets after the fateful fire that followed right after his success at Niagara Falls. In one sense, even physically enduring 2 years of ditch-digging and the associated psychological suffering, the immigrants' plight of indignity, can be considered an extreme experience.

On another level, an extreme of reality may consist of setting a record or achieving the unusual. One of Tesla's most impressive records is the generation of a 130-foot long lightning strike. In the generation of electricity, being able to pump millions of volts into the Earth's surface may be regarded as an extreme achievement.

Contesting of Traditional Ideas

Tesla, like other scientists of the past, had to struggle against tradition. Already as a student, he frequently enraged his professors by questioning the technological status quo. As he began his scientific experiments, he rejected the idea of direct current as the sole means of delivering electrical energy. Later, in the USA, after he had left the Edison Company, he, literally, rebelled against Thomas Edison, himself, who had set out on a name-smearing campaign against Tesla. Using alternating current, Edison conducted demonstrations of electrocuting animals in an attempt to show how unsafe that type of current was. While direct current, then, was the accepted technology, one in which the Edison Company had a huge investment, yet, Tesla contested the prevailing practice and dismissed partisan considerations in struggling to establish his idea of alternating current and its benefits to society.

A Sense of Wonder

Undeniably, even today Tesla's experiments and accomplishments evoke a sense of wonder as they did in his day, astonishing the world with demonstrated marvels of alternating current electricity at the World Exposition in Chicago in 1893, in one illustration sending sparks into the audience, harmlessly. It was only 2 years later that he transmitted electrical power from Niagara Falls to Buffalo City, turning a dream into reality. In 1899, he demonstrated another amazing experiment in wireless electrical energy transmission in Colorado Springs by lighting planted, unwired light bulbs at the touch of his hand. By pumping electricity into the Earth's surface (about 100 million volts), he was able to light 200 lamps from a distance of 25 miles without the use of any wires! It is also astounding that he—one of the greatest geniuses of all time—died destitute and in obscurity, in a New York City hotel room that he shared with a flock of pigeons. He was 87 years old.

Finally, the Tesla story is, in and of itself, a story of wonder: his very nature, his immense accomplishments, his impact on the world. Books written about Tesla, such as *Tesla: Man Out of Time* (Cheney, 1981) and *The Man Who Invented the Twentieth Century: Nikola Tesla, Forgotten Genius of Electricity* (Lomas, 2000), bear witness to his remarkable, heroic qualities and character (see Appendix B for the planning framework used in a study with ninth grade students and Appendix A for the story that provided the material for this framework).

Appendix D

Empirical Evidence of Romantic Understanding of the Idea of 'Alternating Current' Embedded in the Tesla Story (Hadzigeorgiou, Klassen, & Froese-Klassen, 2012)

A. Comments Found Through Content Analysis of Students' Journal Entries

Humanization of Meaning

- It is unbelievable that behind the invention of the alternating current, there was such an effort by a man who suffered a lot because of what happened to him. He lost his job, he became a digger, and his laboratory caught fire.

- The war between Edison and Tesla is something that I did not know, but now I can understand that Tesla's dream and hope are responsible for his victory and for what we all have today because of alternating current.
- Tesla's unbelievable work and passion are the two factors that can explain how alternating current won direct current.
- America showed more interest in Tesla's idea of alternating current than Europe. My father said the same thing is true today. Now if America is more interested in new ideas, science will be more developed there. That is why all the best minds go to America. Tesla I think knew that and that is why he wanted to go there.
- I liked that Tesla had the dream to send electricity to all people on Earth for free. But people wanted to make money and that is why his dream was not realized. I think the fire in his lab after the great triumph at the Niagara Falls was not something that just happened. People who did not want Tesla's dream come true tried to stop him from realizing his dream. Money, money is everywhere.

Heroic Qualities

- Tesla is an admirable man because he wanted to do good to the world. He left his country and worked for another country, wanting to help all people. This is something really admirable, and we should all do like him if we want to help the world.
- I wish I could read and work like Tesla. I really think he is someone special. It is amazing that he could work non-stop for several days. I think he is very special not because he worked all those hours but because he solved the problems he was working on.
- The idea to use a transformer to increase the voltage and decrease the electric current is amazingly clever and shows Tesla's intelligence. As an idea, it is not so difficult. If you want less losses, you need less current, so you must find a way to decrease it. But simple things require intelligence and imagination. And Tesla had a lot of them.
- Transmitting electrical energy to several km from Niagara Falls using electric cables was a big thing. But to transmit electrical energy through the ground and to avoid the use of cables was really excellent and shows Tesla's power of mind.

Sense of Wonder

- The idea of alternating current is a very revolutionary idea. But how did Tesla get it and why did he insist that it was better than the standard current? It is amazing that despite all the problems he faced, he remained faithful to his purpose.
- That we can produce alternating current by using loops of wire moving or rotating inside magnets is very simple indeed. This idea of loop is a very useful idea. In the transformer, making loops of wire and changing the number of loops in the two coils can increase or decrease the electric current. It is just a very clever idea. Simple and very clever. Wow!!

- The transformation of alternating current is something astonishing. I have been thinking about it a lot. Very, very clever. I think Tesla had understood that direct current cannot be transformed, so that's why he insisted on alternating current.
- I am truly amazed by Tesla who fixed the lighting system in a city, in France, by not using a single drawing. He did everything by working it out in his imagination. How did he do this? This is really impressive.
- That Tesla lit 200 or 300 lamps without wires is really astonishing. He sent the electric current through the ground without using any wires. Perhaps we can use it to send electricity to islands without using cables in the sea. I am curious whether islands receive electricity without any wires.

Contesting of Conventional Ideas

- I like the fact that Tesla made his university professors angry. Perhaps if he did not go against what those old professors knew about electricity, there would not have been progress in using alternating current instead of regular current.
- That Tesla did not listen to what his father was saying to him about becoming a priest must have made a big difference in the world as we know it today. I think it all started when Tesla followed his own ideas and feelings, not his father's.
- Tesla went against the Company of Ericson, despite the fact that he was alone and without money. But he won. He taught us a lesson: he who insists wins even when he has to go against problems and difficulties.
- I was thinking that the light we have at home and all those electrical appliances we use [. . .] work because of what Tesla did one century ago. The electric current we use today has a history, behind which is the disagreement/controversy between Tesla and Edison. Who knows, if Tesla had agreed with Edison, whether we would have had electricity in our homes.
- Tesla was a very peculiar person. He went against everything. His father, his teachers, and the American system. But he succeeded. I think people who disagree with the system can be successful. But you need guts. It is not easy. Many times I want to disagree with my father and my teacher, but in the end, I agree with them.

B. Tables Illustrating Time Frame and Intervention Results (Hadzigeorgiou, Klassen, & Froese-Klassen, 2012)

Table D.1 Time frame and research activities for the experimental (romantic) and control (conceptual) croups

Week	Targeted group	Activity
1, 2, 3 4	All students	Teaching fundamentals of direct current electricity (current, voltage, Ohm's law, Joule's law, electric power)
5, 6	All students	Assessment of students' understanding of fundamentals
7, 8, 9	Control group	Teaching intervention: fundamental ideas about AC
	Experimental group	Teaching intervention: Tesla story
10, 18	Both groups	Assessment of students' understanding of AC

Table D.2 Journal entries of the experimental (romantic) group

	Whole class (N=95)	Males (N=55)	Females (N=40)
Number of students who made entries	91 (96%)	53 (96%)	38 (95%)
Total number of entries	275	127	148
Total number of questions asked	293	162	131
Total number of comments	482	190	292

Entries contained both comments and questions and these were counted discretely

Table D.3 Journal entries of the control (conceptual) group

	Whole class (N=102)	Males (N=58)	Females (N=44)
Number of students who made entries	56 (55%)	45 (78%)	11 (25%)
Total number of entries	105	84	31
Total number of questions asked	98	78	20
Total number of comments	177	136	41

Table D.4 Successful responses to test questions on four targeted ideas

	Experimental group (N=95)		Control group (N=102)	
Idea	1st test n (%)	2nd test n (%)	1st test n (%)	2nd test n (%)
Concept of AC	68 (72%)	67 (71%)	42 (41%)	31 (31%)
Production of AC	53 (56%)	48 (51%)	37 (36%)	28 (27%)
Transmission of AC electrical power	65 (68%)	58 (61%)	30 (29%)	22 (22%)
Wireless transmission of electrical power	72 (76%)	70 (74%)	39 (38%)	18 (18%)

Tests were conducted 1 week and 8 weeks after the respective teaching interventions. The same tests were administered to both groups, and the first and second tests were identical. The test items were explicit questions about the targeted ideas

Table D.5 Means and T-test values of student scores on a ten-item test on alternating current

Group	N	M		SD		t	
Experimental	95	8.80	6.45[a]	1.22	1.78[a]	7.64	10.40[a]
Control	102	6.76	4.02[a]	1.83	0.90[a]		

$P<0.01$, df1 = 94, df2 = 101
[a]Delayed posttest

Table D.6 Test items for the assessment of conceptual knowledge

What is an AC?
Draw a graph, showing a DC and an AC
What do we need to describe an AC?
How can we produce an AC?
How do a generator and a motor work?
Why is AC more efficient to transmit electrical power over long distances?
Explain what we can do to an AC if we want to transmit power
Describe the process and the apparatuses (technology) used (and the mathematics we need in order to describe them)
What is the wireless transmission of electricity?
What are its advantages?

Table D.7 Number of romantic characteristics identified in students' journals

Number of romantic characteristics	Number of students
2	12
3	18
4	35
5	30

Table D.8 Romantic characteristics identified in students' journals and their frequency of appearance

Romantic characteristics	Number of students	Frequency of appearance
Humanization of meaning	43 (26M, 17F)[a]	106 (55M, 51F)
Heroic elements	94 (53M, 41F)	221 (157M, 164F)
Wonder	94 (54M, 40F)	258 (131M, 127F)
Extremes of reality	50 (32M, 18F)	72 (52M, 20F)
Contesting of ideas	26 (17M, 9F)	55 (36M, 19F)

[a]M stands for male students and F for female students

Table D.9 Romantic characteristics explicitly associated with science content knowledge and their frequency of appearance

Romantic characteristics	Number of students	Frequency of appearance
Humanization of meaning	28 (19M, 9F)	77 (44M, 33F)
Heroic elements	66 (31M, 35F)	105 (70M, 35F)
Wonder	68 (37M, 31F)	113 (62M, 51F)
Extremes of reality	34 (24M, 20F)	54 (29M, 25F)
Contesting of ideas	10 (6M, 4F)	19 (13M, 6F)

Source: Hadzigeorgiou, Klassen, and Froese-Klassen (2012)

Appendix E

Ideas Selected by a Ninth Grade Teacher Who Attempted to Evoke a Sense of Wonder in Her Students

Force and motion
Newton's third law: two cars in a head-on collision, regardless of how different their masses are, experience the same force (despite the fact that the damage is far greater on the car with the smaller mass) (V)
Newton's first law: there can be motion in straight line and at constant speed, no matter how high the speed, in the absence of a net force. That is, a spaceship can travel in straight line even at thousand miles an hour and yet the resultant force on it is zero (V, D)
The state of rest and that of uniform straight-line motion are equivalent: that is, we cannot tell the difference between being at rest and moving in straight line at constant speed. Any activity taking place in a train compartment, moving in straight line and with constant speed, is like taking place in a train compartment which is at rest (V, D)
Law of universal attraction: any two bodies, regardless of their location, state, size, and shape attract each other. That is, between a student in this class and another student on the other side of the planet, or a rock on Mars, there is always an attraction (V)
An object can participate in more than one motion: this means that the time taken by an object to reach the ground, if dropped from an airplane, is independent of the speed of the airplane (V, D)
All solid objects (of different sizes and weights) reach the ground simultaneously if dropped from the same height (near the surface of the Earth). A peanut, a big orange, and a crumpled piece of paper reach the ground simultaneously (D, H)
The force of gravity is a very weak force, for it is possible for a tiny magnet to hold a paper clip despite the fact that the whole planet exerts a downward force on it (D, H)
Matter
Molecules are so small that the number of molecules in a glass of water is greater than the number of glasses which would hold the water of the whole Mediterranean Sea (V, H)
Matter is 99.99 % empty space (V)
If we could remove all empty space existing in between the atoms in the bodies of all people on the planet, then all the subatomic particles contained in the bodies of these people could pack easily into a ping-pong ball (V)
Although atoms consist of empty space, matter appears solid (our hands cannot through a table) (D)
Light
Light is invisible: we cannot see light; we see only the objects that it hits (V, H)
Light does not require a medium to travel through: yet we know we cannot have water waves without water or sound without air (V)
The immense magnitude of the speed of light: what we see in the sky is the distant past, that is, we see stars-ghosts (V, H)
Light has weight (V)
No matter how fast or slow and in which direction a light source moves, the speed of light is always constant (V)

Source: Hadzigeorgiou (2012)
V, D, H stand for the mode of presentation of ideas
V verbal presentation, *D* demonstration, *H* hands-on activity

Appendix F

'Important Ideas' Remembered by Ninth Grade Students at the End of the School Year

A. Important ideas from the units *force and motion*, *matter*, and *light* remembered at the end of the school year by students in whose class the teacher introduced those units, by evoking a sense of wonder

Idea	%
*Light is invisible	96
*Light is sometimes a stream of particles and sometimes a wave	93
*Matter is empty (or so empty) that if we could remove all empty space existing in between the atoms in the bodies of all people on the planet, all the subatomic particles contained in the bodies of these people could pack easily into a ping-pong ball	93
*Starlight gives us information about the composition, temperature, and velocity of stars	85
*The colour of the sky depends on the way light is scattered (the sky appears blue, red, and black because of the different ways light scatters or not scatter)	82
*The relationship between wavelength, frequency, and the speed of light (mathematical equation)	82
*Matter consists of nuclei (protons, neutrons) and electrons	78
*The speed of a free falling object (near the ground) is independent of its mass (or weight)	78
*When two objects collide, the force exerted by the big body on the small object is equal and opposite to the force exerted by the small object on the big one	78
*There can be motion at very high speeds in a straight line in the absence of a net force (or uniform straight-line motion does not require a net force)	78
*The force of gravity is very small, smaller than the force exerted by a tiny magnet on a paper clip (gravity is the weakest of all four fundamental forces of nature)	78
*Molecules are so small, so there are more water molecules in a glass of water than there are glasses of water in the Mediterranean Sea	74
*Matter is 99.99 % empty space	74
*What we see now if we look at the stars is the distant past	63
*Light can travel in the vacuum	60
*Any two objects regardless of their location, state, size, colour, attract each other	56
*Newton's 2nd law (mathematical formula and verbal expression)	56
*The concept of density (mathematical formula and verbal expression)	44
*Although atoms consist of empty space, matter appears solid (our hand cannot pass through a table)	44
*The Earth accelerates all objects with the same acceleration regardless of their weight	15
*For every action, there is always an opposite and equal reaction	11
*Between two bodies A and B that are or come into contact, and which are at rest or in motion, the forces that one object A exerts on object B is always equal and opposite to the force that object B exerts on object A	11

B. Important ideas from the units *force and motion*, *matter*, and *light* remembered at the end of the school year by students in whose class the teacher introduced those units, according to the textbook and in accordance with the National Curriculum

Idea	%
*For every action, there is always an opposite and equal reaction	47
*Matter consists of nuclei (protons, neutrons) and electrons	40
*The concept of weight (mathematical formula and verbal expression)	37
*Light is a wave	33
*Light travels in the vacuum	33
*Every body continues in its state of rest or in uniform motion unless an unbalanced force starts acting on it	33
*The speed of light is the highest speed that exists	33
*Newton's 2nd law (mathematical formula and verbal expression)	30
*Light travels in a straight line	30
*The concept of density (mathematical formula and verbal expression)	27
*Free fall (equations of motion)	27
*An object can participate in more than one motion	27
*Law of universal attraction	27

Source: Hadzigeorgiou (2012)

Appendix G

'Mystery Narratives' for Introducing Inquiry Science and Problem Solving

1. *The 'mystery egg' story*: In a small community, there was a farm. The farmer, who sold eggs to the local people, played every day a little game. Each time a person came to his farm and bought eggs, the farmer said: 'If you tell me what will happen to this egg I am holding in my hand if I place it in here in this jar—showing the water-filled jar on the shelf behind him—I will give you another five eggs. If you don't, you will give me back two eggs from those you just bought. You see it is easy. You will say whether the egg will float or will sink'. The problem was that no one ever got it right. So the people in the community were talking about a farmer/magician who had a mysterious egg. Can you help the puzzled people to resolve this mystery? (Clue: One day a little girl, who was known for her observational skills and who had been at the farm a couple of times, observed that on the shelf there was not just one but two jars and the farmer never placed the egg in the same jar. The two jars were identical and were both filled with same amount of water or, at least, a liquid that looked like water.)

Appendices

What the Students Investigate

- How the density of water determines the floating behaviour of an egg.
- How the density of a liquid (e.g. water, ethyl alcohol) determines whether or not an object will sink in it
- How liquids of various densities can form layers when poured into a glass or jar
- How the density of a fruit or vegetable determines its floating or sinking behaviour when placed in a jar filled with water

The moral: Science can look like magic but is not. And science can explain 'magic tricks'.

2. *The 'white powder' mystery story*: Dennis, ever since he can remember, loved everything in a white powder form. After he graduated from high school, he went to college and studied science. He decided though that his real dream was to open a shop which would specialize in white powders. So he opened a shop in which one could find anything in white powder form. You name it: salt, sugar, cornstarch, flour, baking soda, cleanser, even plaster of Paris. One day, a truck was unloading small sacks of white powders, and a mysterious customer came in. He looked around, and, without saying a word, he walked away. As he was passing by the truck driver, who was still busy unloading sacks of white powders, he took an empty sack from the back of the truck and emptied the contents of a plastic bag that was hidden in his shoulder bag. The driver did not see him, but Dennis did. However, Dennis did not see the label of the sack that the mystery man used.

Of course, this was quite inexplicable. Why should a man come to the shop, walk and look around, then leave and hide something in a truck, when he knew that all sacks were supposed to be unloaded and carried into Dennis' shop? And why did he walk away anyway?

The mystery did not last very long. The police were there soon after the man left, and explained that a poisonous white substance had been brought to their community by a man, whose description matched the man that Dennis had just seen, and who would probably come during the night to break into the shop and retrieve the sack with the poisonous substance. The police were very concerned that they had to identify the substance very soon. However, they were also puzzled over the fact that the sack containing the poisonous substance could be in any of those sacks that had been unloaded from the truck.

Dennis was certainly concerned about all this, but he knew that his science background could be of some help to the police. So he talked to the officer in charge and asked for some help. The officer agreed and asked his people to open all sacks and take a small sample from each one of them, in order for tests of various kinds to be done. Dennis asked another policewoman to go and get vinegar and iodine. Dennis explained that water and heat along with iodine and vinegar can help them identify the poisonous powder. So he is going to give them a chart containing information on how known substances like sugar, salt, cornstarch, and baking soda behave when they interact with water, heat, vinegar, and iodine and then asks them to do the tests and observe which of the substance does not match the information on the chart.

What the Students Investigate

- How the different powders look under a magnifying glass
- How water, vinegar, and iodine react with the white powders
- How the white powders behave in heat

The moral: There are many kinds of tests that we can run in order to find out about the identity of an unknown, or even 'mysterious', substance.

3. *The water-and-ethyl-alcohol mystery*: Once upon a time, there was a little girl, named Alice, who had been given two 100-ml bottles and was supposed to go and fill one of them with purified water and the other one with purified alcohol. The girl arrived home and her mother, who needed the liquids in order to make some perfume, mixed the two liquids by emptying the contents of both bottles into another 200-ml bottle. However, to her surprise, the total volume of the new liquid was not 200 ml. Because she wanted to follow the instructions for making perfume, she was worried that the 6 or 7 ml of the missing water or alcohol may have an effect on the quality of her perfume. Her daughter though looked relaxed since she had learned in her science class that what is conserved, while mixing substances, is not the volume but the mass. 'Mother I will show you something that you will not forget. I am going to use my brother's marbles to make models of the two liquids. The marbles will represent the molecules of the two liquids...'.

What the Students Investigate

- How molecules of different substances can fit into each other when mixed together
- How mass, during both physical and chemical changes, remains the same (constant)
- How the volume of water increases when it freezes

The moral: Don't confuse mass with volume. Mass is always conserved, while volume is not.

4. *The 'potato house' mystery*: There was a beautiful neighbourhood where people went about their daily business. However, in the spring, the people started to witness a very strange event. Every week a big truck unloaded in the backyard of a house several sacks of potatoes. The people in this strange house had just moved in. They looked nice people, like the rest of the neighbourhood, but their neighbours had started to think strange things about them. The only thing they knew was that the owner of the house was an electrical engineer. One day some kids noticed that the drapes of a window in the basement had not been fully pulled.... So their curiosity made them peer through the drapes of that window. To their great surprise, they saw a whole lab, full of potatoes, wires, different metals, alligator clips, and light bulbs. 'What is going on in here?' all kids wondered! 'Let's take a look. We may find out what this man is up to. Let's see what this man does after he comes home from work'. Let's hope this window is open....

What the Students Investigate

- How a potato can be transformed into a battery
- How potatoes can be connected and produce enough electricity to light a small light bulb
- How the size of a potato influences the amount of electricity it can produce
- Whether water, in which an electrolyte has been added (e.g. table salt), can conduct electricity

The moral: Electricity can be produced in a variety of ways. Creative imagination plays a central role in the process of creating innovative ideas.

5. *The 'magic red cabbage' story*: A young child was sent to her grandfather's house to bring home a few things, among which are cleanser powder and aspirin. Her grandfather explained to her that the aspirin was in powder form (not in tablets) and that she must stick label on one bag, so she knows the contents of each bag. After her grandfather went upstairs to bring the label, and after the label was placed on the bag containing the aspirin, the little girl rode her bicycle and left. When she arrived home, she discovered that the label was missing. One of her friends told her that there is a safe way to test, which powder is aspirin and which is a cleanser. The only thing she had to do is go and buy a red cabbage. Then boil it and use the produced juice to do a very easy test.

What the Students Investigate

- Red cabbage indicator changes colour when it comes into contact with a base or an acid.
- Some substances that are neither bases nor acids (e.g. table salt) do not change colour.
- Whether beet juice can be used as an indicator.

The moral: Science is involved everywhere around us. Things we use in our everyday life can also be used in scientific investigations.

6. *The 'mysterious vase' story*[4]: In a small town, high up on a mountain, a magician used to come before Christmas to entertain the children of the elementary school. This time he came in a brand new car and announced, as he was driving around town, that he had a brand new trick that could amaze and astonish his audience. The people who heard the announcement rushed to the school and joined the young students who were really anxious to watch the new trick. The time came and the magician, after he thanked his audience, placed a mouse with a lit candle in a vase and he sealed it. In a few moments, the flame of the candle went out and the mouse passed out too. He waited a little for the mouse to wake up and then repeated his experiment, this time with another bigger vase. Again the candle flame went out and the mouse passed out, even though this second time it took a little longer for the flame to die out and the mouse to faint. The magician waited again a little, for the mouse to wake up, and repeated his 'experimental trick'

---

[4] Based on John Mayof's experiments.

with the second vase for a third time. Well, this time, to their great surprise, the audience watched the candle burn as if it was outside the sealed vase, and the mouse did not pass out. Many people from the audience were talking about a magic trick that they could not explain. The school kids, however, had had an experience with candles and the way they burn, from their science class, so they decided to solve this mysterious trick. They knew this was not an easy task, but, fortunately, one kid that was sitting very near the magician's table had observed during the show that the magician did not use the same vase as the one he used the second time. In fact, the magician, said the kid to his class mates, reached into his magic kit to take out another vase, identical with one he used for his second 'experimental trick'. And at the bottom of that vase, she could see a twig with leaves on it.

What the Students Investigate

- The relationship between the volume of the vase and the time it takes for the candle to go out
- The effect of the existence of oxygen on burning time
- The factors (i.e. fuel, oxygen, sufficient heat/kindling point) necessary for burning to occur

The moral: Oxygen is absolutely necessary for burning to occur and also central to life. Science can look like magic but is not. And science can explain 'magic tricks'.

Appendix H

A Planning Framework for Teaching a Unit on Light Through Storytelling

- *Introduction to create anticipation*: Humans have always had an insatiable sense for exploration and adventure. But they have also been torn between such sense and their desire for comfort and safety. This tug and pull within the human spirit is embodied by the lighthouse, whose light warned sailors of the dangers that lurk in the sea. From the Pharos of Alexandria, the tallest lighthouse ever built, to the lighthouses that were built in the seventeenth century, the lighthouse has become a symbol of safety and hope. The lighthouse and the work of its keeper to make seafaring safe illustrates the fact that 'light', as a science concept, is inevitably connected to human life. Today we are going to listen to the story of the lighthouse and how it relates to light and the phenomena of reflection and refraction.
- *Central ideas to evoke a sense of wonder*: (a) The heroic lighthouse keepers, who, in isolated and adverse living conditions, had to keep the candles or the kerosene lamps burning day and night, in order to secure the safety of boats and their crewmen. (b) The lighthouse, as a symbol of safety, courage, and hope.

- *Protagonists*: The lighthouses and their keepers around the world, Ida Lewis (lighthouse keeper in Lime Rock Lighthouse, in the harbour of Newport, Rhode island) and Augustine Fresnel (French engineer, commissioned by the French government to work on the problem of making a lens that could in turn make the light visible from afar).
- *Heroic element*: The heroic lighthouse keeper, always on the lookout for ships in distress. Ida Lewis, who had surpassed a half century of service and had tented Lime Rock Lighthouse 18,250 nights. Renowned for her strength at rowing and dauntless courage, she saved the lives of four shipwrecked men in 1858, rescued four others in 1886, and also prevented two soldiers from drowning in 1869. That's why she was the first woman to receive a Congressional gold medal for lifesaving, and was visited, and to be visited President Ulysses Grant.
- *Human values*: Courage, commitment, ingenuity.
- *Mental images*: (a) The housekeepers working day in, day out, to keep the light of the kerosene lanterns on and being busy keeping the lanterns clean from a thick dirty residue on walls and lenses. (b) Ida Lewis on the lookout for ships in distress. (c) Augustine Fresnel working on the solution of the problem of better lighthouse light visibility.
- *Plot of the story*: (a) The voyages for the exploration of the world in the seventeenth century made the need for lighthouses imperative. (b) The work of the lighthouse keepers around the clock and the problems they faced with regard to the visibility of the light from afar. (c) The endeavour, on the part of scientists and engineers, to work on that problem. (d) Augustine Fresnel's invention of a lamp that far outshone other rival lamps. (e) The description of Fresnel's prismatic lens, as a new type of lightweight lens, surrounded by refracting prisms that was capable of directing a concentrated light beam through a large glass magnifier (thus making the light visible from great distances) by French entrepreneurs, who travelled to the USA in order to sell the French invention. (f) Ida Lewis, the keeper of the Lime Rock Lighthouse, in the harbour of Newport, in Rhode island, was able to spot in the sea, due to Fresnel's prismatic lens, shipwrecked men and save them. (g) The contribution of Fresnel's lens to maritime safety. (h) The invention of electricity makes the service of lighthouse keepers redundant, while the lighthouse continues to be a symbol of safety, courage, and hope.
- *Ideas to be learned by the students*: (1) *Physics content*: (a) The concepts of reflection and refraction. (b) Convergent and divergent lenses, prisms, and Fresnel's prismatic lens. (2) *Ideas from the nature of science*: (a) Scientists use both logic and imagination. (b) Science is a human endeavour. (c) Scientific knowledge can be the outcome of the solution of practical problems.

The story about lighthouses is based on Crompton (1999).

Appendix I

An 'Imaginative Starter' for Teaching Current Electricity

Get in a comfortable position and sit relaxed in your seat. When ready, close your eyes.

You are getting smaller and smaller and smaller. You are approaching the size of atoms and molecules. Imagine you are an electron. You are an electron in a copper wire. The copper wire is mostly empty space. You see a few hard, dense copper nuclei. You see other electrons orbiting these nuclei. There is much space between the electrons and the nuclei. Suddenly you feel something pushing you. You are moving very fast in the copper wire. You zoom be nuclei and electrons.

Some electrons are moving in the same direction with you. You are moving with lots of energy. Now the copper changes to nichrome. You are moving swiftly into the nichrome wire. The force is still pushing you ahead. The nichrome wire appears to be more crowded. You keep bumping into things. You bump into other electrons. The pathway in which you are travelling is getting narrow the collisions are getting vicious. The force is still pushing you through the narrow and crowded pathway. You are getting hot. You would like to rest. The force is still pushing you on. You are getting battered by the collisions. You are very hot. There seems to be a red glow in the nichrome pathway. It is unbearably hot.

Now the nichrome wire ends, and the copper wire begins again. There are very few collisions here. You lost nearly all your energy in the nichrome pathway. You are very tired, but thank goodness the battering has ended. You can faintly feel the force, pushing you on. You feel so listless. You stumble on in the spacious copper pathway. You are approaching a mysterious box. You are being pushed into the box. You move through a refreshing liquid. It soothes your bruises. It gives you energy. You begin to move faster. You feel rejuvenated. Your spirit swells. You come out of the box totally refreshed. You are ready for another trip (Source: Stefanich, 2001, p. 275).

References

AAAS. (1990). *Science for all Americans*. Oxford, MA: Oxford University Press.
Abrams, M. H. (1971). *Natural supernaturalism: Tradition and revolution in Romantic literature*. New York: Norton.
Aguirre, I. (2004). Beyond the understanding of visual culture: A pragmatist approach to aesthetic education. *International Journal of Art & Design Education, 23*, 256–269.
Aikenhead, G. (1996). Science education: Border crossing into the subculture of science. *Studies in Science Education, 27*, 1–52.
Aikenhead, G. (2002). Whose scientific knowledge? The colonized and the colonized. In W.-M. Roth & J. Desautels (Eds.), *Science as/for sociopolitical action* (pp. 151–166). New York: Peter Lang.
Alberts, R. (2010). Discovering science through art-based activities. *Learning Disabilities: A Multi-Disciplinary Journal, 16*(2), 79–80.
Alexander, T. (1987). *John Dewey's theory of art, experience and nature: The horizons of feeling*. Albany, NY: State University of New York Press.
Allchin, D. (2015). Listening to whales. *American Biology Teacher, 77*, 220–222.
Allday, J. (2003). Science in science fiction. *Physics Education, 38*, 27–30.
Allen, D. (2012). "Playing" with science. *Primary Science, 121*, 21–24.
Alrutz, M. (2004). Granting science a dramatic license: Exploring a 4th grade science classroom and the possibilities for integrating drama. *Teaching Artist Journal, 2*, 31–39.
Alsop, S. (Ed.). (2005). *Beyond Cartesian dualism*. Dordrecht, The Netherlands: Springer.
Altman, R. (2008). *A theory of narrative*. New York: Columbia University Press.
Amabile, T. M. (1996). *Creativity in context*. Oxford, UK: Westview Press.
Andersson, K., & Gullberg, A. (2014). What is science in preschool and what do teachers have to know to empower children? *Cultural Studies of Science Education, 9*, 275–296.
Andrée, M., & Lager-Nyqvist, Λ. (2013). Spontaneous play and imagination in everyday science classroom practice. *Research in Science Education, 43*, 1735–1750.
Appelbaum, P., & Clark, S. (2001). Science! Fun? A critical analysis of design/content/evaluation. *Journal of Curriculum Studies, 33*, 583–600.
Apple, M. (1999). *Power, meaning, and identity*. New York: Peter Lang.
Arnold, N. (1997). *Fatal forces*. London: Scholastic Children's Books.
Arya, D., & Maul, A. (2012). The role of the scientific discovery narrative in middle school science education. An experimental study. *Journal of Educational Psychology, 104*, 1022–1032.

The original version of this back matter was revised. An erratum to this back matter can be found at DOI 10.1007/978-3-319-29526-8_8

Asay, L., & Orgill, M. (2010). Analysis of essential features of inquiry in articles published in The Science Teacher. *Journal of Science Teacher Education, 21*, 57–79.
Ashley, M. (2000). Science: An unreliable friend to environmental education? *Environmental Education Research, 6*, 269–280.
Ashley, D. (2011). Avenues to inspiration. *Science Scope, 35*, 24–30.
Aubusson, P., Fogwill, S., Barr, R., & Percovic, L. (1997). What happens when students do simulation-role-playing in science? *Research in Science Education, 27*, 565–579.
Ausubel, D., Novak, J., & Hanesian, H. (1978). *Educational psychology: A cognitive view*. New York: Holt, Rinehart and Winston.
Avraamidou, L., & Osborne, J. (2009). The role of narrative in communicating science. *International Journal of Science Education, 31*, 1683–1707.
Bailey, S., & Watson, R. (1998). Establishing basic ecological understanding in younger pupils: A pilot evaluation of a strategy based on drama/role play. *International Journal of Science Education, 20*, 139–152.
Ball, R. (1983). *An essay on Newton's Principia*. New York: Johnson Reprint Corporation.
Banister, F., & Ryan, C. (2001). Developing science concepts through story telling. *School Science Review, 83*, 75–84.
Barnes, D. (1987). *From communication to curriculum*. London: Penguin.
Barron, N. (Ed.). (1987). *Anatomy of wonder: Science fiction*. New York: R. Bowker.
Barrow, L. (2010). Encouraging creativity with scientific inquiry. *Creative Education, 1*, 1–16.
Bass, J., Contant, T., & Carin, A. (2009). *Activities for teaching science as inquiry*. Boston: Pearson/Allyn & Bacon.
Battles, D., & Rhoades, H. (2005). An interdisciplinary approach to art and science: A college course on art and geology. In M. Strokrocki (Ed.), *Interdisciplinary art education: Building bridges to connect disciplines and cultures* (pp. 77–87). Reston, VA: National Art Education Association.
Bauer, H. (1992). *Scientific literacy and the myth of the scientific method*. Chicago: University of Illinois Press.
Begoray, D., & Stinner, A. (2005). Representing science through historical drama. Lord Kelvin and the age of the earth debate. *Science & Education, 14*, 547–571.
Beiser, F. (1992). *Enlightenment, revolution, and romanticism*. Cambridge, MA: Harvard University Press.
Beiser, F. (2003). Romanticism. In R. Curren (Ed.), *A companion to the philosophy of education* (pp. 130–141). New York: Blackwell.
Bell, M. (1991). How primordial is narrative. In C. Nash (Ed.), *Narrative in culture*. London: Routledge.
Bereiter, C. (1994). Implication of postmodernism for science education: A critique. *Educational Psychologist, 29*, 3–12.
Bereiter, C., Scardamalia, M., Cassells, C., & Hewitt, J. (1997). Postmodernism, knowledge building, and elementary science. *The Elementary School Journal, 97*, 329–340.
Berlin, I. (2001). *The roots of romanticism*. Princeton, NJ: Princeton University Press.
Berlyne, D. (1960). *Conflict, arousal, and curiosity*. New York: McGraw-Hill.
Bharucha, J. (2006). *Education as we know it does not accomplish what we believe it does*. Retrieved March 3, 2006, from http://www.edge.org/q2006/q06_10.html
Biesta, G., & Osberg, D. (2007). Beyond re/presentation: A case for updating the epistemology of schooling. *Interchange, 38*, 15–29.
Bloom, B., & Eisner, E. (1971). *Confronting curriculum reform*. Boston: Little, Brown, and Co.
Blown, E., & Bryce, T. (2013). Thought experiments about gravity in the history of science and in research into children's thinking. *Science & Education, 22*, 419–481.
Boden, M. (1998). Creativity and knowledge. In A. Craft, B. Jeffrey, & M. Leibling (Eds.), *Creativity in education* (pp. 95–102). London: Continuum.
Boden, M. (2001). Creativity and knowledge. In A. Craft, B. Jeffrey, & M. Leibling (Eds.), *Creativity in education* (pp. 95–102). London: Continuum.
Boden, M. (2004). *The creative mind: Myths and mechanisms*. London: Routledge.

Bohm, D. (1998). *On creativity*. London: Routledge.
Bortoft, H. (1996). *The wholeness of nature. Goethe's way toward a science of conscious participation in nature*. New York: Lindisfarne Press.
BouJaoude, S., Sowwan, S., & Abd-El-Khalick, F. (2005). The effect of using drama in science teaching on students' conceptions of nature of science. In K. Boersma, M. Goedhart, O. de Jong, & H. Eijkelhof (Eds.), *Research and the quality of science education* (pp. 259–267). Dordrecht, The Netherlands: Springer.
Brake, M., & Thornton, R. (2003). Science fiction in the classroom. *Physics Education, 38*, 31–34.
Braund, M. (1999). Electric drama to improve understanding in science. *School Science Review, 81*(294), 35–41.
Braund, M. (2015). Drama and learning science: An empty space? *British Educational Research Journal, 41*, 102–121.
Brent, D., Sumara, D., & Luce-Kapler, R. (2008). *Engaging minds: Changing teaching in complex times*. New York: Routledge.
Brickhouse, N. (1994). Bringing in the outsiders: The sciences of the future. *Journal of Curriculum Studies, 31*, 131–142.
Brickhouse, N. (2001). Embodying science: A feminist perspective on learning. *Journal of Research in Science Teaching, 38*, 282–295.
Brickhouse, N. (2003). Science for all? Science for girls? Which girls? In R. Cross (Ed.), *A vision for science education* (pp. 93–101). London/New York: Routledge Falmer.
Britton, J. (1970). *Language and learning*. Harmondsworth, UK: Penguin.
Bronowski, J. (1978). *The origins of knowledge and imagination*. New Haven, CT: Yale University Press.
Brooks, J. (2011). *Big science for growing minds: Constructivist classrooms for young thinkers*. New York: Teachers College Press.
Brophy, J. (1987). Synthesis of research on strategies for motivating students to learn. *Educational Leadership, 45*, 40–48.
Brophy, J. (1999). Toward a model of the value aspects of motivation in education. *Educational Psychologist, 34*(2), 75–86.
Brown, A. (1992). Design experiments: Theoretical and methodological challenges in creating complex interventions in classroom settings. *The Journal of Learning Sciences, 2*, 141–175.
Brown, B. (2004). Discursive identities: Assimilation into the culture of science and its implications for minority students. *Journal of Research in Science Teaching, 41*, 810–834.
Brown, J. (1991). *The laboratory of the mind. Thought experiments in the natural sciences*. London: Routledge.
Bruna, C. (2013). Motivating active learning of biochemistry through artistic representation of scientific concepts. *Journal of Biological Education, 47*, 46–51.
Bruner, J. (1966). *Toward a theory of instruction*. New York: Norton.
Bruner, J. (1985). Narrative and paradigmatic modes of thought. In 84th Yearbook of NSSE, *Learning and teaching the ways of knowing*. Chicago: University of Chicago Press.
Bruner, J. (1986). *Actual minds, possible worlds*. Cambridge, MA: Harvard University Press.
Bruner, J. (1990). *Acts of meaning*. Cambridge, MA: Harvard University Press.
Bruner, J. (1991). The narrative construction of reality. *Critical Inquiry, 18*, 1–21.
Bruner, J. (1996). *The culture of education*. Cambridge, MA: Harvard University Press.
Brush, S. (1969). The role of history in the teaching of physics. *The Physics Teacher, 75*, 271–276.
Buczynski, S., Ireland, K., Reed, S., & Lacanienta, A. (2012). Communicating science concepts through art: 21st-century skills in practice. *Science Scope, 35*, 30–35.
Burke, E. (1990). *A philosophical enquiry into the origin of our ideas*. New York: Oxford University Press.
Burns, W. (2003). *Science in the enlightenment*. Santa Barbara, CA: ABC-CLIO.
Butterfield, H. (1965a). *The history of science*. New York: The Free Press.
Butterfield, H. (1965b). *The origins of modern science*. New York: The Free Press.

Bybee, R. (1997). *Achieving scientific literacy: From purposes to practices*. Portsmouth, UK: Heinemann.
Bybee, R. (2005). *Investigating life systems*, BSCS (Science and Technology). Dubuque, IA: Kendall/Hunt.
Bybee, R., & Landes, N. (1990). Science for life and living: An elementary school science program from biological science curriculum study. *The American Biology Teacher, 52*, 92–98.
Cabe Trundle, K., & Sackes, M. (Eds.). (2015). *Research in early childhood science education*. Dordrecht, The Netherlands: Springer.
Caddy, J. (2015). *The role of expressive arts in environmental education*. Retrieved March 19, 2015, from www.morning-earth.org/arts_in_EE.html
Caine, G., & Caine, R. (2001). *The brain, education, and the competitive edge*. Lanham, MD: Scarecrow Education.
Caine, R., Caine, G., McClintic, C., & Klimic, K. (2005). *Brain/mind learning principles in action*. Thousand Oaks, CA: Morgan Press.
Cajigal, A. R., Chamrat, S., Tippins, D., Mueller, M., & Thomson, N. (2011). Beyond the movie screen: An Antarctic adventure. *Science Activities: Classroom Projects and Curriculum Ideas, 48*(3), 71–80.
Cakici, Y., & Bayir, E. (2012). Developing children's views of the nature of science through role play. *International Journal of Science Education, 34*, 1075–1091.
Calandra, A. (1968, December 21). Angels on a pin: A modern parable. *Saturday Review, 235*.
Campbell, A. (2011). Avenues to inspiration. *Science Scope, 35*, 24–30.
Campbell, J. (1990). *Transformation of myth through time*. San Bernardo, CA: The Borgo Press.
Cant, A. (2014). Wonder for sale. In K. Egan, A. Cant, & G. Judson (Eds.), *Wonder-full education: The centrality of wonder in teaching and learning across the curriculum* (pp. 162–177). New York: Routledge.
Capra, F. (1977). *The Tao of physics*. London: Fontana.
Carter, K. (1993). The place of story in the study of teaching and teacher education. *Educational Researcher, 22*(1), 5–12, 18.
Casey, P. (2010). Bringing scientists to life. *Education in Science, 237*, 19.
Cavicchi, E. (2011). Classroom explorations: Pendulums, mirrors, and Galileo's drama. *Interchange, 42*, 21–50.
Cavicchi, E. (2014). Learning science as explorers. *Interchange, 45*, 185–204.
Ceci, S., Ginther, D., Kahn, S., & Williams, W. (2014). Women in academic science: A changing landscape. *Psychological Science, 15*, 75–141.
Chalmers, A. (1990). *Science and its fabrication*. Minneapolis, MN: University of Minnesota Press.
Chalmers, A. (1999). *What is this thing called science?* Milton Keynes, UK: Open University Press.
Chander, S. (2012). Little C creativity: A case for our science classrooms-an Indian perspective. *Gifted Education International, 28*, 192–200.
Chandrasekhar, S. (1987). *Truth and beauty. Aesthetics and motivations in science*. Chicago: University of Chicago Press.
Cheney, M. (1981). *Tesla: Man out of time*. Englewood Cliffs, NJ: Prentice-Hall.
Cheney, M., & Uth, R. (1999). *Tesla: Master of lightning*. New York: Barnes and Noble.
Chevalley, C. (1996). Physics as an art: The German tradition and the symbolic turn philosophy, history of art, and the natural science in the 1920s. In A. Tauber (Ed.), *The elusive synthesis: Aesthetics and science* (pp. 227–250). Boston: Kluwer.
Chin, C. (2007). Teacher questioning in the science classroom. *Journal of Research in Science Teaching, 44*, 815–843.
Chin, C., Brown, D., & Bruce, B. (2002). Student generated questions: A meaningful aspect of learning science. *International Journal of Science Education, 24*, 521–549.
Chin, C., & Chia, L. (2004). Problem-based learning. Using students' questions to drive knowledge construction. *Science Education, 88*, 762–784.

Clandinin, D., & Connelly, F. (1996). Teachers' professional knowledge landscapes: Teacher stories – stories of teachers – school stories – stories of schools. *Educational Researcher, 25*(3), 24–30.

Clavelin, M. (1974). *The natural philosophy of Galileo*. Cambridge, MA: MIT Press.

Clement, J. (2008). *Creative model construction in scientists and students*. Dordrecht, The Netherlands: Springer.

Cleaves, A. (2005). The formation of science choices in secondary school. *International Journal of Science Education, 27*, 471–486.

Clough, M. (2011). The story behind the science: Bringing scientists and science to life in post-secondary science education. *Science & Education, 20*, 701–717.

Cobern, W. (1991). *Worldview theory and science education research* (NARST monograph, Vol. 3). Manhattan, KS: NARST.

Cobern, W. (1996). Worldview theory and conceptual change in science education. *Science Education, 80*, 579–610.

Cobern, W. W. (2000). *Everyday thoughts about nature: An interpretive study of 16 ninth graders' conceptualizations of nature*. Dordrecht, The Netherlands: Kluwer.

Cohn, B. (1978). *Introduction to Newton's Principia*. Cambridge, MA: Cambridge University Press.

Coles, R. (1989). *The call of stories: Teaching and the moral imagination*. Boston: Houghton Mifflin.

Conle, C. (2000). Narrative inquiry: Research tool and medium for professional development. *European Journal of Teacher Education, 23*, 49–63.

Connelly, F., & Clandinin, J. (2000). *Narrative inquiry: Experience and story in qualitative research*. San Francisco: Jossey Bass.

Corni, F., Gilberti, E., & Mariani, C. (2010). *A story as innovative medium for science education in primary school*. Retrieved March 12, 2015, from https://personale.unimore.it/rubrica/pubblicazioni/corni

Costa, V. (1995). When science is "another world": Relationships between the worlds of family, friends, school and science. *Science Education, 79*, 313–333.

Craft, A. (2001). Little C' creativity. In A. Craft, B. Jeffrey, & M. Leibling (Eds.), *Creativity in education*. New York: Continuum International.

Cremin, T., Glauert, E., Craft, A., Compton, A., & Stylianidou, F. (2015). Creative little scientists: Exploring pedagogical synergies between inquiry-based and creative approaches in early years science. *Education, 3–13*(43), 404–419.

Crompton, S. (1999). *The lighthouse book*. New York: Barnes and Noble.

Crowther, G. (2012). Using science songs to enhance learning: An interdisciplinary approach. *Life Sciences Education, 11*, 26–30.

Csikszentmihalyi, M. (1994). The domain of creativity. In D. Feldman, M. Csikszentmihalyi, & H. Gardner (Eds.), *Changing the world: A framework of the study of creativity* (pp. 135–158). Westport, CT: Praeger.

Csikszentmihalyi, M. (1996). *Creativity: Flow and the psychology of discovery and invention*. New York: HarperCollins.

Cummins, J. (2003). Challenging the considerations of difference as deficit: Where are identity, intellect, imagination, and power in the new regime of truth? In P. Trifonas (Ed.), *Pedagogies of difference: Rethinking education for social justice* (pp. 41–60). London: RoutledgeFalmer.

Cunningham, A., & Jardine, N. (Eds.). (1990). *Romanticism and the sciences*. Cambridge, MA: Cambridge University Press.

Dahlin, B. (2001). The primacy of cognition – or of perception? A phenomenological critique of the theoretical bases of science education. *Science & Education, 10*, 453–475.

Dahlin, B. (2013). Poetizing our unknown childhood: Meeting the challenge of social constructivism. The romantic philosophy of childhood and Steiner's spiritual anthropology. *RoSE: Research on Steiner Education, 4*, 16–47.

Darlington, H. (2010). Teaching secondary school science through drama. *School Science Review, 91*(337), 109–113.

Dawkins, R. (1998). *Unweaving the rainbow: Science, delusion, and the appetite for wonder.* New York: Teachers College Press.

DeBoer, G. (1991). *A history of ideas in science education.* New York: Teachers College Press.

Deckert, D. (2001). Science and art: Lessons from Leonardo da Vinci? In G. Burnaford, A. Aprill, & C. Weiss (Eds.), *Renaissance in the classroom: Arts integration and meaningful learning* (pp. 125–139). Mahwah, NJ: Lawrence Erlbaum Associates.

De Cruz, H., & De Smedt, J. (2010). Science as structured imagination. *Journal of Creative Behavior, 44*, 9–44.

De Young, R., & Monroe, M. (1996). Some fundamentals of engaging stories. *Environmental Education Research, 2*, 171–187.

Dennis, M., Duggan, A., & McGregor, D. (2014). Evolution in action. *Primary Science, 131*, 8–10.

Dewey, J. (1934). *Art as experience.* New York: Perigee/Penguin Group.

Dewey, J. (1938). *Experience and education.* New York: Collier Books.

Dewey, J. (1960). *The quest for certainty: A study of the relation of knowledge and action.* New York: Putnam. (Original work published 1929)

Dewey, J. (1966). *Democracy and education.* New York: Macmillan.

Dewey, J. (1998). *How we think. A restatement of the relation of reflective thinking to the educative process.* Boston: Houghton Mifflin. (Original work published 1933)

Dhanapal, S., Kanapathy, R., & Mastan, J. (2014). A study to understand the role of visual arts in the teaching and learning of science. *Asia-Pacific Forum on Science Learning and Teaching, 15*, 2.

Dhingra, K. (2006). Science on television: Storytelling, learning and citizenship. *Studies of Science Education, 42*, 89–123.

Diakidoy, I., & Constantinou, C. (2000–2001) Creativity in physics: Response fluency and task specificity. *Creativity Research Journal, 13*, 401–410.

Dietrich, A. (2004). The cognitive neuroscience of creativity. *Psychonomic Bulletin & Review, 11*, 1011–1026.

Dirac, P. (1963). The evolution of the physicist's picture of nature. *Scientific American, 208*(5), 45–53.

Di Trocchio, F. (1997). *Il genio incompresso.* Milan: Mondadori.

Donald, M. (1991). *The origins of the modern human mind.* Cambridge, MA: Cambridge University Press.

Dorion, K. (2009). Science through drama: A multiple case exploration of the characteristics of drama activities used in secondary science lessons. *International Journal of Science Education, 31*, 2247–2270.

Driver, R. (1983). *The pupil as a scientist?* London: Milton Keynes.

Driver, R. (1991). Culture clash: Children and science. *The New Scientist, 29*, 46–48.

Driver, R., Asoko, H., Leach, J., Mortimer, E., & Scott, P. (1994). Constructing scientific knowledge in the classroom. *Educational Researcher, 23*, 7–12.

Driver, R., & Oldham, V. (1986). A constructivist approach to curriculum development. *Studies in Science Education, 13*, 105–122.

Duckworth, E. (2006). *"The having of wonderful ideas" and other essays on teaching and learning* (3rd ed.). New York: Teachers College Press.

Duit, R., & Treagust, D. (2003). Conceptual change: A powerful framework for improving science teaching and learning. *International Journal of Science Education, 25*, 671–688.

Duschl, R. (1994). Research on the history and philosophy of science. In D. Gabel (Ed.), *Handbook of research on science teaching and learning* (pp. 443–465). New York: McMillan.

Duschl, R., & Grandy, R. (2013). Two views about explicitly teaching the nature of science. *Science & Education, 22*, 2109–2139.

Duschl, R., Schweingruber, H., & Shouse, A. (Eds.). (2007). *Taking science to school: Learning and teaching science in grades K-8.* Washington, DC: National Academies Press.

Dykstra, D. (1992). Studying conceptual change in learning physics. *Science Education, 76*, 615–632.
Dyson, F. (2008). *The scientist as rebel.* New York: New York Review Books.
Eastwell, P. (2002). Poetry: Adding passion to the science curriculum. *Science Education Review, 1*, 2.
Egan, K. (1986). *Teaching as story-telling.* Chicago: University of Chicago Press.
Egan, K. (1988). *Primary understanding.* Chicago: University of Chicago Press.
Egan, K. (1990). *Romantic understanding.* Chicago: University of Chicago Press.
Egan, K. (1992). *Imagination in teaching and learning.* Chicago: University of Chicago Press.
Egan, K. (1997). *The educated mind. How cognitive tools shape our understanding.* Chicago: University of Chicago Press.
Egan, K. (1999). *Children's minds, talking rabbits and clockwork oranges.* New York: Teachers College Press.
Egan, K. (2003). Start with what the student knows or with what the student can imagine? *Phi Delta Kappan, 84*, 443–445.
Egan, K. (2005). *An imaginative approach to teaching.* San Francisco: Jossey-Bass.
Egan, K. (2010). *The future of education. Reimagining our schools from ground-up.* New Haven, CT: Yale University Press.
Egan, K. (2014). Wonder, awe, and teaching techniques. In K. Egan, A. Cant, & G. Judson (Eds.), *Wonder-full education: The centrality of wonder in teaching and learning across the curriculum* (pp. 149–161). New York: Routledge.
Egan, K., Cant, A., & Judson, G. (Eds.). (2014). *Wonder-full education: The centrality of wonder in teaching and learning across the curriculum.* New York: Routledge.
Egan, K., & Nadaner, D. (Eds.). (1988). *Imagination and education.* New York: Teachers College Press.
Egan, K., Stout, M., & Takaya, K. (2007). *Teaching and learning outside the box: Inspiring imagination across the curriculum.* London: Alhouse Press.
Einstein, A. (1931). *Cosmic religion with other opinions and aphorisms.* New York: Covici-Friede.
Einstein, A. (1949). *The world as I see it.* New York: Philosophical Library.
Einstein, A. (1961). *Relativity: The special and the general theory. A popular exposition.* New York: Grown Publishers.
Einstein, A., & Infeld, L. (1966). *The evolution of physics: From early concepts to realitivity and quanta.* New York: Simon and Schuster. (Original work published 1938)
Eiseley, L. (1978). *The star thrower.* New York: Times Books.
Eisner, E. (1985). *The educational imagination.* New York: McMillan.
Eisner, E. (1991). *The enlightened eye.* New York: McMillan.
Elkana, Y. (2000). Science, philosophy of science and science teaching. *Science & Education, 9*, 463–485.
Erten, S., Kiray, A., & Sen-Gumus, B. (2013). Influence of scientific stories on students ideas about science and scientists. *International Journal of Education in Mathematics, Science and Technology, 1*, 122–137.
Esbin, H. (2007/2008). Imagination goes to school. *Education Canada, 48*, 24–26, 28.
Faraday, M. (1978). *The chemical history of a candle.* Marietta, GA: Cherokee Publishing Company. (First published in 1861.)
Feasey, R. (2005). *Creative science. Achieving the wow factor with 5–11 year olds.* London: David Fulton.
Feldman, D. H. (1988). Creativity: Dreams, insights, and transformations. In R. J. Sternberg (Ed.), *The nature of creativity* (pp. 271–297). New York: Cambridge University Press.
Feldman, D., Czikszentmihalyi, M., & Gardner, H. (1994). *Changing the world: A framework for the study of creativity.* Westport, CT: Praeger.
Fenstermacher, G. (1994). The knower and the known: The nature of knowledge in research on teaching. *Review of Research in Education, 20*, 3–56.
Feyerabent, P. (1993). *Against method.* London: Verso.

Feynman, R. (1964). The value of science. In A. Arons & A. Bork (Eds.), *Science and ideas* (pp. 3–12). Englewood Cliffs, NJ: Prentice Hall.
Feynman, R. (1969). What is science? *The Physics Teacher, 7*, 313–320.
Feynman, R. (1989). *What do you care what other people think?* London: Unwin/Hyman.
Feynman, R. (1995). *Six easy pieces*. Reading, MA: Helix Books.
Feynman, R. (2005). *The Feynman lectures on physics. Vol. 1: Mainly mechanics, radiation, and heat*. Boston: Addison Wesley.
Feynman, R. (2015). *The quotable Feynman*. Princeton, NJ: Princeton University Press.
Fisher, P. (2003). *Wonder, the rainbow, and the aesthetics of rare experiences*. Boston: Harvard University Press.
Fleer, M. (2009). Understanding the dialectical relations between everyday concepts and scientific concepts within play-based programs. *Research in Science Education, 39*, 281–306.
Fleer, M. (2012). Imagination, emotions and scientific thinking: What matters in the being and becoming of a teacher of elementary science? *Cultural Studies of Science Education, 7*, 31–39.
Fleer, M. (2013). Affective imagination. in science education: Determining the emotional nature of scientific and technological learning of young children. *Research in Science Education, 43*, 2085–2106.
Fleer, M., & March, S. (2009). Engagement in science, engineering and technology in the early years: A cultural-historical reading. *Review of Science, Mathematics and ICT Education, 3*, 23–47.
Franken, R. (2001). *Human motivation*. Pacific Grove, CA: Brooks/Cole.
Frazier, W., & Murray, K. (2009). Science poetry in two voices: Poetry and the nature of science. *School Science Review, 8*(2), 58–78.
Fredricks, A., Blumenfeld, P., & Paris, A. (2004). School engagement: Potential of the concept, state of the evidence. *Review of Educational Research, 94*, 59–109.
Frontali, C. (2014). History of physical terms: 'Energy'. *Physics Education, 49*, 564–573.
Friedman, M., & Normann, A. (Eds.). (2006). *The Kantian legacy in the nineteenth-century science*. Cambridge, MA: MIT Press.
Fuchs, H. (2015). From stories to scientific models and back: Narrative framing in modern macroscopic physics. *International Journal of Science Education, 37*, 934–957.
Furnham, A., Botey, M., Booth, T., Patel, V., & Lozinskaya, D. F. (2011). Individual differences predictors of creativity in art and science students. *Thinking Skills & Creativity, 6*, 114–121.
Galili, I. (2009). Thought experiments: Determining their meaning. *Science & Education, 18*, 1–23.
Gallagher, S. (1992). *Hermeneutics and education*. New York: SUNY Press.
Gallas, K. (1997). *Sometimes I can be anything: Power, gender and identity in a primary classroom*. New York: Teachers College Press.
Gardner, H. (1991). *The unschooled mind: How children think and how schools should teach*. New York: Basic Books.
Gardner, H. (1983). *Frames of mind. The theory of multiple intelligences*. New York: Basic Books.
Gardner, H. (1993a). *Multiple intelligences. The theory in practice*. New York: Basic Books.
Gardner, H. (1993b). *Creating minds*. New York: Basic Books.
Gardner, H. (1997). *Extraordinary minds: Portraits of four exceptional minds and the extraordinary minds in all of us*. New York: HarperCollins.
Gardner, H. (2010). *Five minds for the future*. Cambridge, MA: Harvard Business School Press.
Gaut, B. (2003). Creativity and imagination. In B. Gaut & P. Livingston (Eds.), *The creation of art* (pp. 268–293). Cambridge, MA: Cambridge University Press.
Getzels, J. W. (1964). Creative thinking, problem-solving, and instruction. In E. R. Hilgard (Ed.), *Theories of learning and instruction* (63rd Yearbook of the National Society for the Study of Education, pp. 240–267). Chicago: University of Chicago Press.
Ghanbari, S. (2015). Learning across disciplines: A collective case study of two university programs that integrate the arts with STEM. *International Journal of Education & the Arts, 16*, 7.
Gilbert, A. (2013). Using the notion of "wonder" to develop positive concepts of science with future primary teachers. *Science Education International, 24*(1), 6–32.

Gilbert, J., & Reiner, M. (2000). Thought experiments in science education: Potential and current realization. *International Journal of Science Education, 22*, 265–283.

Gilbert, J., & Reiner, M. (2004). The symbiotic roles of empirical experimentation and thought experimentation in the learning of physics. *International Journal of Science Education, 26*, 1819–1834. 26.

Gilbert, J., Reiner, M., & Nakhleh, M. (Eds.). (2008). *Visualization: Theory and practice in science education*. Berlin, Germany: Springer.

Gil-Perez, D., & Carrascosa-Alis, J. (1994). Bringing pupils' learning closer to a scientific construction of knowledge: A permanent feature in innovations in science teaching. *Science Education, 78*, 301–315.

Ginsberg, H., & Opper, S. (1969). *Piaget's theory of intellectual development: An introduction*. Englewood Cliffs, NJ: Prentice-Hall.

Girod, M. (2007a). A conceptual overview of the role of beauty and aesthetics in science and science education. *Studies in Science Education, 43*, 38–61.

Girod, M. (2007b). Sublime science. *Science and Children, 44*, 26–29.

Girod, M., Rau, C., & Schepige, A. (2003). Appreciating the beauty of science ideas: Teaching for aesthetic understanding. *Science Education, 87*, 574–587.

Girod, M., Twyman, T., & Wojcikiewicz, S. (2010). Teaching and learning science for transformative, aesthetic experience. *Journal of Science Teacher Education, 21*, 801–824.

Girod, M., & Wong, D. (2002). An aesthetic (Deweyan) perspective on science learning. Case studies of three fourth graders. *The Elementary School Journal, 102*(199–224), 199–224.

Giroux, H. (1992). *Border crossings. Cultural workers and the politics of education*. New York/London: Routledge.

Glasersfeld, V. E. (1989). Cognition, construction of knowledge, and teaching. *Synthese, 80*, 121–140.

Godon, R. (2004). Understanding, personal identity, and education. *Journal of Philosophy of Education, 38*, 589–600.

Goodenough, U. (1997). *The sacred depths nature*. New York: Oxford University Press.

Gooding, F. (Ed.). (1985). *Faraday rediscovered. Essays on the life and work of Michael Faraday*. New York: Stockton.

Goodwin, A. (2001). Wonder in science teaching and learning. *School Science Review, 83*, 69–73.

Good, L. T., & Brophy, J. (1995). *Contemporary educational psychology* (5th ed.). New York: Longman Publishers.

Gordon, C. (2013). *Learning astronomy in the year prior to formal schooling: An intervention study*. Sydney, Australia: Macquarie University.

Goudsmit, R., Clairborne, S., & The Editors of Life. (1967). *Time*. Amsterdam, The Netherlands: Time-Life International.

Gough, A. (2002). Mutualism: A different agenda for science and environmental education. *International Journal of Science Education, 24*, 1201–1215.

Gough, A. (2008). Towards more effective learning for sustainability: Reconceptualising science education. *Transnational Curriculum Inquiry, 5*, 32–50.

Gough, N. (1993). Environmental education, narrative complexity and postmodern science/fiction. *International Journal of Science Education, 5*, 607–625.

Graig-Faxon, A. (1996). Intersections of art and science to create aesthetic perception. In A. Tauber (Ed.), *The elusive synthesis* (pp. 251–266). Dordrecht, The Netherlands: Kluwer.

Green, M., Strange, J., & Brock, T. (Eds.). (2002). *Narrative impact: Social and cognitive foundations*. Mahwah, NJ: Erlbaum.

Greene, M. (1978). *Landscapes of learning*. New York: Teachers College Press.

Greene, M. (1991). Forward. In C. Witherell & N. Noddings (Eds.), *Stories lives tell: Narrative and dialogue in education* (pp. ix–xi). New York: Teachers College Press.

Greene, M. (1988). *The dialectic of freedom*. New York: Teachers College Press.

Greene, M. (2000). Diversity and inclusion. Toward a curriculum for human beings. In F. Parkay & G. Moss (Eds.), *Curriculum planning: A contemporary approach* (pp. 293–300). Boston: Allyn & Bacon.

Gross, P., Levitt, N., & Lewis, M. (Eds.). (1996). *The flight from science and reason*. New York: The Academy of Sciences.

Guilford, J. P. (1950). *Creativity* (American Psychologist, Vol. 5, pp. 444–454). Washington, DC: American Psychological Association.

Guilford, J. P. (1967). *The nature of human intelligence*. New York: McGraw Hill.

Hadzigeorgiou, Y. (1997). Relationships, meaning and the science curriculum. *Curriculum and Teaching, 12*, 83–89.

Hadzigeorgiou, Y. (1999). On problem situations and science learning. *School Science Review, 81*, 43–49.

Hadzigeorgiou, Y. (2001). The role of wonder and «romance» in early childhood science education. *International Journal of Early Years Education, 9*, 63–69.

Hadzigeorgiou, Y. (2002a). The utilization of sensorimotor experiences for introducing young children to molecular motion: A report of a pilot study. *Physics Education, 37*, 239–244.

Hadzigeorgiou, Y. (2002b). From concepts to the great ideas of physics. *Science Education: Theory and Practice, 3*, 9–14 (in Greek).

Hadzigeorgiou, Y. (2005a). Romantic understanding and science education. *Teaching Education, 16*, 23–32.

Hadzigeorgiou, Y. (2005b). *On humanistic science education*. Fulbright project – Part I: Theoretical framework. Unpublished paper, Department of Curriculum & Instruction, University of Northern Iowa, summer 2005. (ED 506504).

Hadzigeorgiou, Y. (2005c). Science, personal relevance, and social responsibility: Integrating the liberal and humanistic traditions of science education. *Educational Practice & Theory, 27*, 82–93.

Hadzigeorgiou, Y. (2006a). Exploring the possibilities for developing romantic understanding through storytelling. Paper presented at the 1st International Conference on Teaching and Learning Science Through Storytelling. Deutsches Museum, Munich, July 4–7, 2006.

Hadzigeorgiou, Y. (2006b). Humanizing the teaching of physics through storytelling: The case of current electricity. *Physics Education, 41*, 42–46.

Hadzigeorgiou, Y. (2007). *Wonder: Why is it important and how can it be evoked in the science classroom?* Paper presented at the 5th international conference on imagination and education. Simon Fraser University, Vancouver, Canada, July 14–17, 2007.

Hadzigeorgiou, Y. (2008). *Reclaiming the value of wonder in science education*. Paper presented at the 2nd summer institute on imaginative education, Delta Vancouver Airport Hotel, Vancouver, Canada, July 7–9, 2008.

Hadzigeorgiou, Y. (2010). *What activities and questions are really challenging for preschool, elementary, and high school students? A study of student engagement in science with implications for curriculum and teaching*. Unpublished paper, University of the Aegean.

Hadzigeorgiou, Y. (2012). Fostering a sense of wonder in the science classroom. *Research in Science Education, 42*, 985–1005.

Hadzigeorgiou, Y. (2014). Reclaiming the value of wonder in science education. In K. Egan, A. Cant, & G. Judson (Eds.), *Wonder-full education: The centrality of wonder in teaching and learning across the curriculum* (pp. 40–66). New York: Routledge.

Hadzigeorgiou, Y. (2015). A critique of science education as socio-political action from the perspective of liberal education. *Science & Education, 24*, 259–280.

Hadzigeorgiou, Y., & Fotinos, N. (2007). Imaginative thinking and the learning of science. *Science Education Review, 6*, 15–22.

Hadzigeorgiou, Y., & Garganourakis, V. (2010). Using Nikola Tesla's story and experiments, as presented in the film "The Prestige", to promote scientific inquiry. *Interchange, 41*, 363–378.

Hadzigeorgiou, Y., & Skoumios, M. (2013). The development of environmental awareness through school science: Problems and possibilities. *International Journal of Environmental & Science Education, 8*, 405–426.

Hadzigeorgiou, Y., Kabouropoulou, M., & Fokialis, P. (2012). Thinking about creativity in science education. *Creative Education, 3*, 603–611.

Hadzigeorgiou, Y., Klassen, S., & Froese-Klassen, C. (2012). Encouraging a 'romantic understanding' of science: The effect of the Nikola Tesla story. *Science & Education, 21*, 1111–1138.

Hadzigeorgiou, Y., & Konsolas, M. (2001). Global problems in the curriculum: Toward a humanistic and constructivist science education. *Curriculum & Teaching, 16*, 29–39.

Hadzigeorgiou, Y., Anastasiou, L., Prevezanou, B., & Konsolas, M. (2009). A study of the effect of preschool children's participation in sensorimotor activities on their understanding of the mechanical equilibrium of a balance beam. *Research in Science Education, 39*(1), 39–55.

Hadzigeorgiou, Y., Prevezanou, B., Kabouropoulou, M., & Konsolas, M. (2010). Teaching about the importance of trees. A study with young children. *Environmental Education Research, 17*, 519–536.

Hadzigeorgiou, Y., & Savage, M. (2001). A study of the effect of sensorimotor activities on the understanding and application of two fundamental physics ideas. *Journal of Elementary Science Education, 31*, 9–23.

Hadzigeorgiou, Y., & Schulz, R. (2014). Romanticism and romantic science: Their contribution to science education. *Science & Education, 23*, 1963–2006.

Hadzigeorgiou, Y., & Stefanich, G. (2001). Imagination in science education. *Contemporary Education, 71*, 23–28.

Hadzigeorgiou, Y., & Stivaktakis, S. (2008). Encouraging involvement with school science. *Journal of Curriculum & Pedagogy, 5*, 138–162.

Hamilton, S. (2002). *A life of discovery. Michael Faraday, a giant of the scientific revolution*. New York: Random House.

Hammer, D. (1995). Student inquiry in a physics class discussion. *Cognition and Instruction, 13*, 401–430.

Hardiman, M. (2012). *The brain-targeted teaching model for 21st-century schools*. Thousand Oaks, CA: Corwin.

Harre, R. (1991). Some narrative conventions of scientific research discourse. In C. Nash (Ed.), *Narrative in culture* (pp. 81–101). London: Routledge.

Haven, K. (2000). *Super simple storytelling*. Englewood, CO: Teacher Idea Press.

Head, J. (1997). *Working with adolescents: Constructing identity*. London: The Falmer Press.

Heidegger, M. (2008). *Being and time* (J. Macquarrie & E. Robinson Trans.). New York: Harper Perennial. (Original work published 1927)

Hein, H. (1996). The art of displaying science: Museum exhibitions. In A. Tauber (Ed.), *The elusive synthesis: Aesthetics and science* (pp. 267–288). Boston: Kluwer.

Heisenberg, W. (1971). *Physics and beyond*. London: Allen & Unwin.

Helm, H., Gilbert, J., & Watts, D. M. (1985). Thought experiments and physics education, Part II. *Physics Education, 20*, 211–217.

Hempel, K. (1945). Studies in the logic of confirmation. *Mind, 54*, 1–26.

Hempel, K. (1966). *Philosophy of natural science*. Cambridge, UK: Pearson.

Henderson, L. (1988). X rays and the quest for invisible reality in the art of Kupka Duchmont, and the cubists. *Art Journal, 47*, 323–340.

Hendrix, R., Eick, C., & Shannon, D. (2012). The integration of creative drama in an inquiry-based elementary program: The effect on student attitude and conceptual learning. *Journal of Science Teacher Education, 23*, 823–846.

Hennessey, B. (1995). Social, environmental and developmental issues and creativity. *Educational Psychology Review, 7*, 163–183.

Herrick, R., & Cording, R. (2013). Using a poetry reading on hemoglobin to enhance subject matter. *Journal of Chemical Education, 90*, 215–218.

Heringman, N. (2003). *Romantic science*. New York: SUNY Press.

Heering, P. (2010). False friends: What makes a story inadequate for science teaching? *Interchange, 41*, 323–333.

Hill, C., & Baumgartner, L. (2009). Stories in science: The backbone of science learning. *Science Teacher, 76*(4), 60–64.

Hirst, P. (1972). Liberal education and the nature of knowledge. In R. Dearden, P. Hirst, & R. Peters (Eds.), *Education and the development of reason* (pp. 391–414). London: Routledge & Kegan Paul.

Hodson, D. (1993). In search of a rationale for multicultural science education. *Science Education, 77*, 685–711.

Hodson, D. (2003). Time for action: Science education for an alternative future. *International Journal of Science Education, 25*(6), 645–670.

Hodson, D. (2004). Going beyond STS: Towards a curriculum for socio-political action. *The Science Education Review, 3*, 2–7.

Holmes, R. (2009). *The age of wonder. How the romantic generation discovered the beauty and terror of science.* New York: Pantheon Books.

Holton, G. (1978). *The scientific imagination. Case studies.* Cambridge, UK: Cambridge University Press.

Holton, G. (1996). *Einstein, history, and other passions.* Reading, MA: Addison-Wesley.

Hong, H.-Y., & Lin-Siegler, X. (2012). How learning about scientists' struggles influences students' interest and learning in physics. *Journal of Educational Psychology, 104*, 469–484.

Hong, M., & Kang, N.-H. (2010). South Korean and the US secondary school science teachers' conceptions of creativity and teaching for creativity. *International Journal of Science and Mathematics Education, 8*, 821–883.

Hove, P. (1996). The face of wonder. *Journal of Curriculum Studies, 28*, 437–462.

Howe, A. (1971). A lost dimension in elementary science education. *Science Education, 55*, 143–146.

Hu, W., & Adey, P. (2002). A scientific creativity test for secondary school students. *International Journal of Science Education, 24*, 389–403.

Huggins, E. (2010). Weighing photons using bathroom scales: A thought experiment. *Physics Teacher, 48*, 287–288.

Hughes, S., Wimmer, J., Towsey, M., Fahmi, M., Winslett, G., Dubler, G., et al. (2014). The greatest shadow on Earth. *Physics Education, 49*, 88–94.

Isabelle, A. (2007). Teaching science using stories: The storyline approach. *Science Scope, 31*, 6–25.

Jackson, P. (1998). *John Dewey and the lessons of art.* New Haven, CT: Yale University Press.

Jakobson, B., & Wickman, P.-O. (2008). The roles of aesthetic experience in elementary school science. *Research in Science Education, 38*, 45–65.

Jardine, D., Clifford, P., & Friesen, S. (2003). *Back to the basics of teaching and learning.* Mahwah, NJ: Lawrence Erlbaum.

Jenkins, E. (1996). The "nature of science" as a curriculum component. *Journal of Curriculum Studies, 28*, 137–150.

Jenkins, E. (1999). School science, citizenship and the public understanding of science. *International Journal of Science Education, 21*, 703–710.

Jenkins, E. (2002). Linking school science education with action. In W.-M. Roth & J. Desautels (Eds.), *Science as/for sociopolitical action* (pp. 17–34). New York: Peter Lang.

Jenkins, E. (2007). School science: A questionable construct? *Journal of Curriculum Studies, 39*, 265–282.

Jenkins, E. (2009). Linking theory to practice: Education for sustainability and learning and teaching. In M. Littledyke, N. Taylor, & C. Eames (Eds.), *Education for sustainability in the primary curriculum: A guide for teachers* (pp. 29–38). South Yarra, Australia: Palgrave Macmillan.

Jenkins, E., & Nelson, N. (2005). Important but not for me: Students' attitudes toward secondary school science in England. *Research in Science and Technological Education, 23*, 41–57.

Johnson, M. (1987). *The body in the mind: The bodily basis of meaning, imagination, and reason.* Chicago: University of Chicago Press.

Johnston, B. (Ed.). (1983). *My inventions: The autobiography of Nikola Tesla.* Williston, VT: Hart Brothers.

Jones, B. (1870). *The life and letters of Faraday* (Vol. 2). London: Longmans, Green and Co.

References

Jonnes, J. (2003). *Empires of light: Edison, Tesla, Westinghouse, and the race to electrify the world.* New York: Random House.

Kalogiannakis, M., & Violintzi, A. (2012). Intervention strategies for changing preschool children's understandings about volcanoes. *Journal of Emergent Science, 4,* 12–18.

Kelley, L. (1972). *Themes in science fiction: A journey into wonder.* New York: McGraw Hill.

Kim, K. (2011). The creativity crisis: The decrease in creativity thinking scores on the Torrance test of creative thinking. *Creativity Research Journal, 23,* 285–295.

Kind, P., & Kind, V. (2007). Creativity in science education: Perspectives and challenges for developing school science. *Studies in Science Education, 43,* 1–37.

King, D., Ritchie, S., Sandhu, M., & Henderson, S. (2015). Emotionally intense science activities. *International Journal of Science Education, 37,* 1886–1914.

Kirikkaya, E. (2011). Grade 4 to 8 primary school students' attitudes towards science: Science enthusiasm. *Educational Research & Review, 6,* 374–382.

Kitchener, K. S. (1983). Cognition, metacognition, and epistemic cognition. A three-level model of cognitive processes. *Human Development, 26,* 222–323.

Kitchener, K. S. (1986). *Piaget's theory of knowledge.* New Haven, CT: Yale University Press.

Klassen, S. (2006a). A theoretical framework for contextual science teaching. *Interchange, 37,* 31–61.

Klassen, S. (2006b). The science thought experiment: How might it be used profitably in the classroom? *Interchange, 37,* 77–96.

Klassen, S. (2009). The construction and analysis of a science story. *Science & Education, 18,* 401–423.

Klassen, S., & Froese-Klassen, C. (2014a). Science teaching with historically based stories: Theoretical and practical perspectives. In M. Matthews (Ed.), *International handbook of research in history and philosophy for science and mathematics education* (pp. 1503–1529). Berlin: Springer.

Klassen, S., & Froese-Klassen, C. (2014b). The role interest in learning science through stories. *Interchange, 45,* 133–151.

Klein, M. (1954). *Mathematics in western culture.* London: George Allen & Unwin.

Kloncher, J. (2013). *Transfiguring the art and science: Knowledge and cultural institutions in the romantic age.* Cambridge, UK: Cambridge University Press.

Klopfer, L., & Cooley, W. (1963). The history of science cases for high schools in the development of student understanding of science and scientists. A report of the HOSC instruction project. *Journal of Research in Science Teaching, 1,* 33–47.

Kokkotas, P., Rizaki, A., & Malamitsa, K. (2010). Storytelling as a strategy for understanding concepts of electricity and electromagnetism. *Interchange, 41,* 379–405.

Kolakowski, L. (1989). *The presence of myth.* Chicago: Chicago University of Chicago Press.

Konicek-Morran, R. (2008). *Everyday science mysteries. Stories for inquiry-based science teaching.* Washington, DC: NSTA Press.

Konicek-Morran, R. (2010). *More everyday science mysteries. Stories for inquiry-based science teaching.* Washington, DC: NSTA Press.

Konicek-Morran, R. (2013). *Everyday physical science mysteries. Stories for inquiry-based science teaching.* Washington, DC: NSTA Press.

Kösem, S. D., & Özdemir, Ö. F. (2014). The nature and role of thought experiments in solving conceptual physics problems. *Science & Education, 23,* 865–895.

Kozoll, R., & Osborne, M. (2004). Finding meaning in science: Life-world, identity, and the self. *Science Education, 88,* 157–181.

Kreps-Frisch, J. (2010). The stories they'd tell: Pre-service elementary teachers writing stories to demonstrate physical science concepts. *Journal of Science Teacher Education, 21,* 703–722.

Kreps-Frisch, J., & Saunders, G. (2008). Using stories in an introductory college biology course. *Journal of Biological Education, 4,* 164–169.

Kroger, J. (1996). Identity, regression, and development. *Journal of Adolescence, 19,* 203–222.

Kubli, F. (2001). Can the theory of narrative help science teachers become better storytellers? *Science & Education, 10,* 595–599.

Kuhn, T. (1970). *The structure of scientific revolution*. Chicago: University of Chicago Press.
Kuhn, T. (1977). *The essential tension*. Chicago: University of Chicago Press.
LaBonty, J., & Danielson, K. E. (2005). Writing poems to gain deeper meaning in science. *Middle School Journal, 36*(5), 30–36.
Lakoff, G., & Johnson, M. (1980). *Metaphors we live by*. Chicago: Chicago University of Chicago Press.
Lakoff, G., & Nunez, R. (2000). *Where mathematics comes from: How the embodied mind brings mathematics into being*. New York: Basic Books.
Lamarque, P. (1991). Narrative as invention: The limits of fictionality. In C. Nash (Ed.), *Narrative in culture*. New York: Routledge.
Latour, B., & Woolgar, S. (1986). *Laboratory life: The construction of scientific facts*. Princeton, NJ: Princeton University Press.
Lederman, N. (1998). The state of science education: Subject matter without context. Editorial. *Electronic Journal of Science Education, 3*, 2.
Lederman, N. (2004). Syntax of nature of science within inquiry and science instruction. In L. Flick & N. Lederman (Eds.), *Scientific inquiry and nature of science* (pp. 301–317). Dordrecht, The Netherlands: Kluwer.
Lederman, N., & Abd-El-Khalick, F. (1998). Avoiding de-natured science: Activities that promote understandings of the nature of science. In W. McComas (Ed.), *The nature of science in science education: Rationales and strategies* (pp. 83–126). Dordrecht, The Netherlands: Kluwer Academic Publishers.
Lehrer, J. (2012). *Imagine: How creativity works*. New York: Houghton Mifflin Harcourt.
Lemke, J. (1990). *Talking science. Language, learning, values*. Norwood, NJ: Ablex.
Lemke, J. (2001). Articulating communities: Sociocultural perspectives on science education. *Journal of Research in Science Teaching, 38*, 296–316.
Liguori, L. (2014). The chocolate shop and atomic orbitals: A new atomic model created by high school students to teach elementary students. *Journal of Chemical Education, 91*, 1742–1744.
Limon, M. (2001). On the cognitive conflict as an instructional strategy for conceptual change. *Learning and Instruction, 11*, 613–623.
Lin, S., Hu, W., Adey, P., & Shen, J. (2003). The influence of CASE on scientific creativity. *Research in Science Education, 33*, 143–162.
Lin, S.-C., & Lin, H.-S. (2014). Primary teachers beliefs about scientific creativity in the classroom context. *International Journal of Science Education, 36*, 1551–1567.
Linfield, R. (2007). Bringing imagination back to science. *Primary Science Review, 100*, 26–27.
Littledyke, M. (2004). Drama and science. *Primary Science Review, 84*, 4–16.
Lodge, D. (1990). Narrative with words. In H. Barlow, C. Blakemore, & M. Weston-Smith (Eds.), *Images and understanding*. Cambridge, MA: Cambridge University Press.
Loewenstein, G. (1994). The psychology of curiosity. A review and reinterpretation. *Psychological Bulletin, 116*, 75–98.
Lomas, R. (2000). *The man who invented the twentieth century: Nikola Tesla, forgotten genius of electricity*. London: Headline.
Longworth, N., & Davies, K. (1996). *Lifelong learning. New visions, new implications. New roles for people, organizations, nations and communities in the 21st century*. London: Kogan Page.
Love, A. (2010). Darwin's "imaginary illustrations": Creatively teaching evolutionary concepts and the nature of science. *American Biology Teacher, 72*, 82–89.
Lunn, M., & Nobel, A. (2008). Revisioning science. Love and passion in the scientific imagination: Art and science. *International Journal of Science Education, 30*, 793–805.
Mach, E. (1896/1976). *Knowledge and error* (T. Cormack & P. Foulkes, Trans.). Dordrecht, The Netherlands: Reidel.
Mandelbrote, S. (2001). *Footprints of the lion: Isaac Newton at work*. Cambridge, UK: Cambridge University Library.
Mandler, J. (1984). *Stories, scripts, and scenes: Aspects of schema theory*. Hillsdale, NJ: Erlbaum.
Manicas, P., & Secord, P. (1983). Implications for psychology of the new philosophy of science. *American Psychologist, 38*, 399–413.

Mannheim, K. (1972). *Ideology and utopia* (L. Wirth & E. Shils, Trans.). London: Routledeg & Kegan Paul. (Original work published 1936)
Marcia, J. (1988). Common processes underlying ego identity, cognitive/moral development, and individuation. In D. Lapsley & F. Power (Eds.), *Self, ego, and identity: Integrative approaches*. New York: Springer.
Marshall, J. (2010). Five ways to integrate: Using strategies from contemporary art. *Art Education, 63*, 13–19.
Martin, B., & Brouwer, W. (1991). The sharing of personal science and the narrative element in science education. *Science Education, 75*, 707–722.
Martin, B., & Brouwer, W. (1993). Exploring personal science. *Science Education, 77*, 441–459.
Martin, R., Sexton, C., Franklin, T., Gerlovich, J., & McElroy, D. (2008). *Teaching science for all children* (5th ed.). New York: Allyn and Bacon.
Marzano, R. (2007). *Art and science of teaching. Though experiment in the classroom*. Alexandria, Egypt: ASCD.
Maslow, A. (1968). *Toward a psychology of being*. New York: Van Nostrand Reinhold.
Maslow, A. (1971). *The farther reaches of human nature*. New York: Viking.
Mathewson, J. (1999). Visual-spatial thinking: An aspect of science overlooked by educators. *Science Education, 83*, 33–54.
Matthews, B. (2005). Emotional development, science and co-education. In S. Alsop (Ed.), *Beyond Cartesian dualism: Encountering affect in the teaching and learning of science* (pp. 173–185). Berlin: Springer.
Matthews, G. (1997). Perplexity in Plato, Aristotle, and Tarski. *Philosophical Studies, 85*, 213–228.
Matthews, M. (1992). History, philosophy, and science teaching. The present rapprochement. *Science & Education, 1*, 11–47.
Matthews, M. (1994). *Science teaching: The role of history and philosophy of science*. New York: Routledge.
Matthews, M. (Ed.). (1998). *Constructivism in science education*. Dordrecht, The Netherlands: Kluwer Academic Publishers.
Matthews, M. (2002). Teaching science. In R. R. Curren (Ed.), *A companion to the philosophy of education* (pp. 342–353). Oxford, UK: Blackwell Publishers.
Matthews, M. (2015). *Science teaching: The contribution of history and philosophy of science*. New York: Routledge.
McAllister, J. (1996). *Beauty and revolution in science*. Ithaca, NY: Cornell University Press.
McComas, W. (1996). Ten myths of science: Reexamining what we think we know about the nature of science. *School Science and Mathematics, 96*, 10–16.
McComas, W. (1998a). *The nature of science in science education: Rationales and strategies*. Dordrecht, The Netherlands: Kluwer Academic Publishers.
McComas, W. (1998b). The principal elements of the nature of science: Dispelling the myths. In W. McComas (Ed.), *The nature of science in science education: Rationales and strategies* (pp. 53–72). Dordrecht, The Netherlands: Kluwer Academic Publishers.
McComas, W. (Ed.). (2002). *The nature of science in science education*. New York/Boston: Kluwer Academic Publishers.
McGregor, D. (2012). Dramatising science learning: Findings from a pilot study to re-invigorate elementary science pedagogy for five- to seven-year olds. *International Journal of Science Education, 34*, 1145–1165.
McGregor, D., & Precious, W. (2012). Dramatic science at key stage 1: Modelling ideas within an Olympics theme. *Primary Science, 123*, 10–13.
McGregor, D., & Precious, W. (2015). *Dramatic science*. London: Routledge.
McKinney, D., & Michalovic, M. (2004). Teaching the stories of scientists and their discoveries. *Science Teacher, 71*, 46–51.
McLean, S. (2009). Stories and cosmogonies. Imagining creativity beyond 'nature' and 'culture'. *Cultural Anthropology, 24*, 213–245.

Medawar, P. (1967). *The art of the soluble: Originality in science.* Middlesex, UK: Penguin Books.
Medawar, P. (1979). *Advice to a young scientist.* New York: Harper & Row.
Medawar, P. (1984a). *Pluto's republic.* Oxford: Oxford University Press.
Medawar, P. (1984b). *The limits of science.* New York: Teachers College Press.
Mellou, E. (1995). Creativity: The imagination condition. *Early Child Development and Care, 114*, 97–106.
Merten, S. (2011). Enhancing science education through art. *Science Scope, 35*(2), 31–35.
Metcalfe, R., Abbot, S., Bray, P., Exley, J., & Wisnia, D. (1984). Teaching science through drama: An empirical investigation. *Research in Science & Technological Education, 2*, 77–81.
Meyer, A. (1970). Karl Friedrich Burdach and his place in the history of neuroanatomy. *Journal of Neurology, Neurosurgery & Psychiatry, 33*(5), 553–561.
Meyer, H. (1971). *A history of electricity and magnetism.* Cambridge, MA: MIT Press.
Midgley, M. (1992). *Science as salvation: A modern myth and its meaning.* London: Routledge.
Midgley, M. (2000a). *Science and poetry.* London: Routledge.
Midgley, M. (2000b). The need for wonder. In R. Stannard (Ed.), *God for the 21st century* (pp. 185–187). Templeton Foundation Press.
Millar, R., & Osborne, J. (1998). *Beyond 2000: Science education for the future.* London: King's College.
Miller, A. (2001). *Einstein, Picasso: Space, time, and the beauty that causes havoc.* New York: Basic Books.
Miller, L. (2007). Solve medical mysteries. *Science Scope, 31*(3), 26–29.
Mills. (2013). A qualitative study: Integrating art and science in the environment. Ph.D. Dissertation, Wayne State University. ERIC document: ED552989.
Miell, D., & Littleton, K. (Eds.). (2004). *Collaborative creativity: Contemporary perspectives.* London: Free Association Books.
Milne, C. (1998). Philosophically correct science stories? Examining the implications of heroic science stories for school science. *Journal of Research in Science Teaching, 35*, 175–187.
Milne, I. (2010). A sense of wonder, arising from aesthetic experiences, should be the starting point for inquiry in primary science. *Science Education International, 21*, 102–115.
Milne, C., Kirch, S., Jhumki Basu, S., Leou, M., & Pamela Fraser-Abder, P. (2008). Understanding conceptual change: Connecting and questioning. *Cultural Studies of Science Education, 3*, 417–434.
Mindell, A. (2000). *Quantum mind: The edge between physics and psychology.* Oakland, CA: Lao Tse Press.
Mintzes, J., Wandersee, J., & Novak, J. (1997). Meaningful learning in science: The human constructivist perspective. In G. Phye (Ed.), *Handbook of academic learning* (pp. 405–451). San Diego, CA: Academic.
Mitchell (Ed.). (1981). *On narrative.* Chicago: University of Chicago Press.
Moje, C. (2007). Developing socially just subject-matter instruction. *Review of Research in Education, 31*, 1–44.
Monk, M., & Dillon, J. (2000). The nature of scientific knowledge. In M. Monk & J. Osborne (Eds.), *Good practice in science teaching: What research has to say* (pp. 72–87). Buckingham, UK: Open University Press.
Monk, M., & Poston, M. (1999). A comparison of music and science education. *Cambridge Journal of Education, 29*, 93–101.
Morais, C. (2015). Storytelling with chemistry and related hands-on activities. Informal learning experiences to prevent "chemophobia" and promote young children's scientific literacy. *Journal of Chemical Education, 92*, 58–65.
Moravcsik, M. (1981). Creativity in science education. *Science Education, 65*, 221–227.
Morton, A. (2013). *Emotion and imagination.* Cambridge, UK: Polity Press.
Mumford, M. (2003). Where have we been, where are we going? Taking stock in creativity research. *Creativity Research Journal, 15*, 107–120.
Murphy, P., Peters, M., & Marginson, S. (2010). *Imagination: Three models of imagination in the age of the knowledge economy.* New York: Peter Lang.

References

Myers, G. (1991). Making rediscovery: Narratives of split genes. In C. Nash (Ed.), *Narrative in culture*. London: Routledge.

Nash, C. (Ed.). (1991). *Narrative in culture*. London: Routledge.

National Research Council. (1996). *National science education standards*. Washington, DC: AAAS.

National Research Council. (2000). *Inquiry and the national science education standards*. Washington, DC: National Academic Press.

National Research Council. (2007). *Taking science to school. Learning and teaching science in grades K-8*. Washington, DC: National Academic Press.

National Research Council. (2009). *Learning science in informal environments: People, places, and pursuits*. Washington, DC: Committee on Learning Science in Informal Environments.

Newton, L., & Newton, D. (2010). Creative thinking and teaching creativity in elementary school science. *Gifted and Talented International, 25*, 111–124.

Nicholas, H., & Ng, W. (2008). Blending creativity, science and drama. *Gifted and Talented International, 23*, 51–60.

Nickerson, R. (1999). Enhancing creativity. In A. Sternberg (Ed.), *Handbook of creativity* (pp. 392–430). Cambridge, MA: Cambridge University Press.

Nikolic, H. (2012). EPR before EPR: A 1930 Einstein-Bohr thought experiment revisited. *European Journal of Physics, 33*, 1089–1097.

Norris, S., Guilbert, S., Smith, M., Hakimelahi, S., & Phillips, L. (2005). A theoretical framework for narrative explanation in science. *Science Education, 89*(4), 535–554.

Norton, J. (1996). Are thought experiments just what you thought. *Canadian Journal of Philosophy, 26*, 333–366.

Novak, J., & Gowin, B. (1984). *Learning how to learn*. New York: Cambridge University Press.

Ødegaard, M. (2003). Dramatic science. A critical review of drama in science education. *Studies in Science Education, 39*, 75–101.

Ogborn, J., Kress, J., Martins, I., & McGillicuddy, K. (1996). *Explaining science in the classroom*. Buckingham, UK: Open University Press.

Olrich, D., Harder, R., Callahan, R., & Gibson, H. (2001). *Teaching strategies*. Boston: Houghton Mifflin Company.

O'Neil, J. (1992). *Prodigal genius: The life of Nikola Tesla*. Chula Vista, CA: Tesla Book Company.

Ong, W. (1971). *Rhetoric, romance and technology*. Ithaka, NY: Cornell University Press.

Opdal, P. M. (2001). Curiosity, wonder and education seen as perspective development. *Studies in Philosophy and Education, 20*, 331–344.

Ormrod, J. (1999). *Human learning*. Upper Saddle River, NJ: Merrill.

Ortony, A., Clore, G., & Collins, A. (1989). *The cognitive structure of emotions*. New York: Cambridge University Press.

Osberg, D., & Biesta, G. (2007). Beyond presence: Epistemological and pedagogical implications of 'strong emergence'. *Interchange, 38*, 31–51.

Osborne, J. (1996). Beyond constructivism. *Science Education, 80*, 53–82.

Osborne, J., Collins, S., Ratcliffe, M., Millar, R., & Duschl, R. (2003). What "ideas-about-science" should be taught in school? A Delphi study of the expert community. *Journal of Research in Science Teaching, 40*, 692–720.

Osborne, J., Simon, S., & Tytler, R. (2009, April 13–17). *Attitudes toward science: An update*. Paper presented at the annual meeting of the American Educational Research Association, San Diego, CA.

Østergaard, E., Dahlin, B., & Hugo, A. (2008). Doing phenomenology in science education: A research review. *Studies in Science Education, 44*, 93–121.

Pantidos, P., Ravanis, K., Valakas, K., & Vitoratos, E. (2014). Incorporating poeticality into the teaching of physics. *Science & Education, 23*, 621–642.

Pantidos, P., Spathi, K., & Vitoratos, E. (2001). The use of drama in science education: The case of 'Blegdamsvej Faust'. *Science & Education, 10*, 107–117.

Papacosta, P. (2008). The mystery in science: A neglected tool in science education. *Science Education International, 19*, 5–8.

Park Rogers, M., & Abell, S. (2007a). Connecting with other disciplines. *Science and Children, 44*, 58–59.

Park Rogers, M., & Abell, S. (2007b). Connecting with other disciplines. *Science and Children, 44*(6), 58–60.

Paul, R., & Elder, L. (2013). *Critical thinking: Tools for taking charge of your professional and personal life*. Upper Saddle River, NJ: Pearson.

Peleg, R., & Baram-Tsabari, A. (2011). Atom surprise: Using theatre in primary science education. *Journal of Science Education and Technology, 20*, 508–524.

Pendrill, A.-M., Ekström, P., Hansson, L., Mars, P., Ouattara, L., & Ulrika, R. (2014). The equivalence principle comes to school—falling objects and other middle school investigations. *Physics Education, 49*, 425–430.

Pera, M. (1993). *The ambiguous frog. The Galvani- Volta controversy ob animal electricity*. Princeton, NJ: Princeton University Press.

Peters, R. (1973). Aims of education: A conceptual enquiry. In R. Peters (Ed.), *The philosophy of education* (pp. 1–35). Oxford, MA: Oxford University Press.

Peters, R. (1988). Democratic values and educational aims. In W. Hare & J. Portelli (Eds.), *Philosophy of education* (pp. 339–357). Calgary, Canada: Detselig Enterprises.

Peters, R. S. (1966). *Ethics and education*. London: Allen and Unwin.

Peters, R. S. (1967). What is an educational process? In R. Peters (Ed.), *The concept of education* (pp. 1–23). London: Routledge & Kegan Paul. New York: The Humanity Press.

Phenix, P. (1964). *Realms of meaning*. New York: McGraw Hill.

Phenix, P. (1982). Promoting personal development through learning. *Teachers College Record, 84*, 301–317.

Piaget, J. (1954). *The child's construction of reality*. New York: Basic Books.

Piaget, J. (1964). Development and learning. In R. Ripple & V. Rockcastle (Eds.), *Piaget rediscovered. A report of the conference on cognitive studies and curriculum development* (pp. 38–46). Ithaca, NY: Cornell University School of Education.

Piaget, J. (1970). *Genetic epistemology*. New York: Norton.

Piaget, J. (1977). *The development of thought. Equilibration of cognitive structures*. Oxford: Basil Blackwell.

Piersol, L. (2014). Our hearts leap up: Awakening wonder within the classroom. In K. Egan, A. Cant, & G. Judson (Eds.), *Wonder-full education: The centrality of wonder in teaching and learning across the curriculum* (pp. 3–21). New York: Routledge.

Pink, D. (2005). *A whole new mind*. New York: Riverhead Trade.

Pintrich, P., Marx, R., & Boyle, R. (1993). Beyond cold conceptual change: The role of motivational beliefs and classroom contextual factors in the process of conceptual change. *Review of Educational Research, 63*, 167–199.

Pintrich, P., & Schunk, D. (1996). *Motivation in education: Theory, research, and applications*. Englewood Cliffs, NJ: Prentice-Hall.

Pittman, K. (1999). Student-generated analogies: Another way of knowing? *Journal of Research in Science Teaching, 36*, 1–22.

Planck, M. (1933). *Where is science going?* (J. Murphy, Trans.). London: Allen & Unwin.

Poggi, S., & Bossi, M. (Eds.). (1994). *Romanticism in science. Science in Europe, 1790–1840*. Dordrecht, The Netherlands: Kluwer Academic Publishers.

Polanyi, N. (1958). *Personal knowledge*. Chicago: University of Chicago Press.

Polanyi, M. (1981). The creative imagination. In D. Dutton & M. Krausz (Eds.), *The concept of creativity is science and art* (pp. 91–108). The Hague, The Netherlands/Boston/London: Martinus Nijhoff Publications.

Polkinghorne, J. (1998). *Beyond science*. Cambridge, MA: Cambridge University Press.

Pongsophon, P., Yutakom, N., & Boujaoude, S. (2010). Promotion of scientific literacy on global warming by process drama. *Asia-Pacific Forum on Science Learning and Teaching, 11*, 1.

Popper, K. (1959). *The logic of scientific discovery*. New York: Routledge.

Popper, K. (1972). *Objective knowledge: An evolutionary approach*. Oxford, MA: Clarendon Press.

Porter, A., & Brophy, J. (1988). Synthesis of research on good teaching. *Educational Leadership, 45*, 74–85.

Posner, G., Strike, K., Hewson, P., & Gertzog, W. (1982). Accommodation of a scientific conception: Towards a theory of conceptual change. *Science Education, 66*, 211–222.

Precious, W., & McGregor, D. (2014). Just imagine: Using drama to support science learning with older primary children. *Primary Science, 132*, 35–37.

Pugh, K. (2002). Teaching for transformative experiences in science: An investigation of the effectiveness of two instructional elements. *Teachers College Record, 104*, 1101–1137.

Pugh, K. (2004). Newton's laws beyond the classroom walls. *Science Education, 88*, 182–196.

Pugh, K. (2011). Transformative experience: An integrative construct in the spirit of Deweyan pragmatism. *Educational Psychologist, 46*, 107–121.

Pugh, K., & Girod, M. (2007). Science, art and experience: Constructing a science pedagogy from Dewey's aesthetics. *Journal of Science Teacher Education, 18*, 9–27.

Pugh, K., Linnenbrink-Garcia, L., Koskey, K., Stewart, V., & Manzey, C. (2010a). Motivation, learning, and transformative experience: A study of deep engagement in science. *Science Education, 94*, 1–28.

Pugh, K., Linnenbrink-Garcia, L., Koskey, K., Stewart, V., & Manzey, C. (2010b). Teaching for transformative experiences and conceptual change: A case study and evaluation of a high school biology teacher's experience. *Cognition and Instruction, 28*, 273–316.

Raffini, J. (1993). *Winners without losers. Structures and strategies for increasing student motivation to learn*. Boston: Allyn and Bacon.

Raymo, C. (1998). *Honey from stone*. Cambridge, MA: Cowley Publications.

Reiner, M. (1998). Thought experiments and collaborative learning in physics. *International Journal of Science Education, 20*, 1043–1058.

Reiner, M., & Burko, L. (2003). On the limitations of thought experiments in physics and the consequences for physics education. *Science & Education, 12*, 365–385.

Reiner, M., & Gilbert, J. (2000). Epistemological resources for thought experimentation in science learning. *International Journal of Science Education, 22*, 489–506.

Rennie, L., & Johnston, D. (2004). The nature of learning and its implications for research on learning from museums. *Science Education, 88*(Supplement), 4–16.

Renzulli, J., Gentry, M., & Reis, S. (2004). A time and place for authentic learning. *Educational Leadership, 62*, 73–77.

Resnick, L. (1983a). Mathematics and science learning: A new conception. *Science, 220*, 477–478.

Resnick, L. (1983b). Toward a cognitive theory of instruction. In S. Paris, G. Olson, & H. Stevenson (Eds.), *Learning and motivation in the classroom* (pp. 5–38). Hillsdale, NJ: Erlbaum.

Reveles, J., Cordova, R., & Kelly, G. (2004). Science literacy and academic identity formation. *Journal of Research in Science Teaching, 41*, 111–144.

Ricchiuto, J. (1996). *Collaborative creativity*. Cleveland, OH: Oakhill Press.

Richards, R. J. (1987). *Darwin and the emergence of evolutionary theories of mind and behavior*. Chicago: Chicago University Press.

Richards, R. J. (1992). *The meaning of evolution: The morphological construction and ideological reconstruction of Darwin's theory*. Chicago: Chicago University Press.

Richards, R. J. (2002). *The romantic conception of life. Science and philosophy in the age of Goethe*. Chicago: University of Chicago Press.

Richards, R. J. (2006). Nature is poetry of mind or how Schelling solved Goethe's Kantian problems. In M. Friedman & A. Nordmann (Eds.), *The Kantian legacy in the nineteenth century science* (pp. 27–50). Cambridge, MA: MIT Press.

Richter, M., & Koppet, K. (2000). *How to increase retention through storytelling*. Retrieved August 13, 2006, from http://www.thestorynet.codarticles~essays/retentionicle.com

Ricoeur, P. (1981). Narrative time. In W. Mitchell (Ed.), *On narrative* (pp. 165–186). Chicago: University of Chicago Press.

Rinne, L., Gregory, E., Yarmolinskaya, J., & Hardiman, M. (2011). Why arts integration improves long-term retention of content. *Mind, Brain, and Education, 5*, 89–96.

Ritz, W. (2007). *A head start on science: Encouraging a sense of wonder.* Arlington, VA: NSTA Press.
Robinson, K. (2001). *Out of our minds. Learning to be creative.* Chichester, UK: Capstone.
Rogers, C. (1961). *On becoming a person.* Boston: Houghton, Mifflin.
Root-Bernstein, R. (1987). Harmony and beauty in medical research. *Journal of Molecular and Cellular Cardiology, 19*, 1043–1051.
Root-Bernstein, R. (1996). The sciences and arts share a common creative aesthetic. In A. Tauber (Ed.), *The elusive synthesis. Aesthetics and science* (pp. 49–82). Boston: Kluwer.
Root-Bernstein, R. (2002). Aesthetic cognition. *International Studies in Philosophy of Science, 16*, 61–77.
Root-Bernstein, R., & Root-Bernstein, M. (2004). Artistic scientists and scientific artists: The link between polymathy and creativity. In R. J. Sternberg, E. L. Grigorenko, & J. L. Singer (Eds.), *Creativity: From potential to realization* (pp. 127–151). Washington, DC: American Psychological Association.
Rose, C., & Nichol, M. J. (1997). *Accelerated learning for the 21st century.* New York: Dell Publishing.
Ross, S. (1981). *Philosophical mysteries.* Albany, NY: SUNY Press.
Roth, W.-M., & Desaultes, J. (Eds.). (2002). *Science education as/for sociopolitical action.* New York: Peter Lang.
Roth, W.-M., & Lee, S. (2004). Science education as/for participation in the community. *Science Education, 88*, 263–291.
Rowlands, S. (2011). Discussion article: Disciplinary boundaries for creativity. *Creative Education, 2*, 47–55.
Rorty, R. (1980). *Philosophy and the mirror of nature.* Princeton, NJ: Princeton University Press.
Runco, M. (2004). Creativity as an extracognitive phenomenon. In L. Shavinina & M. Ferrari (Eds.), *Beyond knowledge: Extracognitive aspects of developing high ability* (pp. 17–25). Mahwah, NJ: Lawrence Erlbaum Associates Publishers.
Saçkes, M., Trundle, K., Bell, R., & O'Connell, A. (2011). The influence of early science experience in kindergarten on children's immediate and later science achievement: Evidence from the early childhood longitudinal study. *Journal of Research in Science Teaching, 48*, 217–235.
Sagan, C. (1979). *Broca's brain: Reflections on the romance of science.* New York: Random House.
Sagan, C. (1980). Science and fiction: A personal view. In J. Williamson (Ed.), *Teaching science fiction: Education for tomorrow* (pp. 1–8). Philadelphia: Owlswick Press.
Sagan, C. (1995). Wonder and skepticism. *The Skeptical Inquirer, 19*, 24–30.
Santayana, G. (1955). *The sense of beauty: being the outline of aesthetic theory.* New York: Dover. (Original work published 1896)
Saul, E. (Ed.). (2004). *Crossing boundaries in literacy and science instruction.* Newark, NJ: IRA & NSTA.
Schakel, P. (2002). *Imagination and the arts in C.S. Lewis: Journeying to Narnia and other worlds.* Columbia, MI: University of Missouri Press.
Schank, R. (1990). *Tell me a story.* New York: Charles Scribner's Sons.
Schank, R. (2004). *Making minds less well educated than our own.* Mahwah, NJ: Lawrence Erlbaum.
Schank, R., & Abelson, R. (1995). Knowledge and memory: The real story. In R. S. Wyer Jr. (Ed.), *Advances in social cognition* (Vol. VIII, pp. 1–85). Hillsdale, NJ: Erlbaum.
Schank, R., & Berman, T. (2002). The pervasive role of stories in knowledge and action. In M. Green, J. Strange, & T. Brock (Eds.), *Narrative impact: Social and cognitive foundations* (pp. 287–314). Mahwah, NJ: Lawrence Erlbaum Associates.
Scheffler, I. (1996). The concept of the educated person. In V. Howard & I. Scheffler (Eds.), *Work, education, and leadership* (pp. 81–100). New York: Peter Lang.
Schiffer, H., & Gueria, A. (2015). Electricity a vital force: Discussing the nature of science through historical narratives. *Science & Education, 24*, 409–434.

Schiller, F. (2006). *Aesthetical and philosophical essays*. Retrieved February 12, 2014, from www.gutenberg.net

Schiller, F. v. (1795/1993). On naïve and sentimental poetry. In W. Hinderer & D. O. Dahlstrom (Eds.), *F. Schiller: Essays* (pp. 179–260). New York: Continuum.

Schmidt, A. (2011). Creativity in science: Tensions between perceptions and practice. *Creative Education, 2*, 435–445.

Schmidt, S. (1980). Science fiction and the science teacher. In J. Williamson (Ed.), *Teaching science fiction: Education for tomorrow* (pp. 110–120). Philadelphia: Owlswick Press.

Schulz, R. (2009). Reforming science education: Part II. Utilizing Kieran Egan's educational metatheory. *Science & Education, 18*, 251–273.

Schulz, R. (2014). Philosophy of education and science education: A vital but underdeveloped relationship. In M. R. Matthews (Ed.), *Handbook of research on history, philosophy and science teaching* (pp. 1259–1315). Berlin: Springer.

Schwartz, R., Lederman, N., & Crawford, B. (2004). Developing views on NOS in an authentic context. An explicit approach to bridging the gap between NOS and scientific inquiry. *Science Education, 88*, 610–640.

Sebrell, W., Haggerty, J., & The Editors of Time-Life Books. (1967). *Food and nutrition*. New York: Time- Life Books.

Seeley, C., & Gallagher, S. (2014). Stories and science: Stirring children's imagination. *Primary Science, 134*, 30–33.

Seifer, M. (1998). *Wizard: The life and times of Nikola Tesla*. New York: Citadel Press.

Selby, D. (1998). *Global education: Towards a quantum model of environmental education*. Retrieved April 11, 2010, from http://www.ec.gc.ca/education/documents/colloquium/selby.htm

Semrud-Clikeman, M. (2012). *Research in brain function and learning*. Washington, DC: American Psychological Association.

Sfard, A., & Prusak, A. (2005). Telling identities: In search of an analytic tool for investigating learning as a culturally shaped activity. *Educational Researcher, 34*, 14–22.

Shapira, O., & Liberman, N. (2009). Why thinking about distant things can make us more creative. *Scientific American*. Retrieved January 12, 2011, from http://www.scientificamerican.com/article.cfm?id=an-easy-way-to-increase-c#comments

Shavinina, L., & Ferrari, M. (2004). Extracognitive facets of developing high ability: Introduction to some important issues. In L. V. Shavinina & M. Ferrari (Eds.), *Beyond knowledge: Extracognitive aspects of developing high ability* (pp. 3–13). Mahwah, NJ: Lawrence Erlbaum Associates Publishers.

Shavinina, L., & Seeratan, K. (2004). Extracognitive phenomena in the intellectual functioning of gifted, creative, and talented individuals. In L. Shavinina & M. Ferrari (Eds.), *Beyond knowledge: Extracognitive aspects of developing high ability* (pp. 73–102). Mahwah, NJ: Lawrence Erlbaum Associates Publishers.

Shea, D., LubinskiI, D., & Benbow, C. (2001). Importance of assessing spatial ability in intellectually talented young adolescents: A 20-year longitudinal study. *Journal of Educational Psychology, 93*(3), 604–714.

Shepard, R. (1988). The imagination of the scientists. In K. Egan & D. Nadaner (Eds.), *Imagination and education* (pp. 153–185). New York: Teachers College Press.

Silverman, M. (1989). Two sides of wonder: Philosophical keys to the motivation of science learning. *Synthese, 80*, 43–46.

Silverman, M. (2003). *A universe of atom, an atom in the universe*. New York: Springer.

Simmons, A. (2006). *The story factor: Inspiration, influence and persuasion through the art of storytelling*. New York: Basic Books.

Simons, S. (2013). "It gets under your skin": Using process drama to explore race and privilege with undergraduate students. ERIC Document: ED 563183.

Simonton, D. (2004). *Creativity in science: Chance, logic, genius, and zeitgeist*. Cambridge, UK: Cambridge University Press.

Siry, C., & Kremer, I. (2011). Children explain the rainbow: Using young children's idea to guide curricula. *Journal of Science Education and Technology, 20*, 643–655.

Sloman, K., & Thompson, R. (2010). An example of large-group drama and cross-year peer assessment for teaching science in higher education. *International Journal of Science Education, 32*, 1877–1893.

Snelders, H. (1990). Oersted's discovery of electromagnetism. In A. Cunningham & N. Jardine (Eds.), *Romanticism and the sciences* (pp. 228–240). Cambridge, MA: Cambridge University Press.

Snow, C. (1959). *The two cultures and the scientific revolution*. London: Cambridge University Press.

Sofronieva, T. (2014). Erwin Schrodinger's poetry. *Science & Education, 23*, 655–672.

Solomon, J. (2002). Science stories and science texts: What can they do for our students? *Studies in Science Education, 37*, 85–106.

Solomon, J., & Aikenhead, G. (Eds.). (1994). *STS education: International perspectives on reform*. New York: Teachers College Press.

Sorensen, R. (1992). *Thought experiments*. Oxford, MA: Oxford University Press.

Spencer, H. (1880). *Education: Intellectual, moral and physical*. New York: Appleton & Company.

Spier-Dance, L., Mayer-Smith, J., Dance, N., & Khan, S. (2005). The role of student-generated analogies in promoting conceptual understanding for undergraduate chemistry students. *Research in Science and Technological Education, 23*, 163–178.

Statham, M. (2014). Using visualisation to enhance learning. *Primary Science, 132*, 31–34.

Steele, A., & Ashworth, E. (2013). Walking the integration talk: An ArtSci project. *Canadian Journal for the Scholarship of Teaching and Learning, 4*, 2.

Stefanich, G. (Ed.). (2001). *Science teaching in inclusive classrooms*. Cedar Falls, IA: Wolverton.

Stefanich, G., & Hadzigeorgiou, Y. (2001). Models and applications. In G. Stefanich (Ed.), *Science teaching in inclusive classrooms* (pp. 61–90). Cedar Falls, IA: Wolverton.

Sternberg, R. (Ed.). (1999). *Handbook of creativity*. Cambridge, MA: Cambridge University Press.

Sternberg, R. (2006). Creating a vision of creativity: The first 25 years. *Psychology of Aesthetics, Creativity, and the Arts, 1*, 2–12.

Sternberg, R., & Lubart, T. (1999). The concept of creativity: Prospects and paradigms. In R. Sternberg (Ed.), *Handbook of creativity* (pp. 3–15). London: Cambridge University Press.

Stevens, S., Warshofsky, F., & The Editors of Life. (1966). *Sound and hearing*. The Netherlands: Time-Life International.

Stinner, A. (1993). Science, humanities and society – The Snow-Leavis controversy. *Interchange, 20*, 16–23.

Stinner, A. (1995). Contextual settings, science stories and large context problems: Toward a more humanistic science education. *Science Education, 79*, 555–581.

Stinner, A., & Teichmann, J. (2003). Lord Kelvin and the age-of-the-Earth debate: A dramatization. *Science & Education, 12*, 213–228.

Stinner, A., & Williams, H. (1993). Conceptual change, history, and science stories. *Interchange, 24*, 87–103.

Stolberg, T. (2008). Whither the sense of wonder f pre-service primary teachers' when teaching science? A preliminary study of personal experience. *Teaching and Teacher Education, 24*, 1958–1964.

Stone, R. (2006). Curiosity as the thief of wonder. *Kronoscope, 6*, 205–229.

Strike, K., & Posner, G. (1992). A revisionist theory of conceptual change. In R. Duschl & R. Hamilton (Eds.), *Philosophy of science, cognitive psychology, and educational theory and practice* (pp. 147–176). New York: State University of New York Press.

Stutler, S. (2011). From "The Twilight Zone" to "Avatar": Science fiction engages the intellect, touches the emotions, and fuels the imagination of gifted learners. *Gifted Child Today, 34*(2), 45–49.

Sutton, C. (1996). Beliefs about science and beliefs about language. *International Journal of Science Education, 18*, 1–18.

Sutton-Smith, B. (1988). In search of the imagination. In K. Egan & D. Nadaner (Eds.), *Imagination and education* (pp. 3–29). New York: Teachers College Press.

References

Swimme, B. (1988). The cosmic creation story. In D. Griffin (Ed.), *The reenchantment of science* (pp. 47–56). New York: SUNY Press.
Taber, K. (2013). *Modelling learners and learning in science education.* Dordrecht, The Netherlands: Springer.
Tanriseven, I. (2013). The effect of school practices on teacher candidates' sense of efficacy relating to use of drama in education. *Educational Sciences: Theory and Practice, 13*, 402–412.
Tauber, A. (Ed.). (1996). *The elusive synthesis: Aesthetics and science.* Boston: Kluwer.
Taylor, J. (1998). *Poetic knowledge. The recovery of education.* New York: SUNY Press.
Taylor, J. (2003). Probing the limits of reality: The metaphysics in science fiction. *Physics Education, 38*, 20–26.
Tegmark, M. (2015). *Our mathematical universe.* London, UK: Penguin Books.
Tesla, N. (1982). *My inventions.* New York: Ben Johnson. (original work published 1919)
Tesla Society. (2011). *Tesla Memorial Society of New York.* Available: http://www.teslasociety.com
The Royal Society. (2008). *Exploring the relationship between socioeconomic status and participation and attainment in science education.* London: The Royal Society.
The Royal Bank Letter. (1989). The importance of teaching. *The Royal Bank of Canada, 70*(5), 1–4.
Thomas, L. (1995). *Late night thoughts on listening to Mahler's ninth symphony.* New York: Penguin.
Tolstory, I. (1990). *The knowledge and the power: Reflection on the history of science.* London: Canongate.
Tomas, L., Jackson, C., & Carlisle, K. (2014). The "wonder of science challenge" project. *Teaching Science, 60*(2), 48–57.
Toulmin, S. (1976). *Knowing and acting. An invitation to philosophy.* New York: McMillan.
Toulmin, S. (1982). *Cosmopolis. The hidden agenda of modernity.* Chicago: University of Chicago Press.
Treagust, D., & Duit, R. (2008). Conceptual change: A discussion of theoretical, methodological and practical challenges for science education. *Cultural Studies of Science Education, 3*, 297–328.
Trefil, J. (2003). *The nature of science: An A-Z guide to the laws and principles governing the universe.* Boston: Houghton-Mifflin.
Truby, J. (2007). *The anatomy of story: 22 steps to becoming master storyteller.* New York: Faber & Faber.
Tytler, R., Prain, V., Hubber, P., & Waldrip, B. (Eds.). (2013). *Constructing representations to learn science.* Rotterdam, The Netherlands: Sense Publishers.
Van't Hoff, J. (1967). *Imagination in science.* New York: Springer.
Varelas, M., Pappas, C., Tucker-Raymond, E., Kane, J., Hankes, J., Ortiz, I., et al. (2010). Drama activities as ideational resources for primary-grade children in urban science classes. *Journal of Research in Science Teaching, 47*, 3012–3325.
Van Manen, M. (1990). *Researching lived experience.* Albany, NY: SUNY Press.
van Eijck, M., & Roth, W. M. (2013). *Imagination of science in education: From epics to novelization.* Dordrecht, The Netherlands: Springer.
Velentzas, A., & Halkia, K. (2011). The 'Heisenberg's microscope' as an example of using thought experiments in teaching physics theories to students of the upper secondary school. *Research in Science Education, 41*, 525–539.
Velentzas, A., & Halkia, K. (2013). From Earth to Heaven: Using "Newton's cannon" thought experiment for teaching satellite physics. *Science & Education, 22*, 2621–2640.
Velentzas, A., Halkia, K., & Skordoulis, C. (2007). Thought experiments in the theory of relativity and in quantum mechanics: Their presence in textbooks and in popular science books. *Science & Education, 16*, 353–370.
Verhoven, C. (1972). *The philosophy of wonder.* New York: McMillan.
Vernon, P. (Ed.). (1970). *Creativity: Selected readings.* Harmondsworth, UK: Penguin.
Vosniadou, S. (2008). *International handbook of research on conceptual change.* New York: Routledge.

Vygotsky, L. S. (1991). Imagination and creativity. *Soviet Psychology, 29,* 73–89. (Original work published 1930)

Vygotsky, L. S. (2004). Imagination and creativity in childhood. *Journal of Russian and East European Psychology, 42*(1), 7–97. (Original work published 1930)

Wang, M., Eccles, J., & Kenny, S. (2013). Not lack of abilities but lack of choice: Individual and gender differences in choice of careers in science, technology, engineering, and mathematics. *Psychological Science, 14,* 1–6.

Wardekker, W. (1998). Scientific concepts and reflection. *Mind, Culture, and Activity, 5,* 143–153.

Warnock, M. (1976). *Imagination.* London: Faber.

Waterman, A. (1992). Identity as an aspect of optimal physiological functioning. In G. Adams, T. Gullotta, & R. Montemayor (Eds.), *Adolescent identity formation* (pp. 121–176). Newbury Park, CA: Sage.

Waterman, A. (2004). Finding someone to be: Studies on the role of intrinsic motivation in identity formation. *Identity, 4,* 209–228.

Watts, M. (2001). Science and poetry: Passion v. prescription in school science? *International Journal of Science Education, 23,* 197–208.

Watson, P. (2010). *The German genius. Europe's third renaissance, the second scientific revolution, and the twentieth century.* New York: Harper Perennial.

Weaver, G. (1998). Strategies in K-12 science instruction to promote conceptual change. *Science Education, 82,* 455–472.

Weiping, H., Adey, P., Jiliang, S., & Chondge, L. (2004). The comparisons of the development of creativity between English and Chinese adolescents. *Acta Psychologica Sinca, 36,* 718–731.

Weisberg, R. (1993). *Creativity: Beyond the myth of genius.* New York: Freeman.

Weisskopf, V. (1979). *Knowledge and wonder.* Harvard, MA: MIT Press.

Wells, G., & Haaf, M. (2013). Investigating art objects through collaborative student research projects in an undergraduate chemistry and art course. *Journal of Chemical Education, 90,* 1616–1621.

Wenger, E. (1998). *Communities of practice: Learning, meaning, and identity.* Cambridge, UK: Cambridge University Press.

White, H. (1981). The value of narrativity in the representation of reality. In W. Mitchell (Ed.), *On narrative* (pp. 1–23). Chicago: University of Chicago Press.

White, M. (1997). *Isaac Newton – The last sorcerer.* London: Fourth Estate.

Whitehead, A. (1957). *The aims of education.* New York: The Free Press (Original work published 1929).

Wickmann, P. (2006). *Aesthetic experience in science education: Learning and meaning making as situated talk and action.* Mahwah, NJ: Lawrence Erlbaum.

Williams, C., Stanisstreet, M., Spall, K., Boyes, E., & Dickson, D. (2003). Why aren't secondary school pupils interested in physics? *Physics Education, 38,* 324–329.

Wilson, E. (1986). *Biophilia.* Cambridge, MA: Harvard University Press.

Wilson, E. (1998). *Consilience.* New York: Knopf.

Wilson, M., & The Editors of Life. (1965). *Energy.* The Netherlands: Time- Life International.

Witz, K. (1996). Science with values and values for science education. *Journal of Curriculum Studies, 28,* 597–612.

Wong, D. (2002). To appreciating variation between scientists: A perspective for seeing for seeing science's vitality. *Science Education, 86,* 386–400.

Wong, D., Pugh, K., & Dewey Ideas Group at Michigan State University. (2001). Learning science: A Deweyan perspective. *Journal of Research in Science Teaching, 38,* 317–336.

Woolgar, S. (1993). *Science: The very idea.* London: Routledge.

Yager, R. (1996). *Science/technology/society as reform in science education.* Albany, NY: State University of New York.

Yager, R. (2000). What the future vision for science education should be like for the first 25 years of the new millennium. *School Science & Mathematics, 100,* 327–341.

Yokoi, C., & Yee, B. (2011). The art and science of notebooks. *Science and Children, 49,* 42–46.

References

Zembylas, M. (2002). Emotions in teaching science: A case study of a teacher's views. *Research in Science Education, 34*, 342–354.

Zembylas, M. (2005). Emotions and science teaching: Present research and future agenda. In S. Alsop (Ed.), *Beyond Cartesian dualism: Encountering affect in the teaching and learning of science* (pp. 123–132). Berlin: Springer.

Zenasni, F., Besancon, M., & Lubart, T. (2011). Creativity and tolerance of ambiguity: An empirical study. *Journal of Creative Behavior, 42*, 61–73.

Ziman, J. (2000). *Real science: What it is, and what it means*. Cambridge, UK: Cambridge University Press.

Zimmerman, S. (2009). Faster than light: Two curiosities to stimulate interest. *Physics Education, 44*, 509–510.

Zuckerman, J. (1998). Science supervisors' stories: A way to communicate pedagogical values. *Science Educator, 7*, 20–25.

Zuckerman, J. (1999). Student science teachers constructing practical knowledge from inservice science supervisors' stories: A way to communicate pedagogical values. *Journal of Science Teacher Education, 10*, 235–245.

Printed in Great Britain
by Amazon